621.2
H995d / 1999

10-23-00

HAWKEYE COMMUNITY COLLEGE
7944 1009 8401 8

HYDRAULICS

The fundamentals of service and theory of operation
in off-road vehicles, trucks and autom

621.2 H995d 1999
Schink, Robert J
Hydraul
fundam
theory
hydraul
off-roa
and aut

16.95

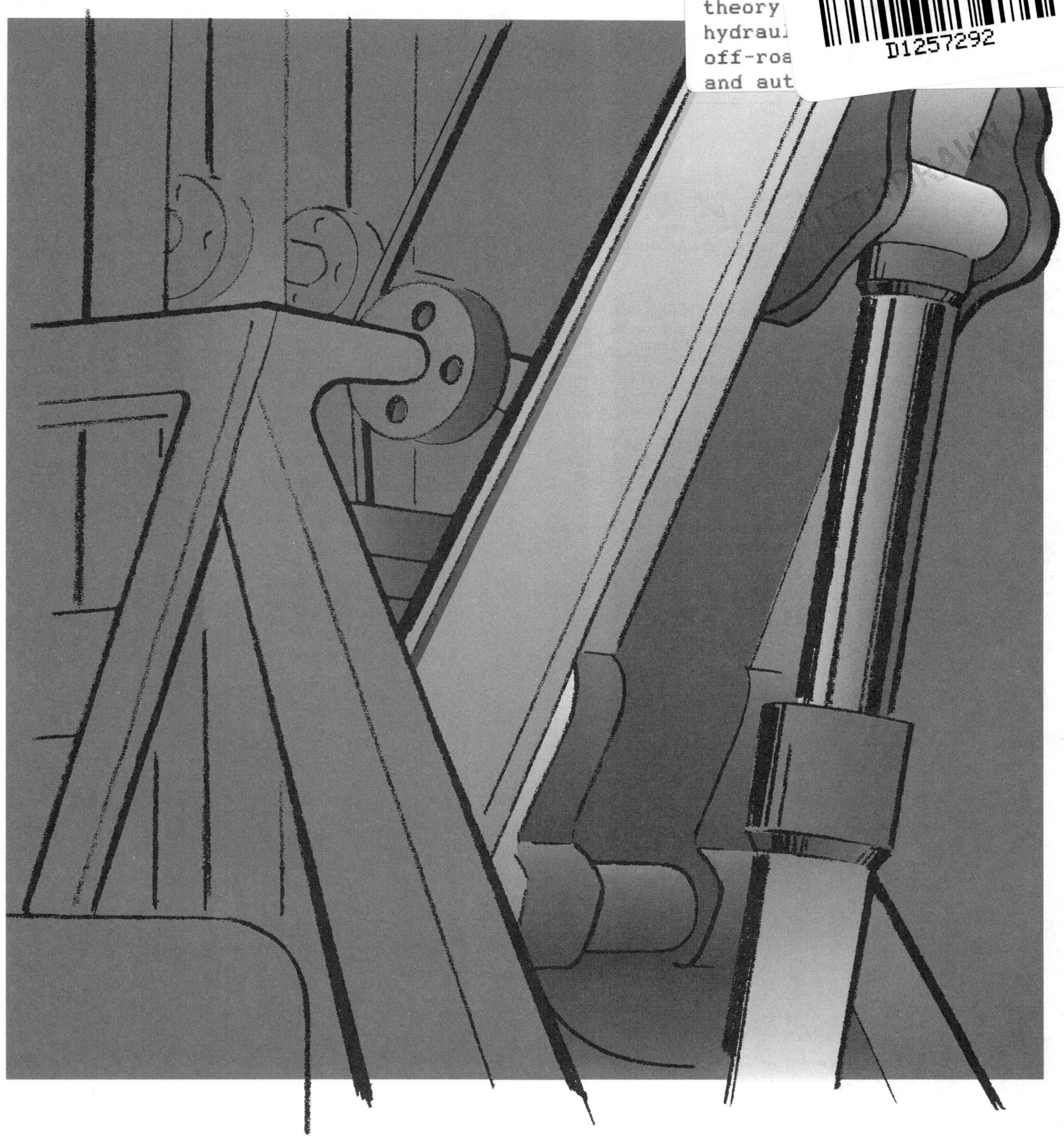

050857

FUNDAMENTALS OF SERVICE

PUBLISHER

DEERE AND COMPANY
JOHN DEERE PUBLISHING
Almon TIAC Bldg, Suite 104
1300 - 19th Street
East Moline, IL 61244
http://www.deere.com/aboutus/pub/jdpub
1-800-522-7448

EDITORIAL STAFF
CINDY CALLOWAY

Fundamentals of Service (FOS) is a series of manuals created by Deere & Company. Each book in the series is conceived, researched, outlined, edited, and published by Deere & Company, John Deere Publishing. Authors are selected to provide a basic technical manuscript that could be edited and rewritten by staff editors.

TO THE READER:
PURPOSE OF THIS MANUAL: The main purpose of this manual is to train readers so that they can understand and service hydraulics systems with speed and skill. Starting with "how it works", we build up to "why it fails" and "what to do about it." This manual is also an excellent reference for trained mechanics who want to refresh their memory on hydraulics. It has been written in a simple form using many illustrations so that it can be easily understood.

APPLICATION OF HYDRAULICS IN THIS MANUAL: "Hydraulics" is a broad field. It covers any study of fluids in motion or at rest. But in this manual, the prime interest is on oil hydraulics as it is commonly used in agricultural and industrial applications.

HOW TO USE THIS MANUAL: This manual can be used by anyone - experienced, mechanics and shop trainees, as well as vocational students and interested laymen. By starting with the basics, build your knowledge step by step. Chapter 1 covers the basics - "how it works". Chapter 2 provides important information on safety. Chapter 3 teaches you to read hydraulic diagrams. Chapters 4 through 13 go into detail about the working parts of hydraulics circuits. Chapters 14 and 15 return to the complete circuit in terms of general maintenance, why it fails, and "how to remedy these failures."

AUTHOR OF THIS EDITION: Robert J. Schink is retired from a career of various Technical Product Support positions involving agricultural and industrial equipment. After receiving and Agricultural Engineering degree from the University of Wisconsin he spent 42 years working with John Deere and John Deere dealers throughout the world. Those positions involved technical problem solving on machines in the field as well as the training of field personnel. In that time he was involved in the hydraulic development, from the simplistic systems of the 1950's to the sophisticated systems used in today's equipment.

ACKNOWLEDGEMENTS: John Deere gratefully acknowledges help from the following groups: Aeroquip Corporation, American Oil Company, Cessna Aircraft Company, Char-Lynn Company, Commercial Shearing and Stamping Company, Greer Hydraulics, Inc., Gresen Manufacturing Company, H7dreco Division of New York Air Brake Company, National Fluid Power Association, The Nuday Company, Owatonna Tool Company, Sun Oil Company, Sundstrand Corp., Texaco Inc., Vickers Inc.

FOR MORE INFORMATION: This book is one of many books published on agricultural and related subjects. For more information or to request a FREE CATALOG, call 1-800-522-7448 or send your request to address above or

Visit Us on the Internet--
http://www.deere.com/aboutus/pub/jdpub/

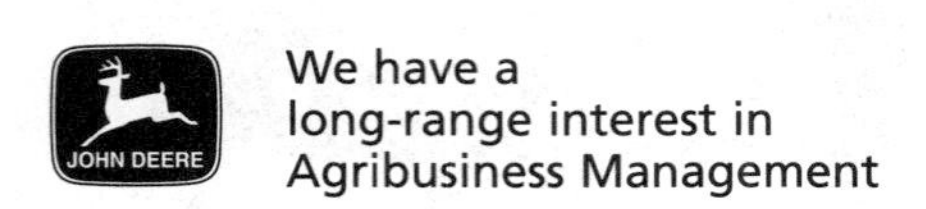

Copyright © 1967, 1972, 1979, 1987, 1992, 1999/Deere & Company, John Deere Publishing, Moline, IL/Sixth Edition./All rights reserved.

This material is the property of Deere & Company, John Deere Publishing, all use and/or reproduction not specifically authorized by Deere & Company, John Deere Publishing is prohibited.

FOSI006NC

ISBN 0-86691-265-7

CONTENTS

1 HYDRAULICS - HOW IT WORKS

2 SAFETY RULES FOR HYDRAULICS

3 SYMBOLS USED IN FLUID POWER DIAGRAMS

4 HYDRAULIC PUMPS

5 HYDRAULIC VALVES

6 HYDRAULIC CYLINDERS

7 HYDRAULIC MOTORS

8 HYDRAULIC ACCUMULATORS

9 HYDRAULIC FILTERS

10 RESERVOIRS AND OIL COOLERS

11 HOSES PIPES, TUBES AND COUPLERS

12 HYDRAULIC SEALS

13 HYDRAULIC FLUIDS

14 GENERAL MAINTENANCE

15 DIAGNOSIS AND TESTING OF HYDRAULIC SYSTEMS

APPENDIX

HOW TO READ THE OIL FLOW DIAGRAMS

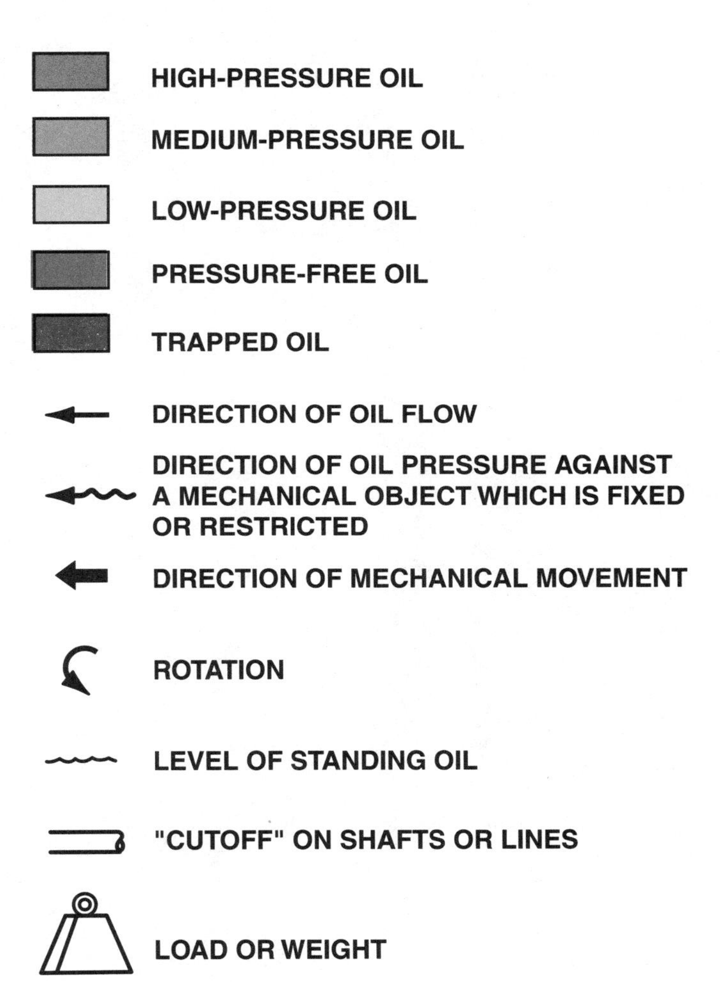

HYDRAULICS - HOW IT WORKS

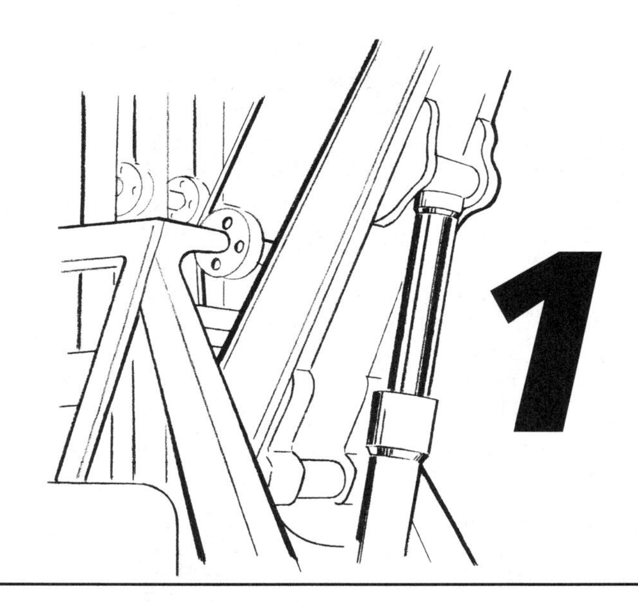

BASIC PRINCIPLES OF HYDRAULICS

The basic principles of hydraulics are few and simple:

- **Liquids have no shape of their own.**
- **Liquids are practically incompressible.**
- **Liquids transmit applied pressure in all directions, and act with equal force at right angles to all surfaces.**
- **Liquids under pressure follow the path of least resistance.**
- **Pressure can be created only by a resistance to flow.**
- **Flow across an orifice results in a pressure drop that is directly proportionate to the flow and inversely proportionate to the area of the orifice opening.**
- **Hydraulic systems <u>can</u> provide great increases in work force.**
- **Energy put into a hydraulic system in the form of flow under pressure will result in either work or heat.**

Fig. 1 - Liquids Have No Shape of Their Own

LIQUIDS HAVE NO SHAPE OF THEIR OWN. Liquids take the shape of any container (Fig. 1). Therefore fluid will flow in any direction and into a passage of any size or shape.

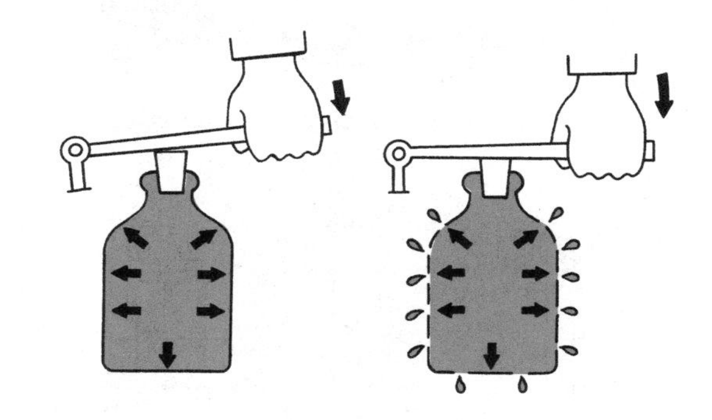

Fig. 2 - Liquids Are Practically Incompressible

LIQUIDS ARE PRACTICALLY INCOMPRESSIBLE. For safety reasons, we obviously wouldn't perform the experiment shown in Fig. 2. We will fill the container with a liquid and insert a cork. If we were then to push down on the cork with enough force, the container would bulge or break because the liquid would not compress.

NOTE: Liquids do compress slightly under pressure, but from a practical point, the amount is negligible.

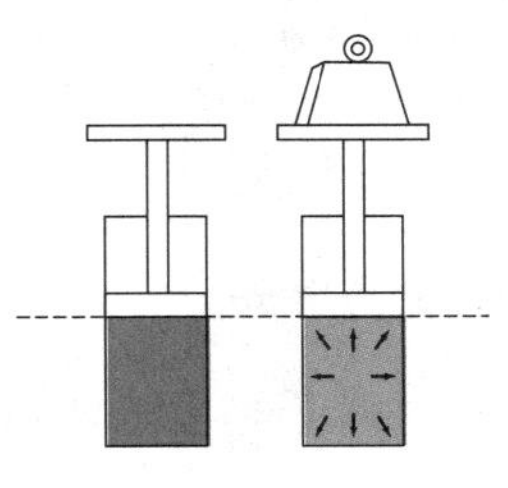

Fig. 3 - Liquids Transmit Applied Pressure Equally in All Directions

LIQUIDS TRANSMIT APPLIED PRESSURE IN ALL DIRECTIONS. The experiment in Fig. 2 shattered the glass jar and also showed how liquids transmit pressure in all directions when they are put under compression. This is very important in a hydraulic system. In Fig. 3, when a weight is added a force is applied to the liquid. The pressure is the same at every point on the cylinder and piston. The force acts at right angles to all surfaces.

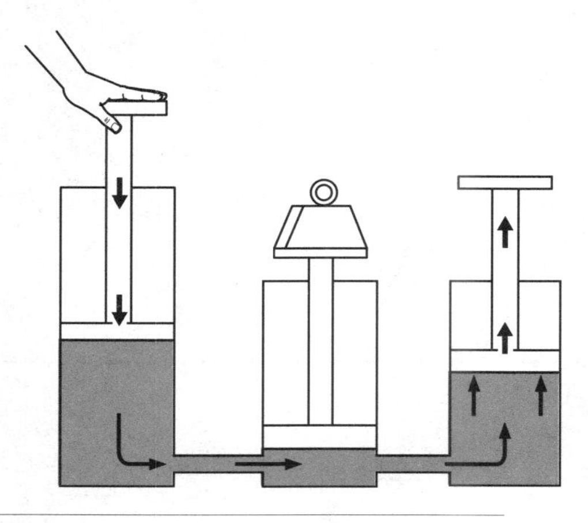

Fig. 4 - Liquids Follow the Path of Least Resistance

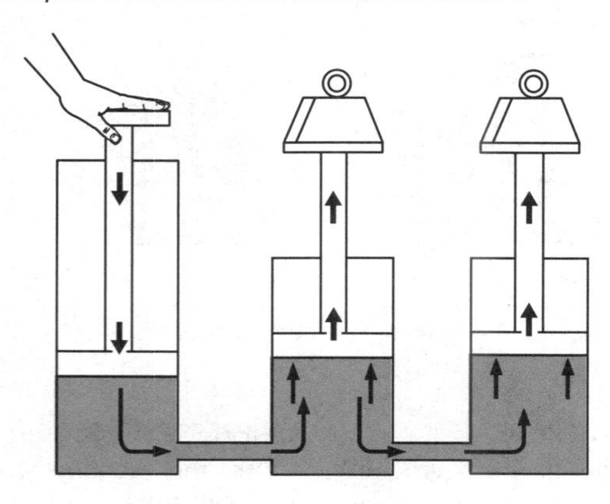

Fig. 5 - Paths of Equal Resistance

LIQUIDS UNDER PRESSURE FOLLOW THE PATH OF LEAST RESISTANCE. Refer to the three cylinders in Fig. 4. One cylinder has a weight on the piston. If a weight is applied to one of the unweighted pistons, the fluid would become pressurized and that pressure would act on the other two cylinders. The unweighted cylinder would extend while the weighted cylinder would remain stationary. The weighted cylinder would extend only after the unweighted one reached the end of its stroke.

If equal weights are added to both cylinders (Fig. 5), the pistons will rise at the same time as force is applied to the third cylinder. Both cylinders offer the same resistance to fluid flow.

PRESSURE CAN BE CREATED ONLY BY RESISTANCE TO FLOW. In Fig. 4, pressure built in the system as the piston was pushed down would be negligible as it could not exceed that necessary to raise the unweighted cylinder. When the unweighted cylinder reached the end of its stroke, the pressure would then be limited to that necessary to raise the weighted cylinder. When the second cylinder reaches the end of its stroke, the pressure is limited only by the amount of force exerted on the first cylinder.

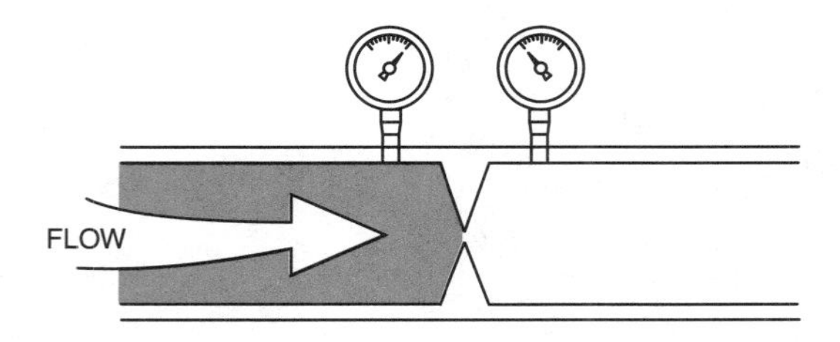

Fig. 6 - Flow Across an Orifice

FLOW ACROSS AN ORIFICE CAUSES A PRESSURE DROP

Flow across an orifice results in a pressure drop which is proportional to the square of the flow and inversely proportional to the square of the area of the orifice opening.

In Fig 6:

1. With a constant flow	1. Changing the orifice size will change the pressure drop across the orifice. *Increase opening — Decrease pressure drop* *Decrease opening — Increase pressure drop*
2. With constant inlet pressure (Variable displacement Pump)	2. Changing the orifice size will change the flow. *Increase opening — Increase flow* *Decrease opening — Decrease flow*
3. With constant inlet pressure (Variable displacement Pump)	3. Changing the outlet pressure will change flow. (It changes differential pressure) *Increased outlet pressure — Decreased flow* *Decreased outlet pressure — Increased flow*
4. With a fixed orifice size	4. Changing the inlet pressure will change the flow. *Increase pressure — Increase flow* *Decrease pressure — Decrease flow*
5. With a fixed orifice size	5. Changing the flow will change the pressure drop across the orifice. *Increased flow — Increased pressure drop* *Decreased flow — Decreased pressure drop*

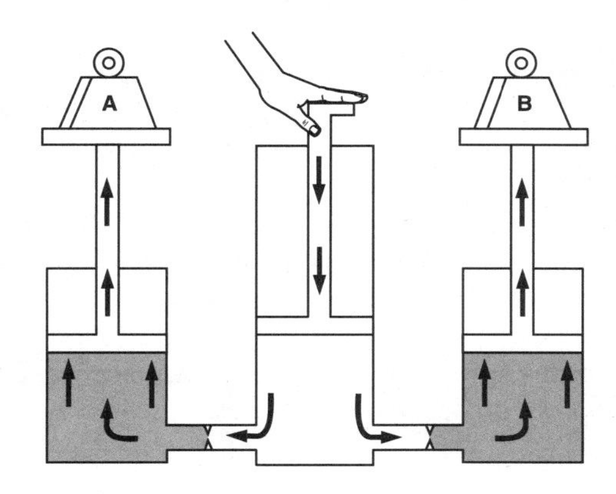

Fig. 7 - Orifices Control Flow and Pressure Drops

In Fig. 7, we will connect three cylinders. Equal sized orifices will be installed in the connecting lines. Equal weights A and B will be placed on the outside cylinders. These weights will be such that it will require 50 psi (345 kPa) in the cylinders to raise them.

On the center cylinder, we will push with a force that will create 100 psi (690 kPa) in the cylinder. The restriction of the orifices will allow the pressure to rise to100 psi (690kPa) in the center cylinder. Pressure will drop to a level necessary to lift the load (50 psi-345 kPa) as it passes through the orifices.

ENERGY PUT INTO A HYDRAULIC SYSTEM CANNOT BE DESTROYED. Energy is put into a hydraulic system in the form of fluid flow under pressure. This energy can accomplish work by moving a load with a cylinder or motor. Any oil that loses pressure without accomplishing work is turned to heat. This heat (loss of energy) can be the result of a line restriction.

In the example in Fig. 7, oil was pressurized to 100 psi (690 kPa). It required only 50 psi (345 kPa) to raise the load. Therefore, 50% of the energy created by pushing on the center cylinder went to work and 50% went to heat. This heating occurred when the oil dropped pressure as it passed through the orifices.

This type of restriction can be an essential part of the hydraulic system operation. However, it can also be undesirable when caused by hoses and lines which are too small, when fittings are restricted, when there is internal leakage in the system, or when oil is unnecessarily metered to a hydraulic function.

LIQUIDS CAN PROVIDE A GREAT INCREASE IN WORK FORCE. In Fig. 8, we have cylinders of different sizes connected with a tube. The piston in Cylinder A has an area of one square inch (6.45 cm2), but the piston in Cylinder B has an area of ten square inches (64.5 cm^2).

This time we'll place a 10-pound (45 N)(4.5 kg) weight on Cylinder B. Because the area of the piston is ten square inches (64.5 cm^2), the ten pound (45 N)(4.5 kg) weight will cause a force of one pound (4.5 N)(0.45 kg) on each square inch (6.45 cm^2) of the surface contacted by the oil. Or a pressure of one pound per square inch (1 psi.)(7 kPa).

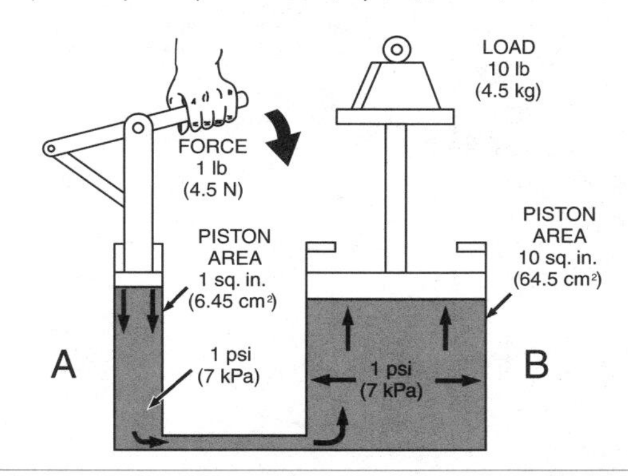

Fig. 8 - Liquids Can Provide a Great Increase in Work Force

We'll again push down on Cylinder A. It will require a force of one pound (4.5 N)(0.45 kg) to build the 1 psi (7 kPa) to move the weight up. Thus, a force of one pound (4.5N)(0.45 kg) is able to lift a10 pound (45 N)(4.5 kg) weight. It must be noted that for each inch (1 in)(2.45 cm) of travel of Piston A, Piston B will travel only 1/10 (0.1) inches (0.245 cm).

PRESSURE, AREA, DISPLACEMENT, AND FLOW

Pressure is force per unit of **Area**. U.S. customary measurement for pressure is pounds per square inch of surface area (psi). The metric unit of measurement for pressure is the Pascal (Pa.) or Newton per square meter (N/m2). The unit is often given in kilopascals (kPa) which is equal to 1000 Pascals. 1 kPa would be 1 kg acting on an area of 100 cm^2.1 psi is about 7000 Pa or 7 kPa.

NOTE: The force exerted by a piston can be determined by multiplying the piston area by the pressure applied.

Displacement is a term used for volume. It is piston area multiplied by stroke length, when referring to a cylinder (Fig. 9) or the quantity of fluid delivered in one cycle when referring to hydraulic pumps (See Chapter 4).

Two hydraulic cylinders can have the same displacement, even if they have different dimensions (Fig. 10). Displacement is usually rated in cubic inches (cu. in.) cubic centimeters (cms.), liters (l) or milliliters (mL).

Flow is the volume of fluid moved per unit of time. The flow rate for fluid flowing to the double-acting lift cylinders on a loader (Fig. 13) determines the speed the operator lifts the load. Flow is measured in gallons per minute (gpm) or liters per minute (L/min).

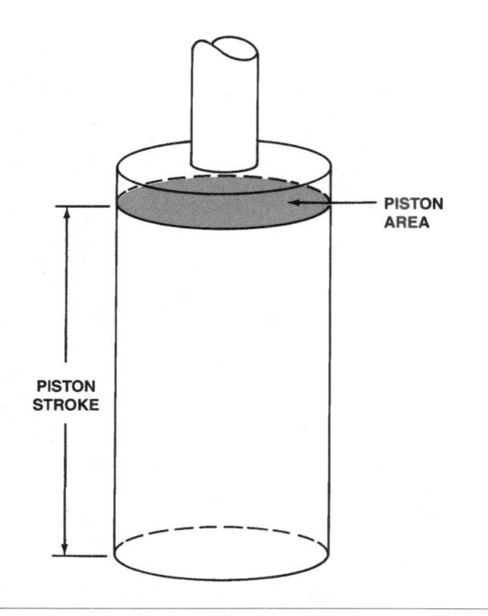

Fig. 9 - Displacement is Stroke Multiplied by Piston Area

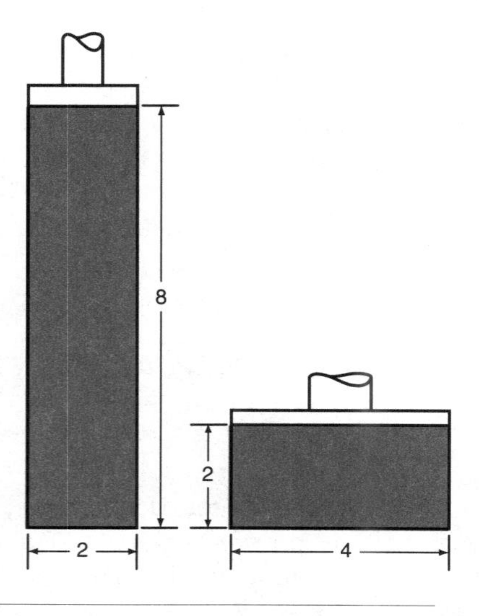

Fig. 10 - Both Cylinders Have the Same Displacement

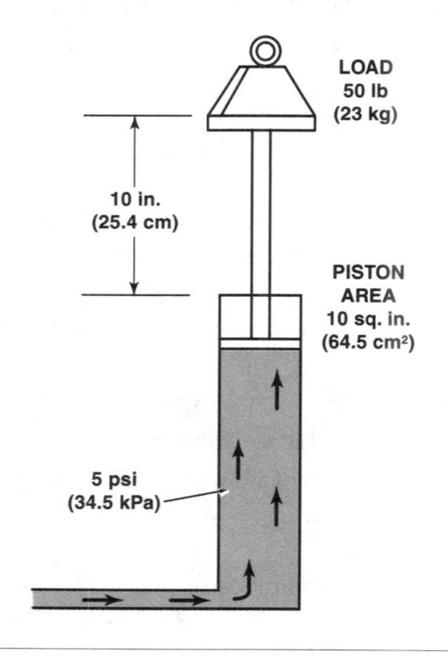

Fig. 11 - Force is Pressure Multiplied by Area
Work is Force Multiplied by Distance Traveled

FORCE, WORK, AND POWER

Force is pressure multiplied by the area to which the pressure is applied. In Fig. 11, the system is exerting 50-lb (222.5 N) of force or 5 psi times 10 square inches (34.5 kPa times 64.5 cm^2).

Work is force multiplied by distance. In Fig. 11, the piston is raising a load using 50 Lb (23 kg) over a distance of 10 inches (25.4 cm). 500 Lb-in. (56.5 N·m) of work is performed.

In some of the previous examples, force was applied, but no movement occurred. There was, therefore, no work accomplished.

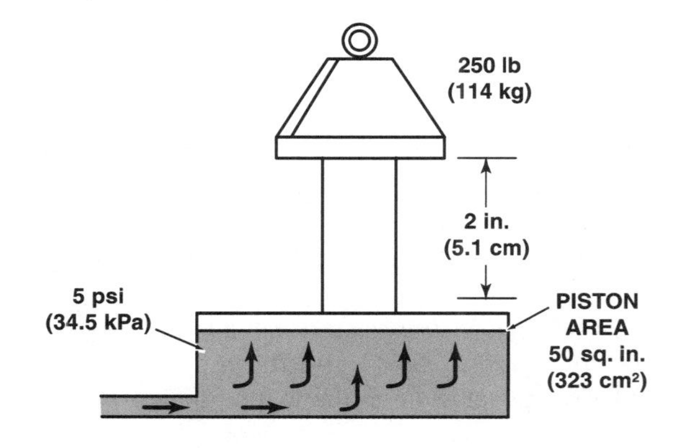

Fig. 12 - Force Increases if Cylinder Area Increases

In Fig. 12, a larger cylinder, with five times the piston area as the one shown in Fig. 11, is raising a 250-Lb (114-kg) load a distance of 2 inches (5.1 cm). With the same pressure and volume of oil input, the force will be 5 times as great, however, the work performed is equal because the load is raised only 2 in. (5.1 cm.). So 500 Lb-in. (56.5 Nm) of work is performed in both examples.

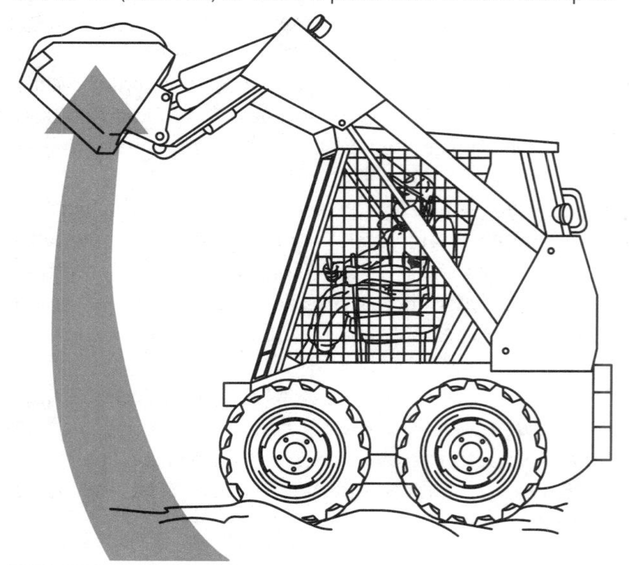

Fig. 13 - Time is a Factor

The term **power** adds the dimension of time to work. If one machine can do the same amount of work as another in less time, it has developed more power.

Power equals work per unit time.

ADDITIONAL CONCEPTS NEEDED TO UNDERSTAND A HYDRAULIC SYSTEM

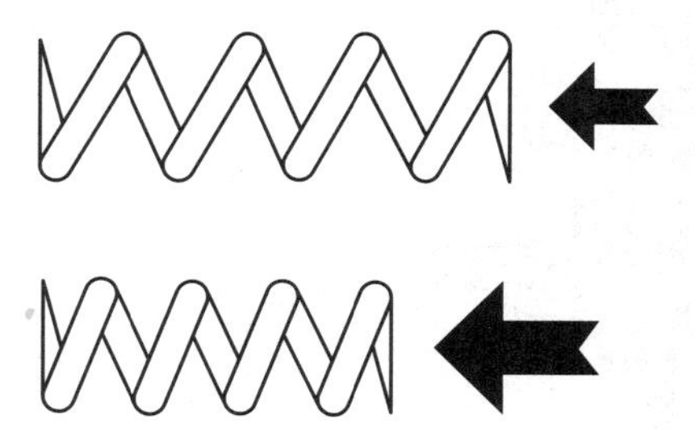

Fig. 14 - Spring Force

The Spring requires more and more force as the spring is compressed. The spring will always return to a specific length when a specific force is applied to it. These characteristics will be used in flow and pressure regulation on many of the hydraulic components covered in this manual.

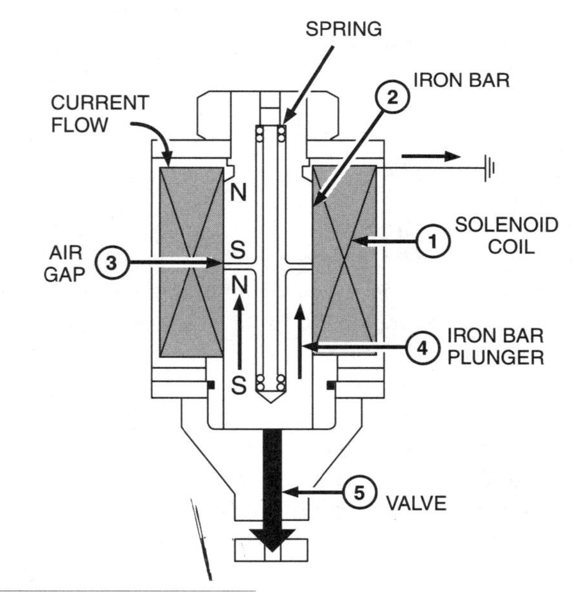

Fig. 15 - Electrical Solenoid

An Electrical Solenoid can be used to control hydraulic systems. It consists of an electrical winding (coil) and an armature (steel bar)(Fig. 15). As an electrical current is passed through the windings, it creates a magnetic field. The field magnetizes the iron bars and causes the plunger (armature) to compress the spring and move toward the stationary bar.

The solenoid in Fig. 16 is attached to a two position, three way valve. Full system voltage is applied to the windings. The magnetic field moves the armature against a spring force to open the valve. The valve sends oil to the circuit and blocks the return flow.

When the solenoid is deenergized, the spring will return the valve to the closed position, blocking pressure oil and dumping the circuit oil.

The strength of the magnetic field is determined by the current flowing through the windings. If a partial voltage were applied to the windings in Fig. 15, a weaker magnetic field will be developed. This would move the armature against the spring with less force, partially compressing the spring and partially opening the valve. As the voltage is increased, the stronger field will compress the spring further and open the valve further. In this way, we can have infinite positions of the valve.

The voltage to the solenoid can be controlled by an Electronic Control Unit (ECU) to provide infinite control of flows and pressures in hydraulic systems.

Reversing the current flow in the solenoid would force the armature in the opposite direction. This is used in some equipment where operation of a three-position valve is required.

HOW A HYDRAULIC SYSTEM WORKS

Let's build up a hydraulic system, piece by piece. The heart of the hydraulic system is the PUMP and the CYLINDER or MOTOR with LINES to connect them.

1. The PUMP which moves the oil (Converts linear or rotary energy into fluid energy).

2. The CYLINDER uses the moving oil to do work (Transfers the fluid energy back into a linear or rotating force to do work).

3. The LINES carry the fluid from one hydraulic component to another.

In Fig. 17, when you apply force to the lever, the hand pump forces oil into the cylinder. The pressure of this oil pushes up on the piston and lifts the weight. The pump has converted a mechanical force to hydraulic power, the line has carried it to the cylinder and the cylinder has converted the hydraulic power back to a linear force.

The load has moved up, but if the hand pressure is released, the oil would be forced back to the pump by the load.

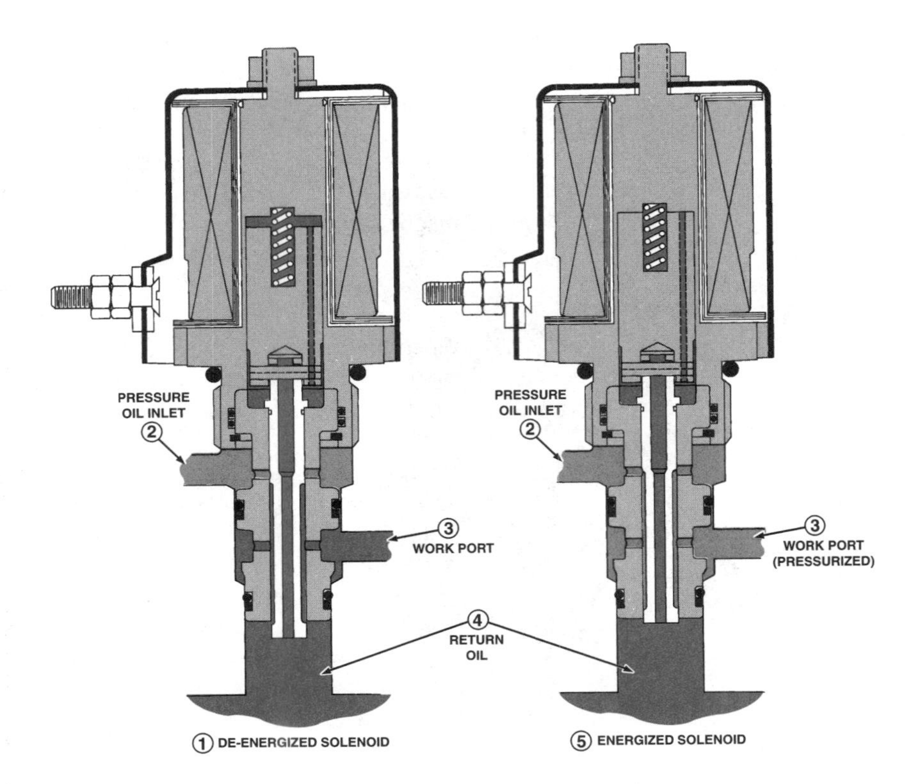

Fig. 16 - Solenoid Valve

We must therefore add a CHECK VALVE to trap the oil in the cylinder.

4. The CHECK VALVE (Fig. 18) allows fluid to pass freely into the cylinder as the fluid pushes the ball off its seat. When the flow stops the spring pushes the ball on the seat trapping the fluid in the cylinder.

So that the pump can have a supply of fluid to continue to move the load, we add a RESERVOIR.

5. The RESERVOIR is a vented container, which contains fluid to be forced by gravity or atmospheric pressure into the pump piston when retracted.

To prevent fluid from being forced back to the reservoir on the power stroke, a second check valve is installed.

Notice that the pump is smaller than the cylinder. This means that with each stroke of the pump, the piston will move only a small amount. The load lifted by the cylinder is much greater than the force applied to the pump piston.

If you want to lift the weight faster, then you must work the pump faster, increasing the volume of oil to the cylinder.

To lower the cylinder, we would need to add a line with a shutoff valve connecting the cylinder to the reservoir.

The system we have just described is a system that might be found on a hydraulic jack or a hydraulic press. To meet the hydraulic requirements in most other applications, however, we must provide a greater quantity of oil at a more consistent rate and be able to have better control of the oil movement.

To make a more usable system, let's add some new features as shown in Figs. 19 and 20.

In Fig. 19, the hand pump has been replaced with a gear-type pump. This pump will supply a continuous flow of oil. The gear pump is one of many types of pumps that transform the rotary force of a motor or engine to hydraulic energy. For more on pumps, see Chapter 4.

The cylinder has also been changed. It has two lines connected to it; one at the top and one at the bottom. The cylinder cavity is sealed above as well as below the piston. These features make this a **double-acting cylinder**.

Fluid from the bottom line supplies fluid to lift the cylinder piston. Fluid from the top line supplies fluid to push the piston back down.

The cylinder in Fig 17 is a **single-acting cylinder**. It has a hydraulic line to only one end of the cylinder and can push the load in only one direction. The only way to get the load back down is to release the oil so that the load will lower by gravity.

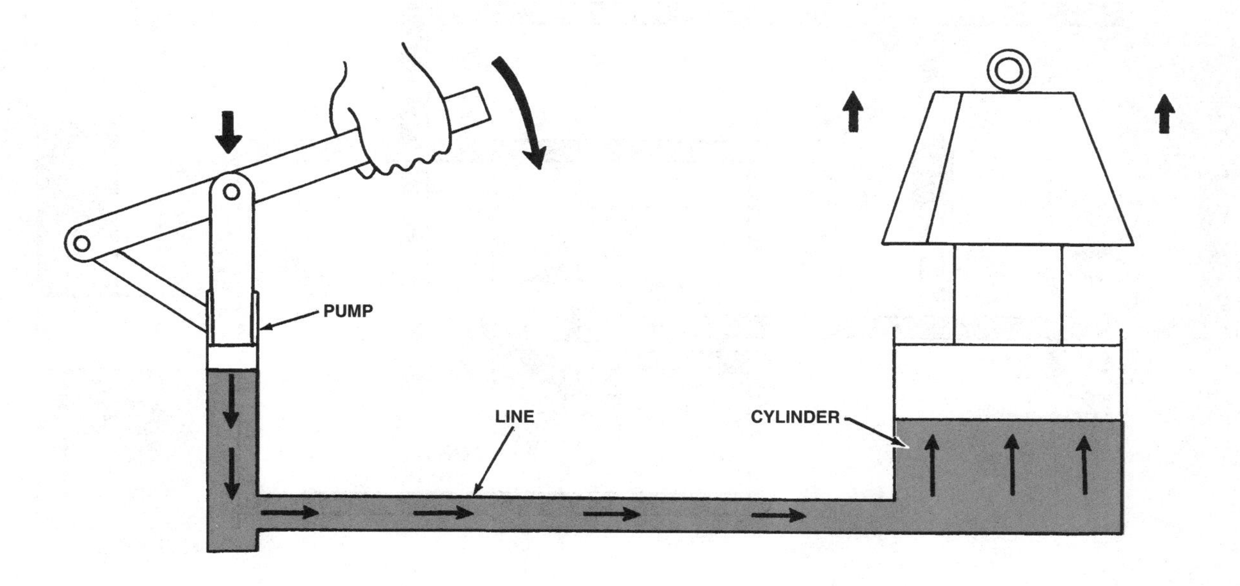

Fig. 17 - A Basic Hydraulic System

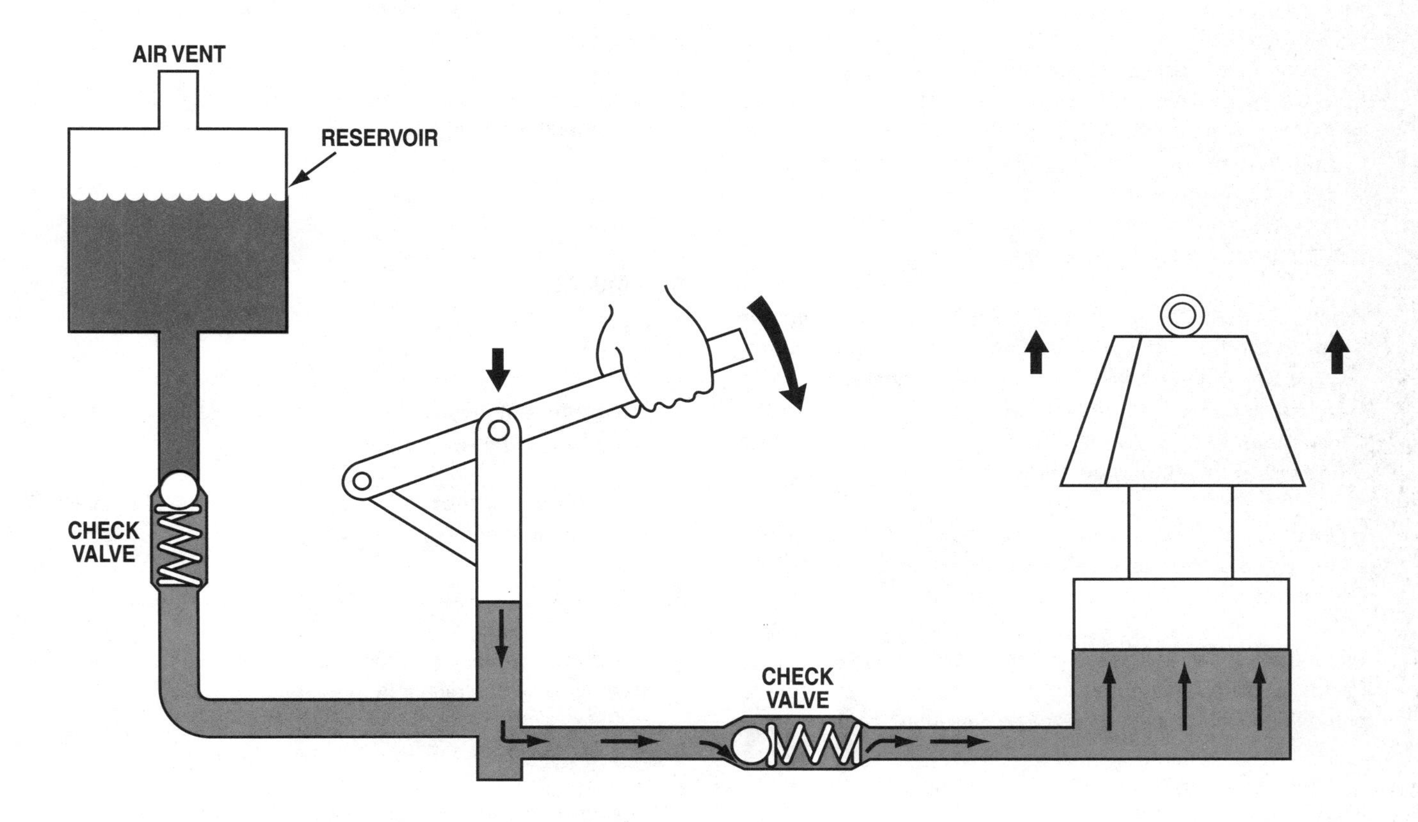

Fig. 18 - Reservoir and Check Valves Added

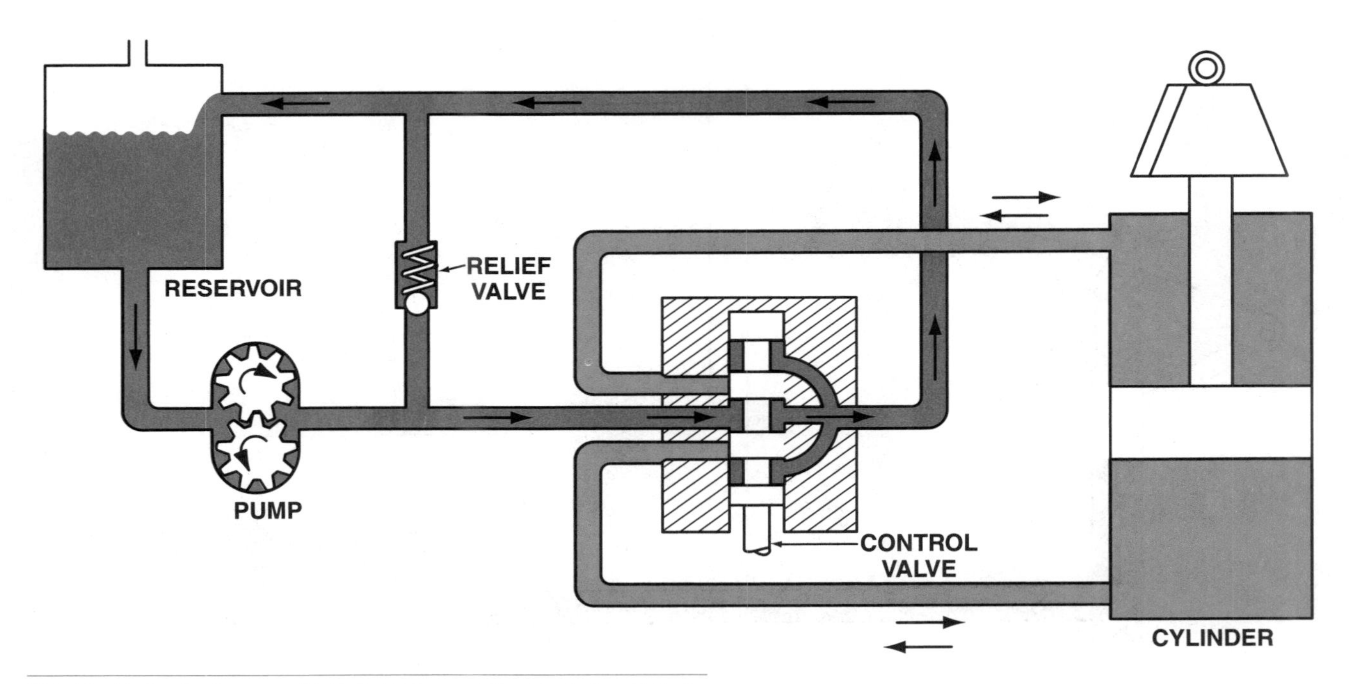

Fig. 19 - Hydraulic System with Relief Valve and Double-Acting Cylinder

To direct the pump fluid to both ends of the cylinder and to the reservoir, we need to add a CONTROL VALVE.

6. The CONTROL VALVE directs the oil. This allows the operator to control the constant supply of oil from the pump to and from the hydraulic cylinder. When the control valve is in the neutral position shown in Fig. 19, the flow of oil from the pump goes directly through the valve to a line that carries the oil back to the reservoir. At the same time, the valve has trapped oil on both sides of the hydraulic cylinder, thus preventing its movement in either direction.

When the control valve is moved down (Fig. 20), the pump oil is directed to the cavity on the bottom of the cylinder piston, pushing up on the piston and raising the weight. At the same time, the line at the top of the cylinder is connected to the return passage, thus allowing the oil forced from the topside of the piston to be returned to reservoir.

When the control valve is moved up (not shown), pump oil is directed to the top of the cylinder, lowering the piston and the weight. Oil from the bottom of the cylinder is returned to the reservoir.

When the cylinder reaches the end of its stroke, there is nowhere for the fluid from the pump to go. The pump will continue to pump fluid until something fails (hose, driveshaft, cylinder, etc.).

To limit the pressure in a hydraulic system, we will add a RELIEF VALVE.

7. The RELIEF VALVE limits the maximum pressure in the system. When the load is too great or when the cylinder bottoms, the pressure will rise. That pressure acts against the ball which is held on its seat by the spring. When the pressure is high enough to force the ball off its seat, it allows fluid to flow to the reservoir.

This completes our basic hydraulic system. We have used 7 components to do this. The modern hydraulic system will use many if not all of these.

SUMMARY

To summarize:

1. The **pump** = generates flow - converts mechanical energy to hydraulic energy.
2. The **cylinder** or **motor** = converts hydraulic energy back to mechanical energy.
3. The **lines** = carry the fluid to the hydraulic components.
4. The **check valve** = controls the flow of fluid, allowing it to go in only one direction in a line.
5. The **reservoir** = contains a reserve of fluid for the system.
6. The **control valve** = directs the flow of fluids to the proper components.
7. The **relief valve** = protects the system from high pressures.

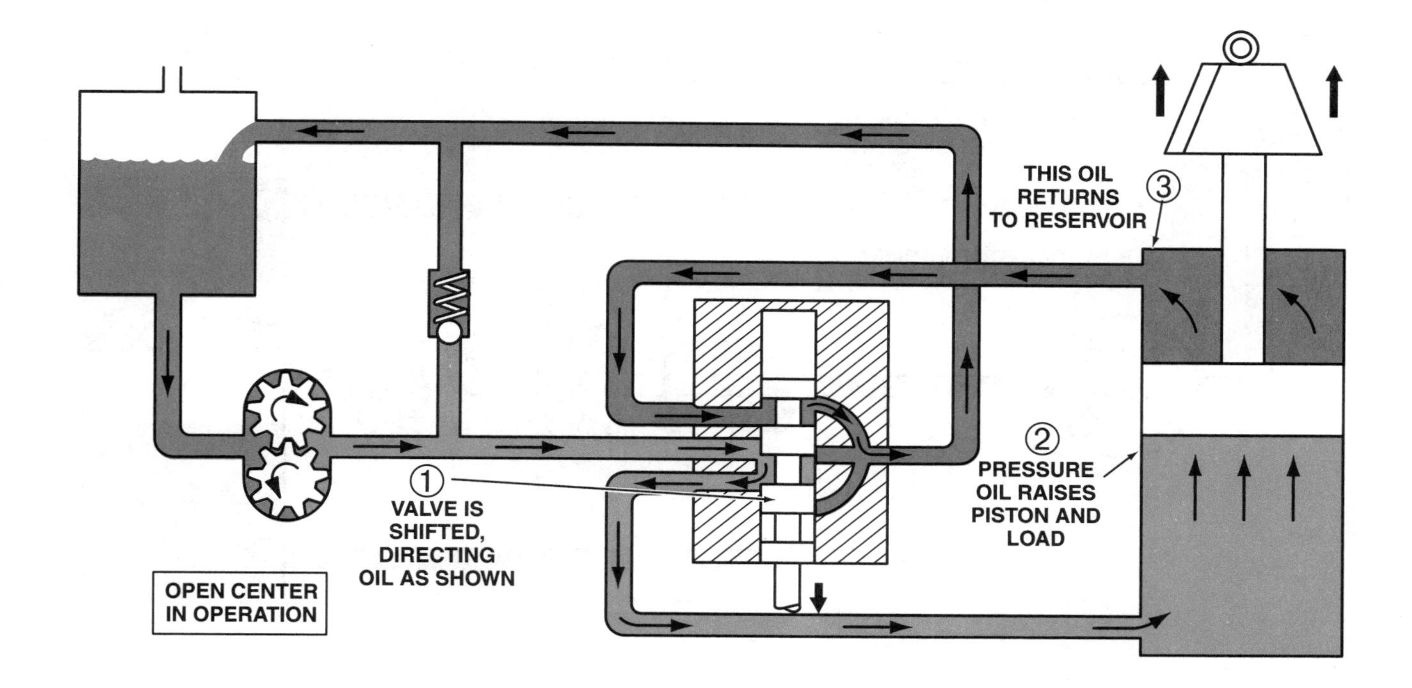

Fig. 20 - Hydraulic System in Operation

For more details on how these parts operate, go to Chapter 4 - Pumps, Chapter 5 - Valves, Chapter 6 - Cylinders, Chapter 10 - Reservoirs and Oil Coolers, and Chapter 11 - Hoses, Pipes, Tubes and Couplers

THE PROS AND CONS OF HYDRAULICS

As you have seen in the simple hydraulic system we have just developed, the purpose is to transmit power from a source (engine or motor) to the location where this power is required for work.

To look at the advantages and disadvantages of the hydraulic system, let's compare it to the other common methods of transferring this power. These would be mechanical (shafts, gears, belts, chains, or cables) or electrical.

ADVANTAGES

1. FLEXIBILITY—The mechanical method of power transmission requires that the positions of the engine and work site remain relatively constant. With the flexibility of hydraulic lines, power can be sent to almost any location.

2. MULTIPLICATION OF FORCE—In the hydraulic system, very small forces can be used to move very large loads simply by changing cylinder sizes.

3. SIMPLICITY—The hydraulic system has fewer moving parts, fewer points of wear. And it lubricates itself.

4. COMPACTNESS—Compare the size of a small hydraulic motor with an electric motor of equal horsepower. Then imagine the size of the gears shafts and levers that would be required to create the forces which can be attained with a small hydraulic press. The hydraulic system can handle more horsepower for its size than either of the other systems.

5. ECONOMY—As the result of simplicity and compactness, cost for the amount of power transmitted is relatively low. Also, power and frictional losses are comparatively small.

6. SAFETY—There are fewer moving parts such as gears, chains, belt and electrical contacts than in other systems. Controlling overloads with relief valves is much simpler than the troublesome overload devices used on the other systems.

7. DURABILITY—With proper care and fluids, they can outlast the other systems.

DISADVANTAGES

1. EFFICIENCY—While the efficiency of the hydraulic system is much better than the electrical system, it is not as efficient as most mechanical systems of transmitting power.

2. NEED FOR CLEANLINESS—Failure to use proper cleanliness and maintenance practices can result in damage from rust, corrosion, dirt, heat and breakdown of fluids.

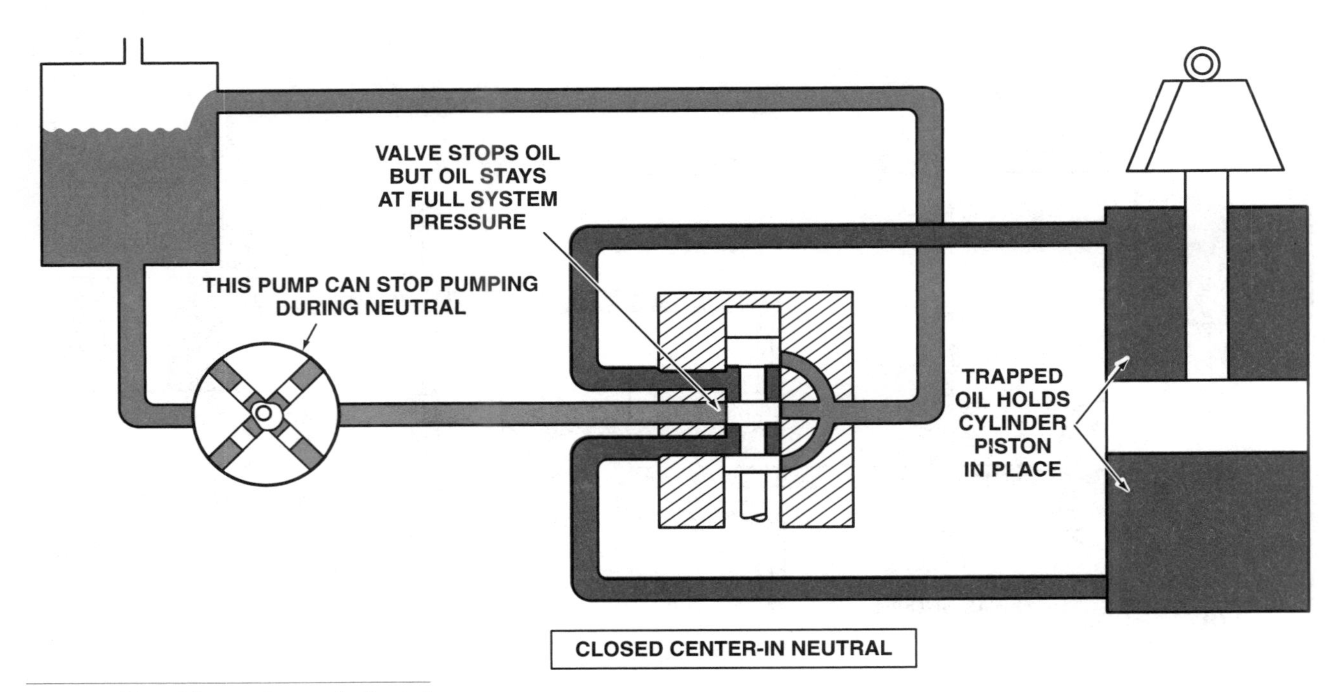

Fig. 21 – Closed-Center System in Neutral

OPEN-CENTER and CLOSED-CENTER HYDRAULIC SYSTEMS

There are many types of systems used on today's equipment. They fall into two major types determined by the type of control valves they use.

- **•Open-Center Systems**
- **•Closed-Center Systems**

The simple hydraulic system, which we developed earlier in this chapter (Fig. 19), is what we call an OPEN-CENTER SYSTEM. This system requires that the control valve spool be open in the center to allow pump flow to pass through the valve and return to the reservoir when the valve is in neutral. The pump we have used supplies a constant flow of oil and the oil must have a path for return to the reservoir when it is not required to operate a function.

In the CLOSED-CENTER SYSTEM, the control valve, on the other hand, blocks pump flow when the valve is in neutral. There is no passage to the reservoir while in neutral. It, therefore, requires a pump or a system that supplies oil only when needed. This can be done by using a pump that shuts itself off (takes break) when oil is not required to operate a function.

The open-center system is shown in neutral position in Fig. 19. It is shown in the operating position in Fig 20. Note that in neutral, pump oil flow is directed through the valves **open center** to the reservoir

CLOSED-CENTER SYSTEM

Let's look at a closed-center system with a variable displacement pump.

In neutral, Fig 21, the pump pumps oil until pressure rises to a predetermined level. Then a pressure-regulating valve allows the pump to shut itself off and to maintain this pressure to the valve.

When the control valve is operated as shown in Fig. 22, oil from the pump is sent to the bottom of the cylinder.

The drop in pressure caused by connecting the pump pressure line to the bottom of the cylinder causes the pump to go back to work, pumping oil to the bottom of the piston and raising the load.

When the valve was moved, the top of the piston was connected to a return line, thus allowing return oil forced from the piston to be sent to the reservoir or back to the pump.

When the valve is returned to neutral, oil is again trapped on both sides of the cylinder and the pressure passage from the pump is dead-ended. At this time, the pump again takes a break.

Moving the spool down (not shown), directs oil to the top of the piston, moving the load downward. The oil from the bottom of the piston is sent into the return line.

With the closed-center system, if the load exceeds the predetermined standby pressure or if the piston reaches the end of its stroke, the pressure build-up simply tells the pump to take a break, thus eliminating the need for relief valves to protect the system.

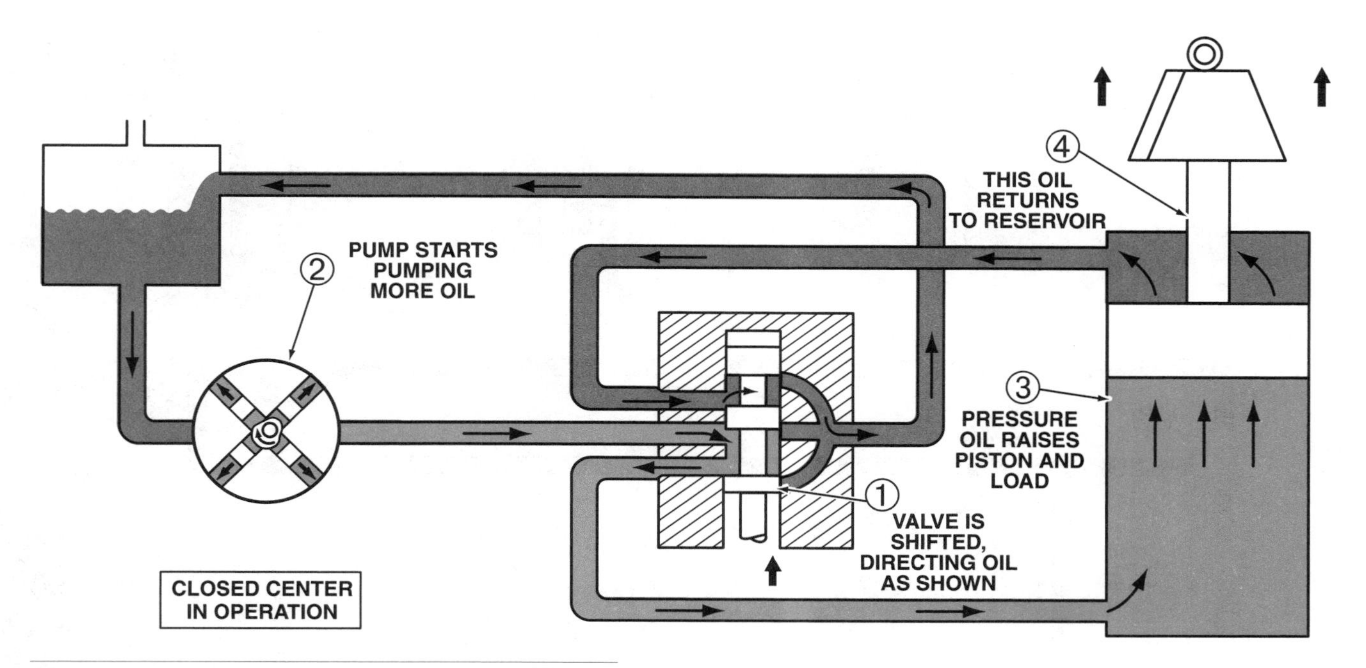

Fig. 22 - Closed-Center System in Operation - Raising a Load

We have now built the simplest of open- and closed-center systems. Most hydraulic systems, however, require their pump to operate more than one function.

Let's look at how this is done and compare the advantages and disadvantages of each system.

VARIATIONS ON OPEN- AND CLOSED-CENTER SYSTEMS

To operate several functions at once, hydraulic systems have the following connections:

OPEN-CENTER SYSTEMS

- *Fixed Displacement Pump with Open-Center Valves Connection in Series*
- *Fixed Displacement Pump with Open-Center Valves Connection in Series Parallel*
- *Fixed Displacement Pump with Flow Divider and Open-Center Valves*
- *Variable-Displacement Pump with Open-Center Valves*

CLOSED-CENTER SYSTEMS

- *Fixed Displacement Pump with Accumulator and Closed-Center Valves*
- *Fixed Displacement Pump with Priority Valve and Closed-Center Valves*
- *Pressure Sensing Variable Displacement Pump with Closed-Center Valves*
- *Load Sensing Variable Displacement Pump with Closed-Center Valves*

CLOSED-LOOP SYSTEMS

Let's discuss each of these systems.

FIXED DISPLACEMENT PUMP WITH OPEN-CENTER VALVES CONNECTED IN SERIES

A multiple valve system with valves connected in series is shown in Fig 23. With all the valves in the neutral position, oil from the pump passes through each of the valves in series and returns it to the reservoir as shown by the arrows.

When a control valve is operated, incoming oil is diverted to the cylinder that the valve serves. Return oil from the cylinder is directed to the return line and on to the next valve. This system is satisfactory as long as only one valve is operated at a time. In this case, full pump flow and pressure are available to that cylinder.

However, if two or more valves are operated at the same time as in Fig. 24, all the oil goes to the first cylinder. Return oil from that cylinder is routed to the second valve. When this happens, the system pressure will be the sum of the pressures required to operate each of the activated cylinders.

In Fig. 24, if the pressure required to move the load on No. 1 cylinder is 900 psi(6205 kPa) and 1100 psi(7585kPa) is required for load on No. 2, the total system pressure would be 2000 psi (13,789 kPa). The return pressure from No. 1 cylinder would be 1100 psi (7584 kPa). Therefore, it would require 2000 psi(13,789 kPa) to overcome this pressure and supply the 900 psi (6205 kPa) required to move the No. 1 load.

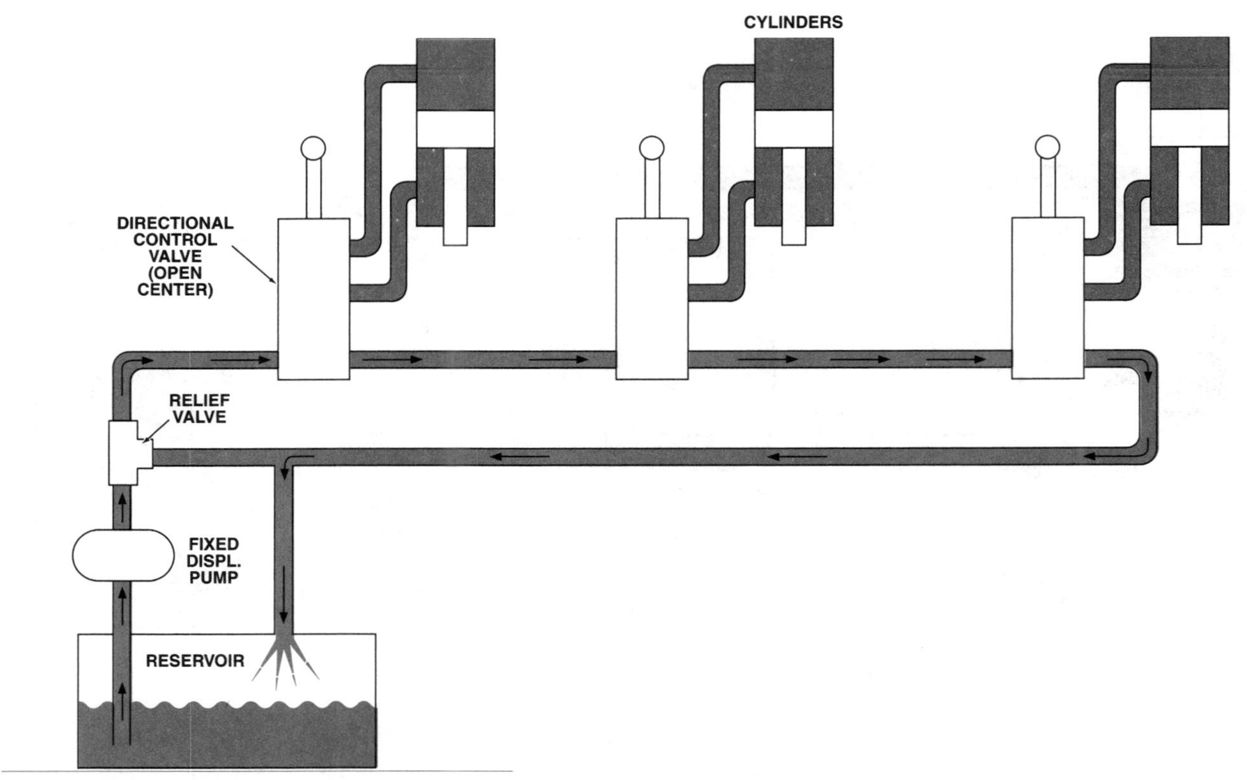

Fig. 23 - Open-Center System with Series Connection

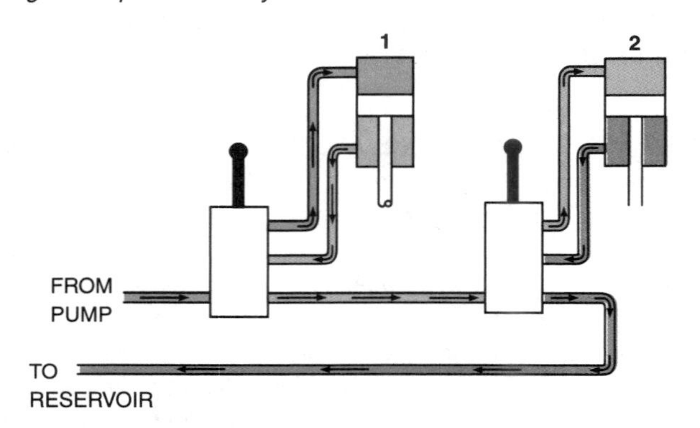

Fig. 24 - Series Valves Activated

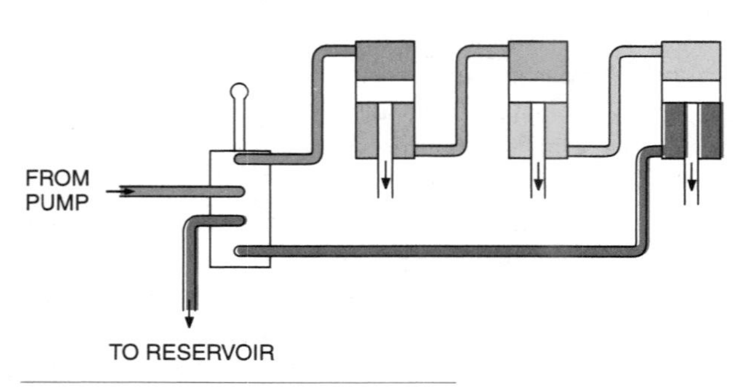

Fig. 25 - Cylinders Connected in Series

A more common application of the series hookups is shown in Fig. 25. A single valve operates several cylinders that are connected in series. Like the example used above, the system pressure is the sum of the pressures required to move each of the individual loads.

To keep the cylinders synchronized, the cylinders would bypass oil when fully extended. This would refill all cylinders in the event there was any leakage. This type of hookup is one that might be used where several functions need to be synchronized, such as raising a disc and its wing sections.

FIXED DISPLACEMENT PUMP WITH OPEN-CENTER VALVES WITH SERIES-PARALLEL CONNECTION

The system shown in Fig. 26 is the most common one used on open-center hydraulic systems. All of the valves in this system are commonly found in a single assembly. The individual valve sections are either bolted together (Stack Valve) or cast into a single housing (Unibody).

Oil from the pump is routed through the control valve in series. When all the valves are in neutral, this passage is open and allows oil to return to the reservoir. When any of the valves are activated, this passage is blocked.

There is also a parallel passage that connects to the valve inlet. This passage runs past all the valve spools and dead ends at the last section.

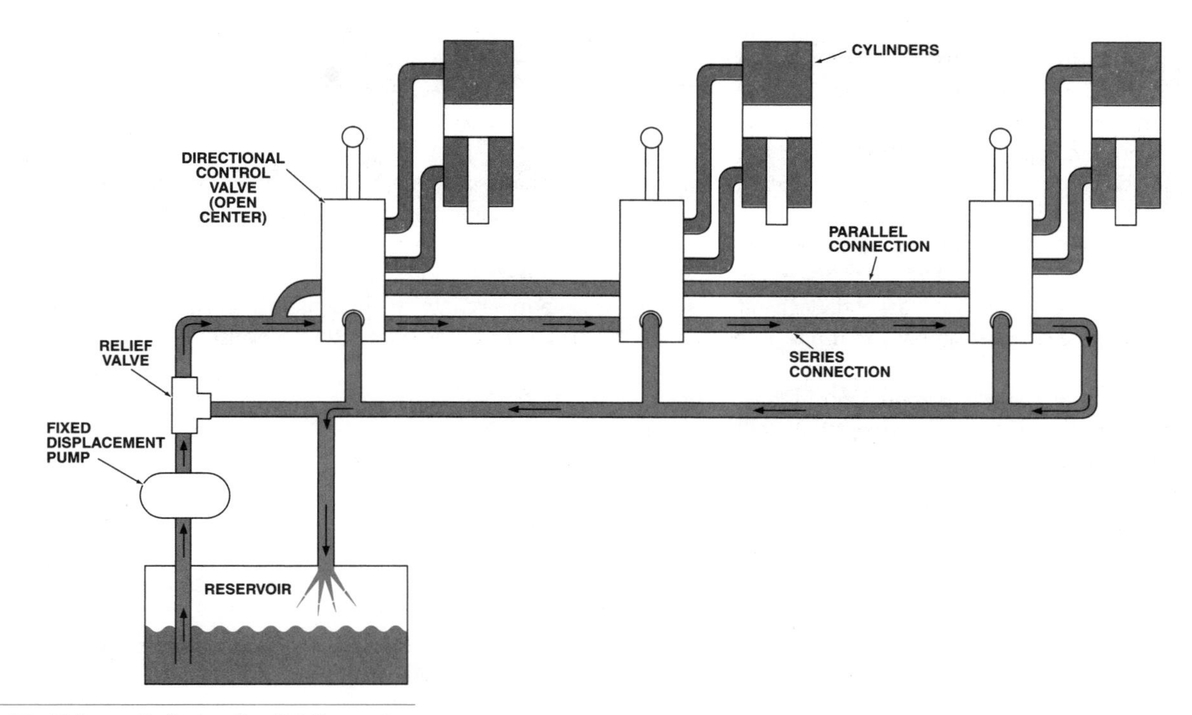

Fig. 26 - Valves with Series-Parallel Connection

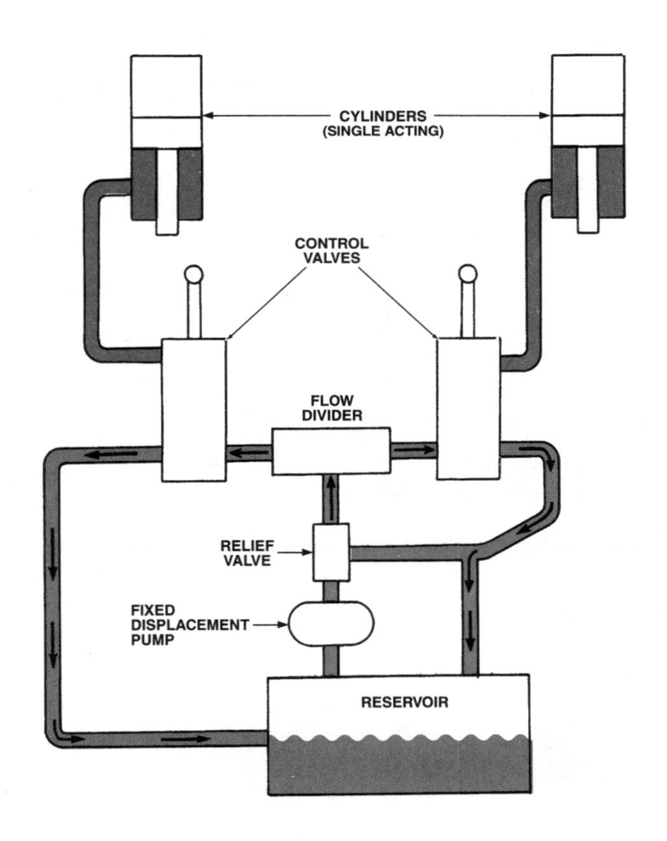

Fig. 27 - Open-Center System with Flow Divider

When any of the valves are operated, the neutral passage is blocked and the oil from the parallel passage is routed to a cylinder port of that valve. Oil in the parallel port is available equally to all of the valve spools.

When two or more valves are operated, the cylinder, which requires the least pressure, will move first and then the one requiring the next highest pressure.

In operation, the operator can meter the oil to the functions so that they will all move at the same time.

This valve can also be used on a closed-center system by blocking the neutral (series connection) passage.

FIXED DISPLACEMENT PUMP WITH FLOW DIVIDER AND OPEN-CENTER VALVES

Fig. 27 shows a flow divider used with an open-center system. The proportional flow divider splits the pump output flow and sends it proportionally to two different circuits. The percentage is designed into the valve and the flow will always go to each valve in those proportions regardless of the activity of the other valve. Those proportions can be 50%-50%, 25%-75%, etc.

In this type of system, the pump will always have to pump all the oil against the highest pressure needed in either circuit. If we had a 25%-75% divider and the "25%" circuit was operated and required 1500 psi (10,500 kPa), the pump would have to pump all the oil against that pressure even though only 25% is needed. The other 75% of the oil would lose it's pressure as it went through the flow valve. Because this pressure loss did not accomplish work, 75% of the energy put into pumping the oil would go into heat and only 25% into work.

A priority type flow divider can also be used in this system. This valve directs a specific flow of oil to the primary circuit before any is allowed to go to the secondary circuit. In some cases, the secondary circuit is simply a return to resérvoir.

This system is used primarily where pump output varies greatly because of engine speeds and a constant flow of oil is required for the function, i.e. automotive power steering.

The same inefficiencies exist as described in the proportional dividers. For this reason, they are used primarily on functions that are not cycled frequently.

VARIABLE DISPLACEMENT PUMP WITH OPEN-CENTER VALVES

These systems are being used on machines like excavators on which the total engine output is used through the hydraulic system.

This system uses a variable displacement pump, which does not go completely out of stroke. Because there is always some oil being pumped, it is connected to a multiple section series-parallel open-center valve.

In Fig. 28, we show a typical system using a bent-axis axial piston pump and a multiple section series-parallel open-center valve. The valve spools are often operated hydraulically by pilot controllers.

When no valve is operated, the pump is held in the minimum flow position. Oil flows through the valves neutral passage and back to the reservoir. When any valve is operated, the controller allows the pump to increase its output. It then determines how much oil the hydraulic system needs and regulates the pump output flow to satisfy the requirements of the system. To control the flow, the controller receives the following signals:

1. Valve inlet pressure.

2. The highest work port pressure in the valve.

3. Reservoir return flow from the neutral passage of the valve.

4. The position of the valve spools (senses the highest Pilot Controller operating pressure).

Some systems may also sense engine speed and the output of a second hydraulic system.

Most systems use pressure lines to send the sensing pressures to the pump controller, however, some use electronic sensors and send signals to an Electronic Control Unit (ECU). The ECU

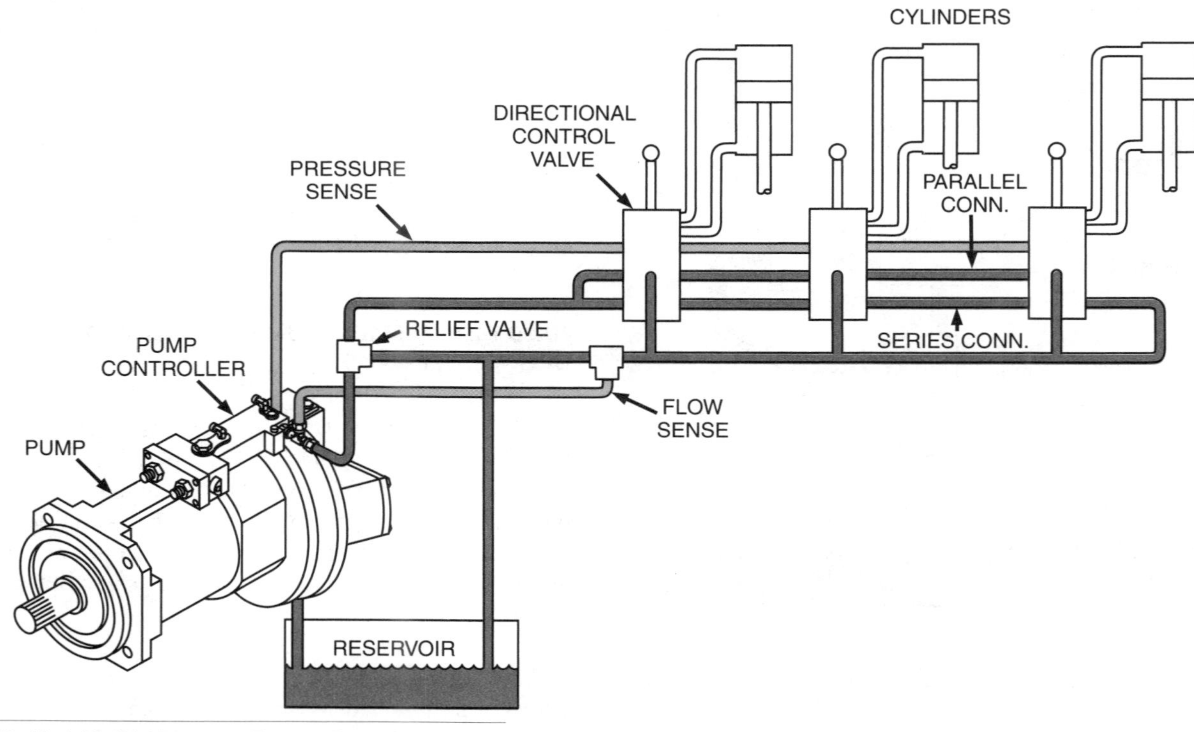

Fig. 28 - Variable Displacement Pump - Open-Center Valve

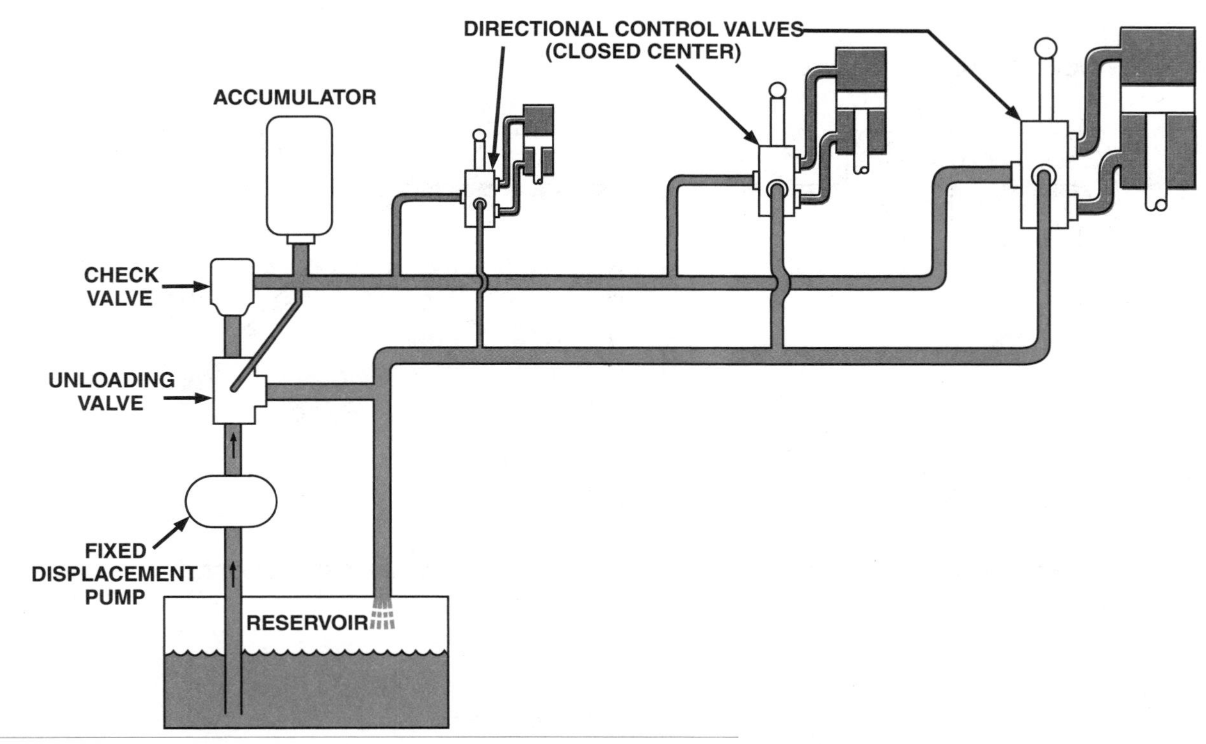

Fig. 29 - Closed-Center System with Fixed Displacement Pump and Accumulator

controls a variable electric solenoid that in turn operates a valve controlling the pump output.

These systems often have a load-limiting feature. On machines, such as excavators and backhoes, it is desirable to use the largest possible pump so that maximum production can be obtained at normal operating pressures.

With this high flow, the engine can be overloaded when full pressure and full flow are required. The load limiting valve senses the system operating pressure and slightly limits the pump output flow as maximum pressure is reached so the engine is not overloaded.

CLOSED-CENTER SYSTEMS

CLOSED-CENTER SYSTEM WITH FIXED DISPLACEMENT PUMP AND ACCUMULATOR

This system is shown in Fig. 29. A small fixed displacement pump charges an accumulator. When the accumulator is charged to full pressure, an unloading valve diverts the pump flow back to the reservoir. The check valve traps pressure oil in the accumulator circuit.

When a control valve is operated, the accumulator discharges its oil and actuates the cylinder. As pressure in the accumulator begins to drop, pump flow is again directed by the unloading valve to recharge the accumulator.

This system, using a small capacity pump, is effective only when operating oil is needed for short periods of time. Typical uses for this system would be for brakes, differential locks, rockshafts, etc. It is not practical when a steady flow of oil is needed.

FIXED DISPLACEMENT PUMP WITH PRIORITY VALVE AND CLOSED-CENTER VALVE

This system is shown in Fig. 30. The output of the fixed displacement pump is directed to a priority valve. The priority valve divides that flow to a primary circuit, which has a closed-center valve and a secondary circuit, which has an open-center valve.

In our example, the priority valve restricts oil flow to the open-center loader circuit to maintain specific pressures in the closed-center steering circuit. Oil not needed for steering is allowed to go to the secondary circuit(s).

A sensing line connected to the primary (steering) valve work port(s) will let the priority valve know what pressure is needed to move the primary load. The priority valve will restrict secondary flow enough to always maintain a higher pressure to the primary valve than it takes to move the load.

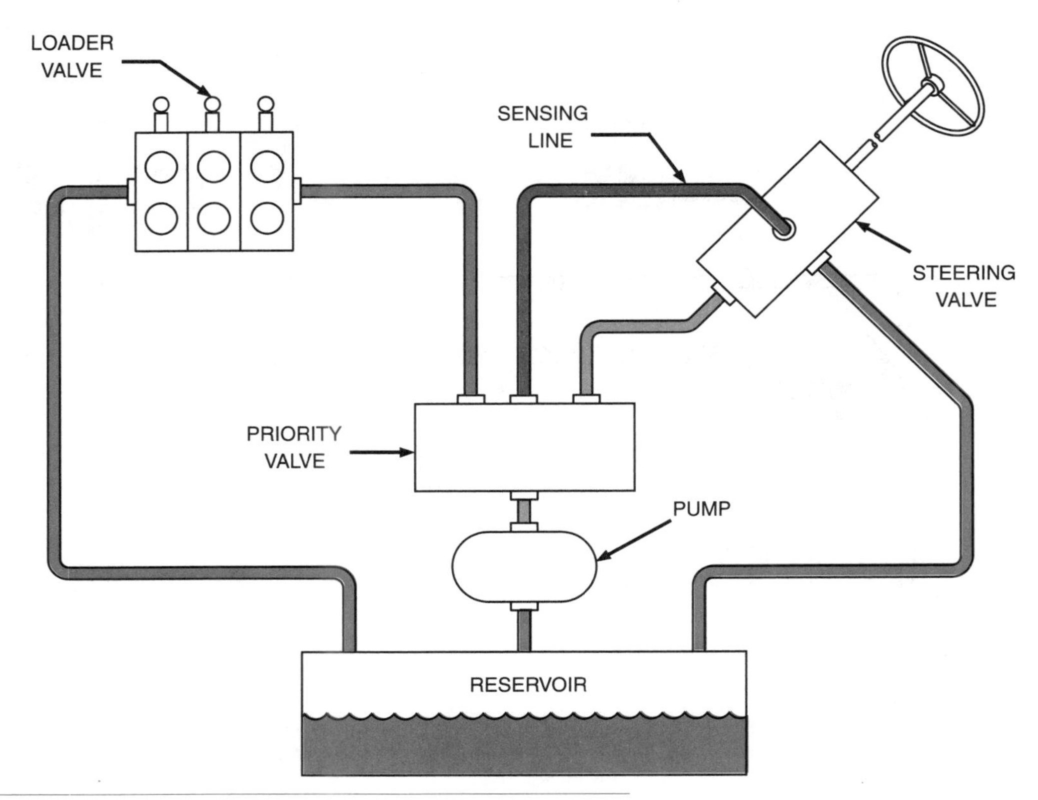

Fig. 30 - Fixed Displacement Pump with Priority Valve and Closed-Center Valve

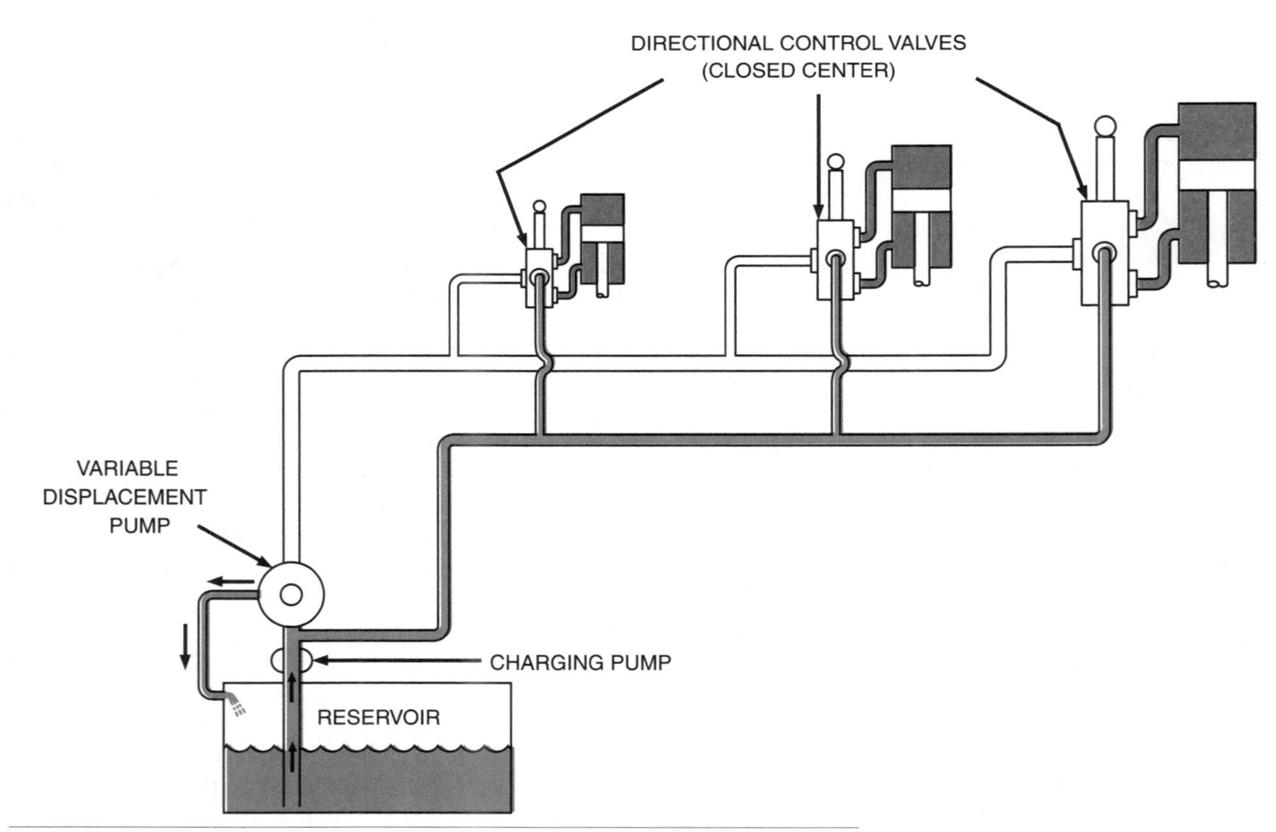

Fig. 31 - Pressure Compensated Variable Displacement Pump with Closed-Center Valves

In our example, with the steering in neutral, the sensing pressure would be 0.0. A spring in the priority valve is adjusted to maintain a standby pressure to the steering. For our example, we'll use a standby pressure of 100-psi (700 kPa).

When the steering valve is activated, the valve working pressure will be directed, by the sensing line, to the spring area of the priority valve. The priority spool will now have to move against this working pressure, plus the spring force, before oil will go to the loader circuit. This means that the valve will always maintain a pressure to the steering valve which is 100 psi (700 kPa) greater than the pressure required to steer the wheels.

When the loader is being used, the priority valve will reduce the higher pressure to maintain the proper pressure to the steering (See Chapter 5 for valve operation).

PRESSURE SENSING VARIABLE DISPLACEMENT PUMP WITH CLOSED-CENTER VALVES

This system is shown in Fig. 31. The variable displacement pump supplies oil to the valves that are closed, or "Blocked" in the neutral position. The pump will pump oil until the system pressure reaches "Standby" which is the pressure at which the pump shuts itself off. This is the maximum pressure that occurs in the system and could be compared to the relief valve setting of the open-center systems. In this system, the pump senses its outlet pressure and always tries to maintain that pressure ("Standby").

When a valve is actuated, flow will begin. This will cause a drop in system pressure. The pump senses this drop and goes into "Stroke", pumping enough oil to try to maintain standby pressure. That flow will satisfy the needs of the machine function. Return oil from the cylinder is directed by the valve to the return line, which carries the oil to the reservoir, or to the inlet of the pump.

When the valve is returned to the neutral position or the cylinder reaches the end of its travel, pressure builds to standby and the pump stops pumping.

When the pressure required to move the load exceeds standby, the pump will go out of stroke and maintain that pressure on the cylinder.

Because the pump may be mounted above or away from the reservoir, it is often necessary to use a charge pump to supply oil to the hydraulic pump. Fig. 31 shows a charge pump supplying reservoir oil to the main pump. The charge pump has much less capacity than the variable displacement pump so return flow from the valves is returned to the main pump inlet. The charge pump, therefore, needs only to supply the system with makeup oil. On machines, however, this oil is usually used to operate a hydraulically engaged transmission or reversor, supply cooler oil flow and to pressure lubricate transmissions.

LOAD SENSING VARIABLE DISPLACEMENT PUMP WITH CLOSED-CENTER VALVES

This system shown in Fig. 32 has all of the advantages of the pressure sensing system. It has a standby pressure setting that limits the maximum pressure of the system. In addition, it has a second stroke control valve that controls the operating pressure at slightly more than required to move the load.

The spring area of this low-pressure valve is connected to the control valve with a sensing line. The sensing passage of the control valve, through check valves, senses the highest workport pressure.

When there is no hydraulic function operated, there is no sensing pressure in the spring area of the low-pressure stroke control valve. The pump will maintain a pressure established by the spring force. For our example, we'll use 300 psi (2070 kPa) for this low standby pressure. There is no flow at this time and a pressure of 300-psi (2070 kPa) is maintained at the closed control valve.

When a valve(s) is operated, the pressure in the sensing passage and line will be that of the highest pressure needed to lift the load(s). This pressure is fed to the spring side of the low stroke control valve.

In order to destroke the pump, pressure will have to overcome the spring force (300-psi)(2070 kPa) plus the pilot working pressure. The pump will maintain a pump flow which is adequate to keep the pump outlet pressure 300 psi (2070 kPa) higher than the highest pressure needed to lift the load.

For example, if valves were activated, one to a load requiring 1500 psi (10,340 kPa), the other 1200 psi (8275 kPa). The sensing line and spring area of the low pressure stroke control valve would read 1500 psi (10,340 kPa). The pump would then go into stroke and maintain enough flow to keep the pump outlet pressure at 1800 psi (12,410 kPa)

The high-pressure stroke control valve functions only when the system pressure reaches the maximum pressure established for the system.

When this system is used on machines, such as backhoes, which use full engine power through the hydraulic system, the pump can be equipped with a load-limiting valve. This allows the use of a large pump to supply plenty of oil for normal operating pressures. When high pressures are required, the load limiter reduces the maximum pump flow so that the engine is not overloaded. This feature was also described in the "Variable Displacement Pump with Open-Center Valves".

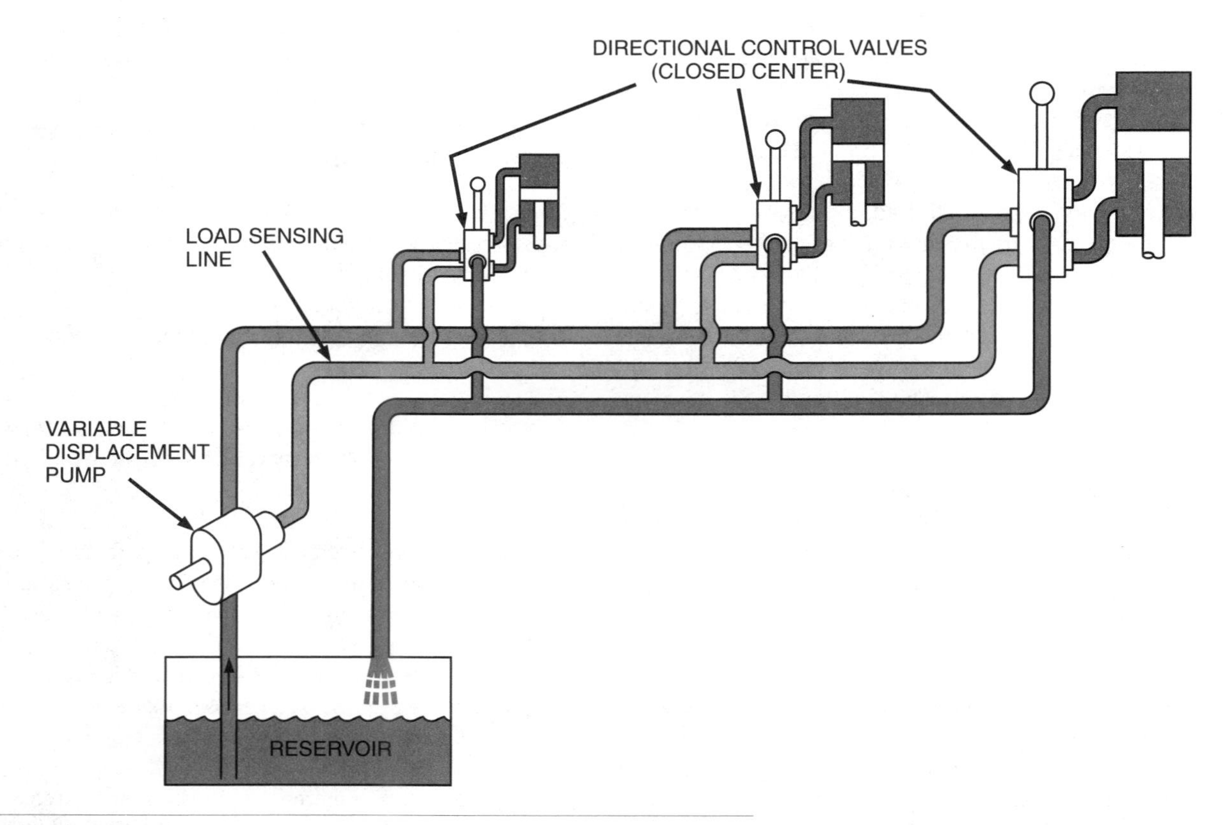

Fig. 32 - Flow Compensated Variable Displacement Pump with Closed-Center Valves

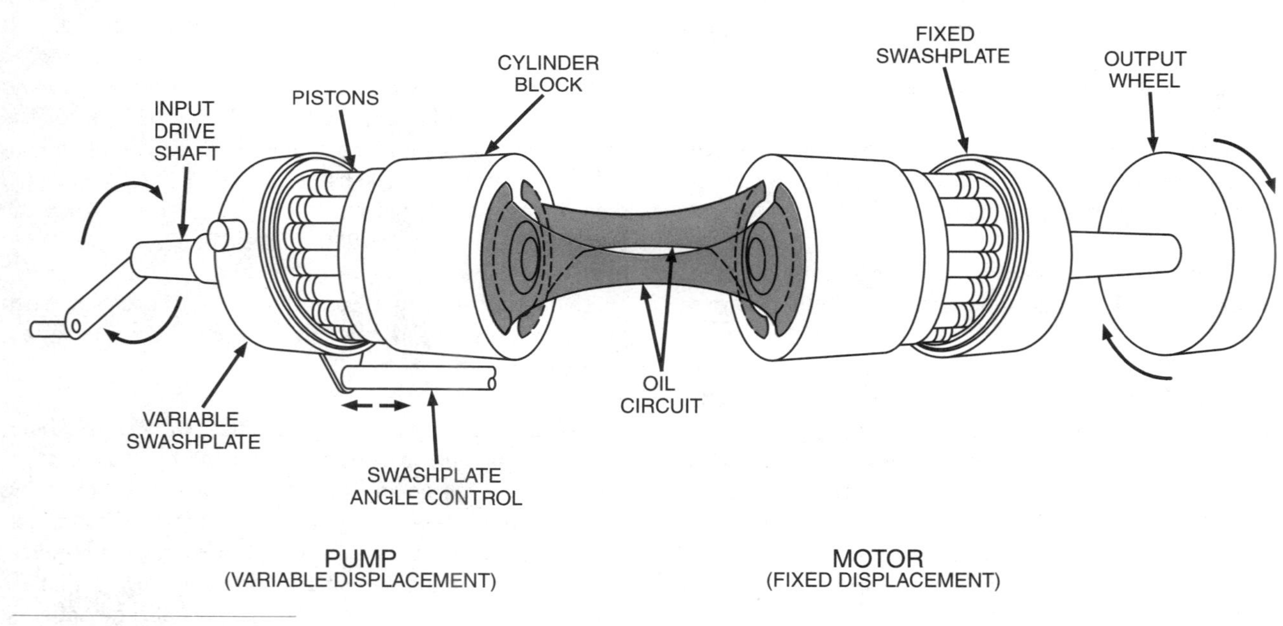

Fig. 33 - Closed Loop System

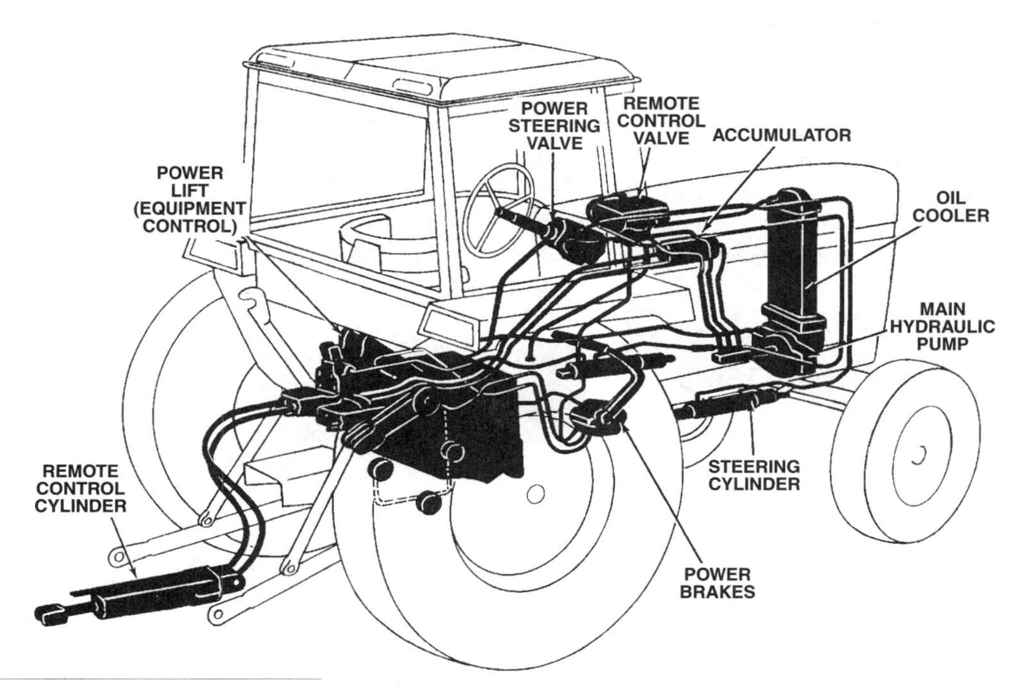

Fig. 34 - A Modern Tractor with Hydraulic System

SOME ADVANTAGES OF CLOSED-CENTER SYSTEMS

1. There is no requirement for relief valves in a basic closed-center system because the pump simply shuts itself off when standby pressure is reached. This prevents heat build-up in systems where relief pressure is frequently reached.

2. The size of lines, valves, and cylinders can be tailored to the flow requirements of each function. Components in the open-center system must be sized to accept the total output flow of the pump.

3. By using a larger pump, reserve flow is available to insure full hydraulic speed at low engine rpm with no loss of efficiency because the pump supplies only the oil required by the function(s) being operated.

4. On functions such as brakes and differential locks that require force but very little or no flow, this system is very efficient. By holding the valve open, standby pressure is constantly applied to the piston with no loss of efficiency because the pump has returned to standby. To accomplish this in an open-center system, pressure could be maintained only by pumping oil past a relief valve.

We saw earlier that the open-center system is the simplest and least expensive for hydraulic systems that have only a few functions or where the entire pump output can be used in any of the valves. But today's machines need more versatile hydraulic systems as more functions with varying demands for each function are added.

To meet those varying demands, the open-center system requires the use of flow dividers to proportion the oil flow. The use of these inefficient flow dividers results in reduced useable power and creates heat build-up in the system.

The trend has been to use the closed-center systems to better meet the needs of today's varying flow applications.

CLOSED LOOP SYSTEMS

The closed loop system shown in Fig 33 uses no valves to direct the high-pressure oil. A pump and a motor are connected together by hoses, lines or passages.

In this illustration, a variable displacement pump is used with a fixed displacement motor. This means that pump flow can vary from zero, when the swashplate has no angle, to maximum capacity when the swash plate is fully tilted. A swashplate angle control operates the swashplate. The operation of the pumps and motors are covered in Chapters 4 & 7 respectively.

The pump can be equipped with a reversible swashplate that means it can be tilted in either direction. When the swashplate is reversed, oil flow is reversed. The return passage becomes the pressure passage and the pressure passage becomes the return passage. The direction of the motor is reversed.

This system can use any combination of fixed, variable and reversible pumps with fixed or variable displacement motors.

THE USES OF HYDRAULICS

Hydraulics are used at many points on a single machine. The tractor in Fig. 34 uses hydraulics to steer, brake, control mounted equipment, and supply oil for remote operation of tools. A single hydraulic system serves to power all these functions.

Let's briefly discuss the major uses of hydraulics.

HYDRAULIC STEERING SYSTEMS

Three major types of steering are used for today's machines:

1. MANUAL STEERING

2. POWER STEERING

   a. Hydraulic steering with mechanical drag link

   b. Hydrostatic steering

   c. Metering pump steering

3. HYDRAULIC ASSIST STEERING

1. **MANUAL STEERING** - The steering wheel is linked directly to the turning wheels and the operator does all the work of steering. No hydraulics are used—only mechanical effort.

2. **POWER STEERING** - These systems are divided into three major categories.

A. HYDRAULIC STEERING WITH MECHANICAL DRAG LINK

Fig. 35 illustrates hydraulic steering with a mechanical drag link. We are showing it with the open-center spool valve, however, it is equally adaptable to the closed-center system. A rotary valve could also be used on either open or closed-center systems. A rotary valve could be mechanically connected to the wheel linkage (This would not contain the follow-up motor described in Fig. 37). A poppet valve could also be used with a closed-center system.

Operation is shown during a right turn. In the right turn, the operator turns the steering wheel as shown. Because of the resistance in turning the front wheels, the shaft is forced up out of the worm nut. This shifts the spool valve and the steering shaft up, which directs oil to the cylinder at the front wheels. This cylinder rotates a rack and pinion device that turns the front wheels. Oil from the other side of the steering cylinder is returned through the spool valve to the reservoir as shown.

As long as the steering wheel is turned, oil will continue to move the wheels. As soon as the steering wheel motion is stopped, the hydraulic pressure will turn the wheels slightly further to the right, moving the steering linkage forward and pulling the valve back to the neutral position.

When turning to the left, the valve spool is pulled down as the shaft is threaded into the worm nut. This sends oil to the other side of the steering cylinder turning the wheels to the left.

B. HYDROSTATIC STEERING

Hydrostatic steering has no mechanical connection between the steering valve and the steering cylinders. Basically the operation is the same as that just described except that we have a hydraulic "drag link" instead of a mechanical one.

Fig. 36 shows a hydrostatic steering system used with a poppet valve on a closed-center hydraulic system. Operation is shown during a right turn.

When the operator turns the steering wheel to the right, the steering shaft, which is threaded through the steering valve piston, attempts to pull this piston upward. Because oil is trapped in the circuit at this time, the shaft instead moves the collar downward, rotating the pivot lever and opening a pressure and a return valve.

When the valves open, pressure oil enters the steering valve cylinder, forcing the piston upward. This pushes the oil out of the valve cylinder and into the right-hand steering cylinder, turning the front wheels to the right.

As the wheels turn, oil is forced out of the left-hand steering cylinder and returns through the open return valve to the reservoir or pump. This will continue as long as the operator turns the wheel.

When the operator stops turning the steering wheel, the steering shaft is moved upward by the steering valve cylinder, pulling the collar upward and centering the pivot lever, thus closing the valves.

For simplicity, a second pivot lever and a second set of pressure and return valves are not shown.

During the left turn, the steering wheel shaft moves upward as the steering wheel is turned. This moves the collar up to open the second set of valves. Oil is sent directly to the left steering cylinder. Oil from the right cylinder will go to the top of the valve piston, moving the piston, shaft and pivot lever down to neutralize the valves. Oil from the bottom of the steering piston is forced through the open return valve back to the reservoir or pump.

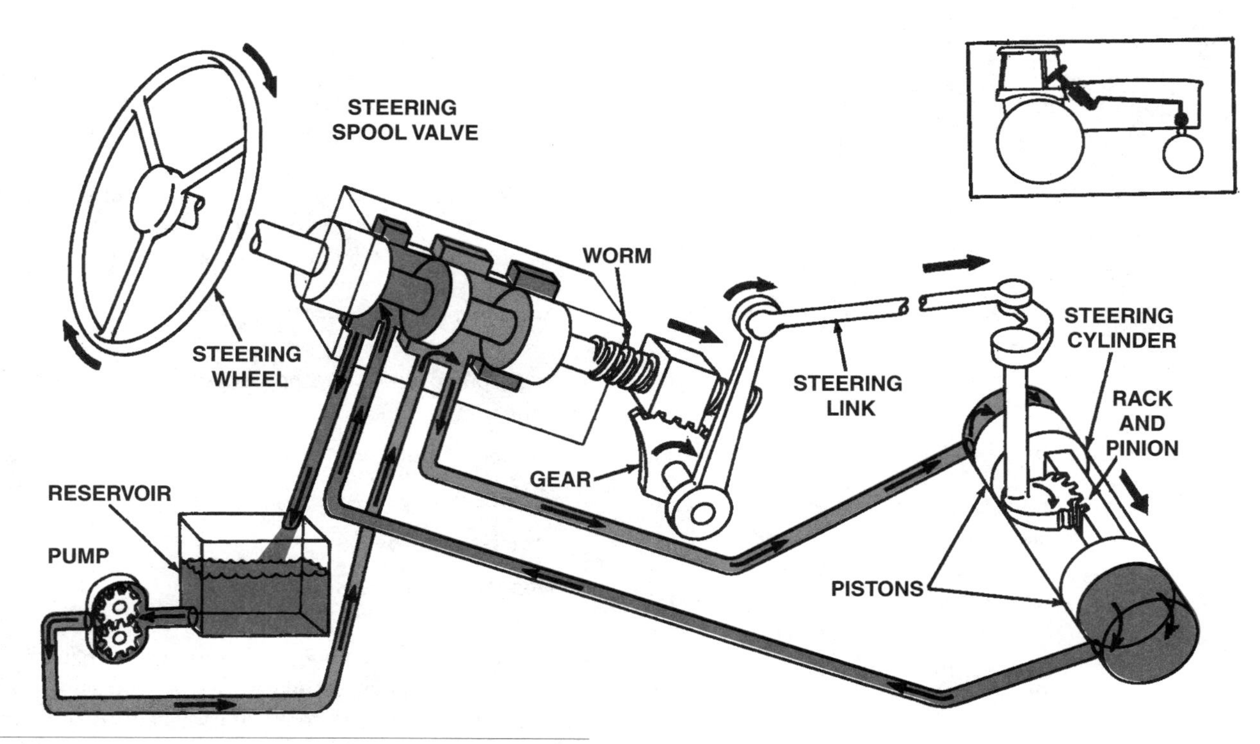

Fig. 35 - Hydraulic Steering with Mechanical Drag Link (Right turn)

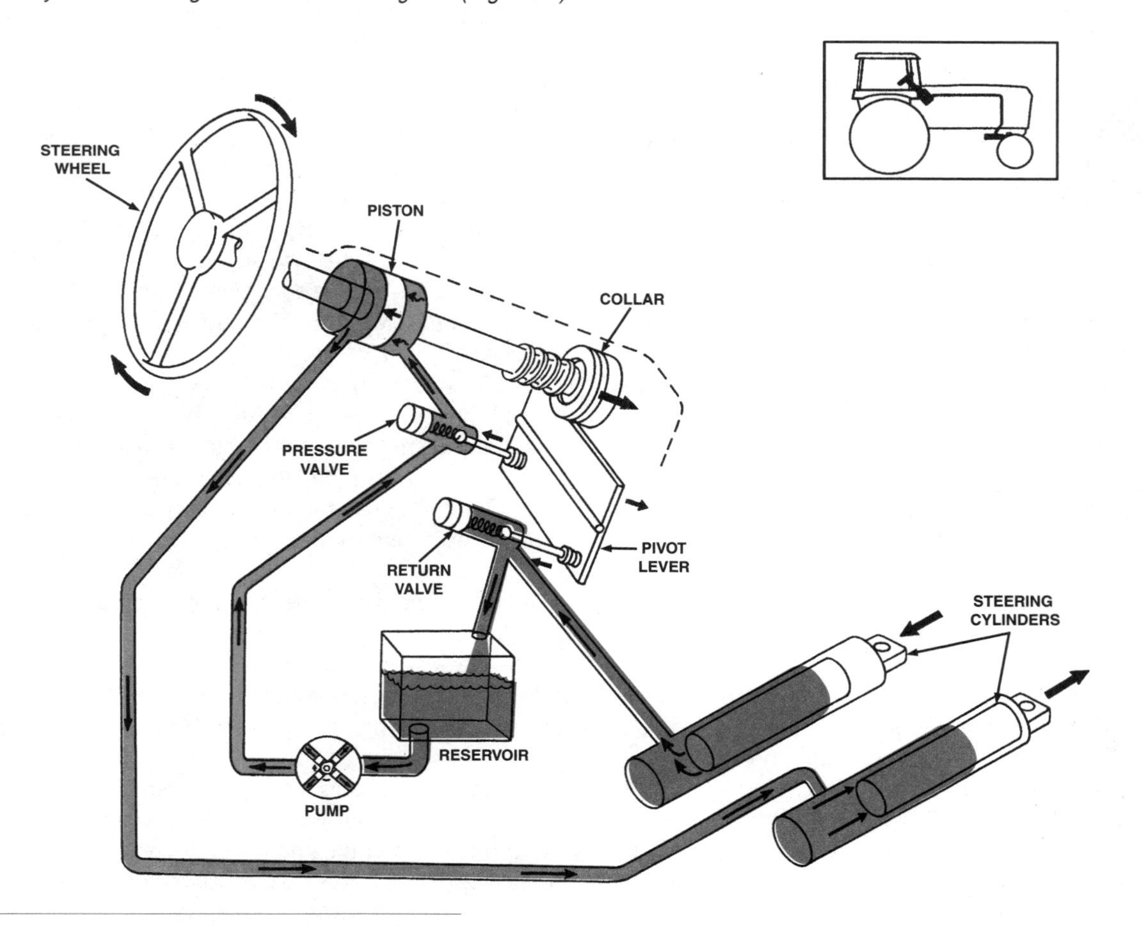

Fig. 36 - Poppet Type Hydrostatic Steering (Right Turn)

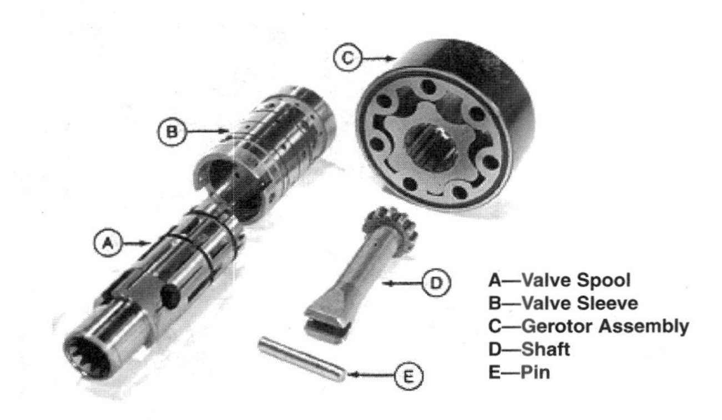

Fig. 37 - Hydrostatic Steering with Rotary Valve

The trapped oil and the follow-up piston in the steering valve provides the hydraulic drag link connection between the steering valve and the wheels. When the steering wheel has turned to the full right, the piston is at the top, the right turn cylinder is fully extended and the wheels are in the full turn position.

When in full left position the piston is at the bottom and the left turn cylinder is fully extended.

Anywhere the steering wheel is stopped in its rotation, the wheels will be in a corresponding position. This is called "Position Responsive Steering".

Manual steering is accomplished when the collar is bottomed in the housing. Turning the steering wheel further moves the piston up or down pressurizing the oil and sending it to the right or left steering cylinders.

Fig. 37 shows the components of a rotary steering valve. The steering wheel is connected to the valve spool with a splined shaft. The inner gear of the gerotor assembly is connected to the valve sleeve by the shaft and pin.

When the steering wheel is turned, the valve spool is turned inside the valve sleeve. Trapped oil in the gerotor assembly prevents the sleeve from turning. This aligns passages to send pressure oil to the gerotor and oil returning from the gerotor to a work port connected to the steering cylinder. It also connects the return passage from the steering cylinders to the reservoir return.

As the gerotor turns it also turns the sleeve trying to catch up to the spool to neutralize the valve. This causes steering as long as the steering wheel is turned and stops when the desired steering position is reached.

The pin goes through a hole in the sleeve and a slot in the valve spool. The slot allows only about 8 degrees of rotation between the valve and sleeve. When no hydraulic power is available, turning the steering wheel turns the valve, sleeve and the gerotor. The gerotor acts as a pump, pressurizing oil and sending it to the right or left steering cylinders, thus providing manual steering capability.

This valve can be designed to be used in an open-center or closed-center system. It can also be equipped with a sensing line to be used on a flow or load sensing system.

C. METERING PUMP POWER STEERING

Metering pump power steering consists of four assemblies (Fig. 38):

- **metering pump**
- **steering valve**
- **steering motor**
- **feedback cylinders**

As with hydrostatic steering, there are no mechanical connections between the steering valve and the wheels being turned. However, the metering pump steering also has a hydrostatic connection between the steering wheel and the valve, which also ties in to a follow-up system independent of the working pressure.

Note: Indications of directions refer to those as seen from the operator's seat. Side A and side B will help identify direction of movement in the steering valve housing.

Fig. 38 shows the metering pump power steering operating during a right turn. When the operator turns the wheel to the right, the gears in the metering pump direct oil in the trapped oil circuit to the steering valve housing and to the left end of the feedback piston.

This oil (under some pressure) moves the steering valve toward side B. The movement of the steering valve opens the pressure oil circuit to the left end of the steering piston. Oil from the right end of the piston flows back to the steering valve oil gallery and to the reservoir.

Oil from the right end of the feedback piston cylinder is forced out, by piston movement, and returns through the steering valve housing to the metering pump. This movement of the steering and feedback pistons from left to right causes the spindle to rotate clockwise and turn the front wheels to the right.

When the operator stops turning the steering wheel, the gears in the metering pump stop directing oil to the steering valve. Circuit pressure, from the right end of the feedback piston to the steering valve, (caused by movement of the feedback piston) acts against the side B of the steering valve. The valve

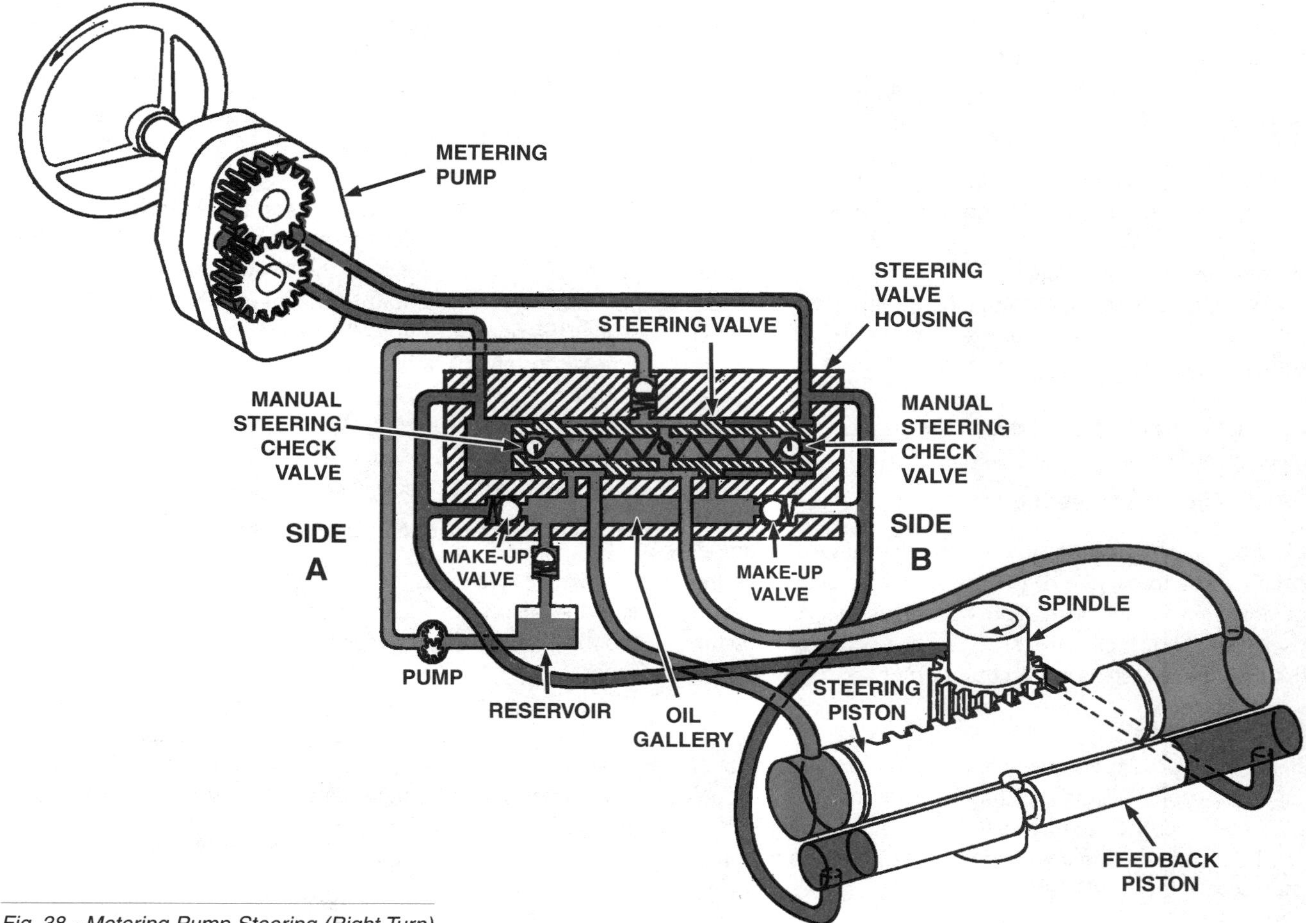

Fig. 38 - Metering Pump Steering (Right Turn)

moves toward side A, closing the pressure oil passage from the main hydraulic pump, and stops the turning movement. The valve becomes centered and traps oil in passages to both sides of the steering piston. The trapped oil holds the wheels in position until the operator again turns the steering wheel.

If oil is lost from the control circuit, pressure in the control circuit drops. The reduced pressure allows oil in the return circuit to unseat the make-up valve, filling the control circuit.

On articulated tractors, hydraulic cylinders that control steering replace the steering and feedback pistons.

Manual Turn with Metering Pump Steering

When there is no inlet pressure oil to the steering valve housing, the machine may be steered manually. Without pressure oil, the inlet check valve is seated preventing oil in the steering system from entering the hydraulic system pressure circuit.

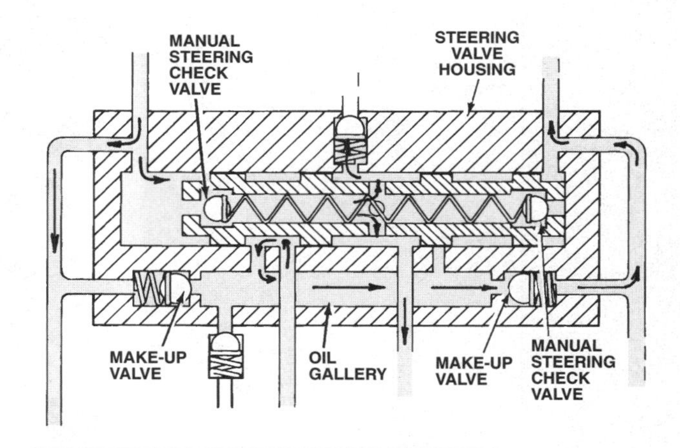

Fig. 39 - Metering Pump Steering (Manual Turn)

When the operator turns the wheel to the right (Fig. 39), oil in the trapped oil circuit is again directed to side A of the steering valve and the left end of the feedback piston. Enough pressure is exerted on side A of the steering valve to unseat the manual steering check valve in the hollow steering valve.

Oil then passes through the steering valve to the left end of the steering piston. The force of the oil on the feedback piston and steering piston moves both pistons to the right, turning the steering spindle clockwise, and turning the front wheels to the right.

Oil from the right end of the feedback piston cylinder returns through the steering valve housing to the metering pump.

Oil from the right end of the steering piston opens the make-up valve on side B and joins with oil returning from the feedback piston cylinder. This insures a recirculating oil supply in the steering circuit.

When not turning, oil is trapped in the circuits and holds the wheels in position.

3. Hydraulic Assist Steering

In hydraulic assist steering systems, steering force is an amplification of the force used by the operator. It is used where the operator must maintain a high degree of feel, such as in crawler steering. In these systems, the amount of pressure built up in the system is in direct proportion to the effort used by the operator.

Fig. 40 is a hydraulic assist system used on a crawler tractor. It is an open-center system that has a flowdivider to divide the pump flow evenly to the two valves. In neutral (lower valve), the oil enters the piston area and then goes through the open piston seat and out to return.

When the steering lever or pedal are operated, the steering valve is pulled forward restricting oil flow at the piston seat. This will cause pressure to build behind the piston moving it forward to put a force on the steering control arm. With more restriction, the pressure will rise and more force exerted on the control arm.

The land on the steering valve is slightly larger than the seat. This means that whatever pressure is built behind the piston will have a negative force on the steering valve. The operator will have to overcome that force. The higher the pressure, the higher the operator effort. So, while the force on the control arm is considerably more than the effort of the operator, it is in direct proportion to it.

If hydraulic power is lost, the valve bottoms on the piston seat and there is a mechanical linkage to the control arm.

Protection Against Failure of Power Steering

If hydraulic steering is lost in steering with a mechanical drag link, the solid link takes over and the operator can still steer the machine mechanically, but with more effort.

The same protection is also provided in most hydrostatic power steering. This is done by pressurizing trapped oil in the steering valve and using the steering valve piston, the gerotor or metering pump to build pressure and direct it to the steering cylinders.

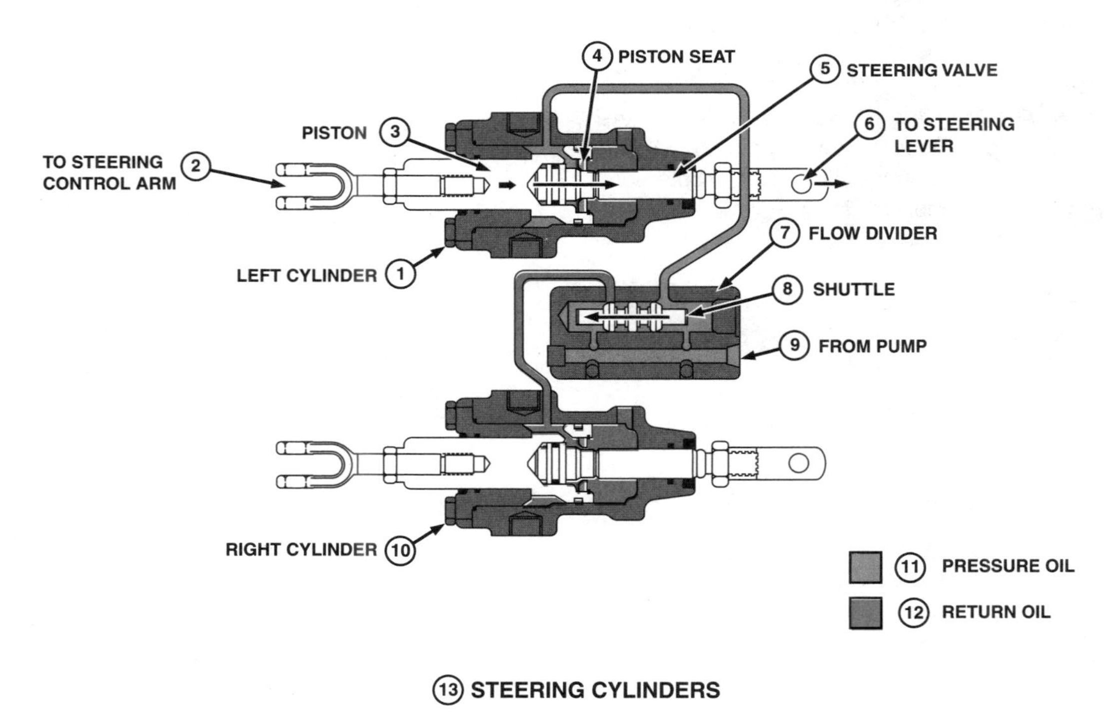

Fig. 40 - Crawler Hydraulic Assist Steering

HYDRAULIC AND POWER BRAKE SYSTEMS

Three major types of brakes are used to turn or stop farm and industrial machines:

1. Manual brakes

2. Hydraulic brakes

3. Power brakes

1. MANUAL BRAKES. When the operator applies the brakes, a mechanical linkage connects the lever or pedal to linkage which forces friction surfaces against a drum or disc to slow or stop a vehicle.

2. HYDRAULIC BRAKES. When the operator applies the brakes, a column of trapped oil is forced through a tube to a cylinder. The cylinder then forces the friction surfaces against a drum or disc. The force exerted on the lever or pedal generates all pressure in the system.

3. POWER BRAKES. When operator applies the brakes, he/she simply directs pressure oil to the brake cylinder to slow or stop the vehicle.

On some machines, two types of brakes may be used. For example, the power brakes for stopping may be backed up by a manual brake for parking. They may or may not use the same braking mechanism.

Most modern machines have parking brakes that are spring engaged and require hydraulic power to disengage them. In this case, pressurized oil is directed to the park brake to keep it disengaged while operating. To engage the brake, a valve is closed not allowing pressurized oil to reach the brake piston. In case of engine or hydraulic failure, the springs will automatically engage the brake.

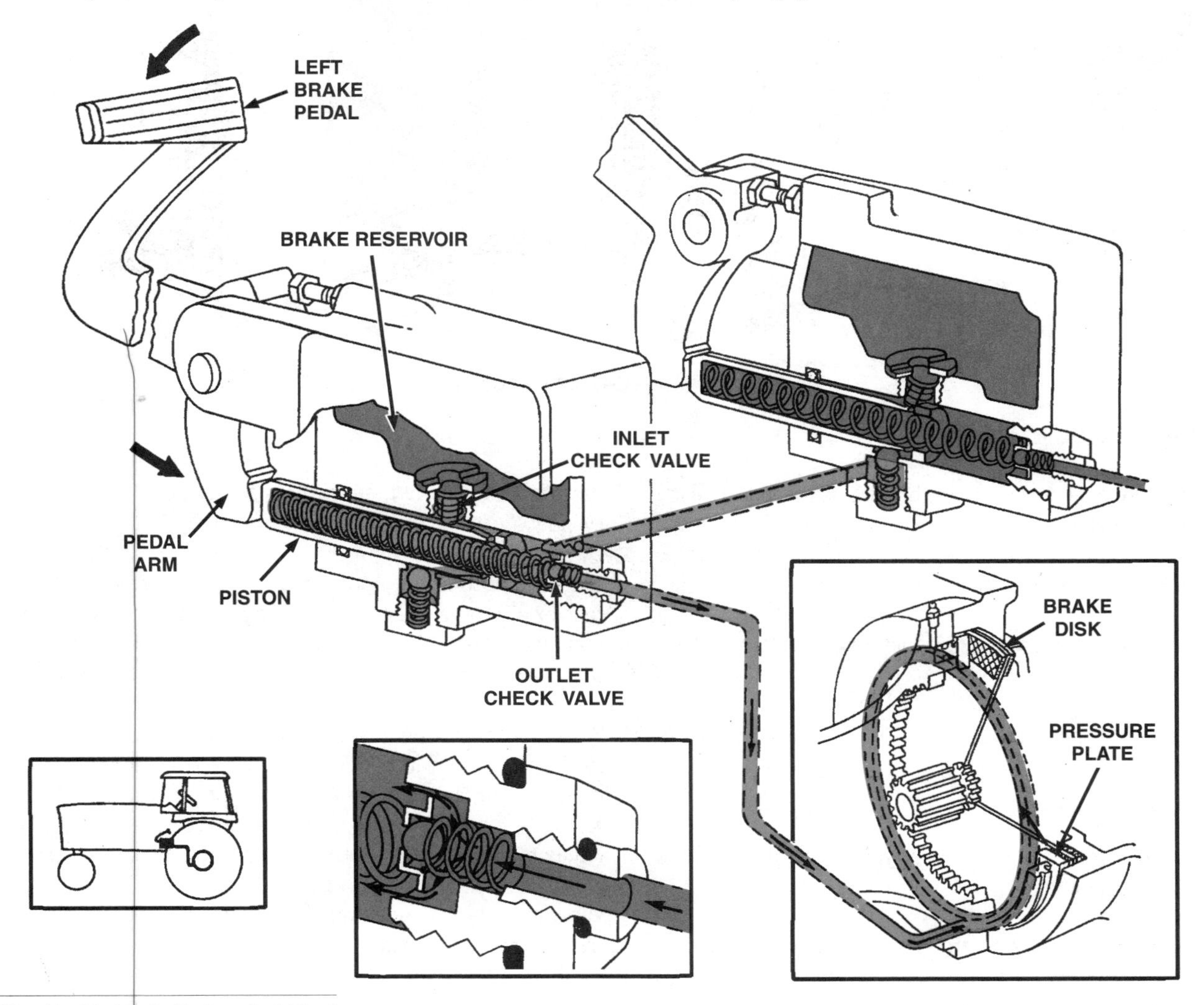

Fig. 41 - Hydraulic Brakes (Left Turn)

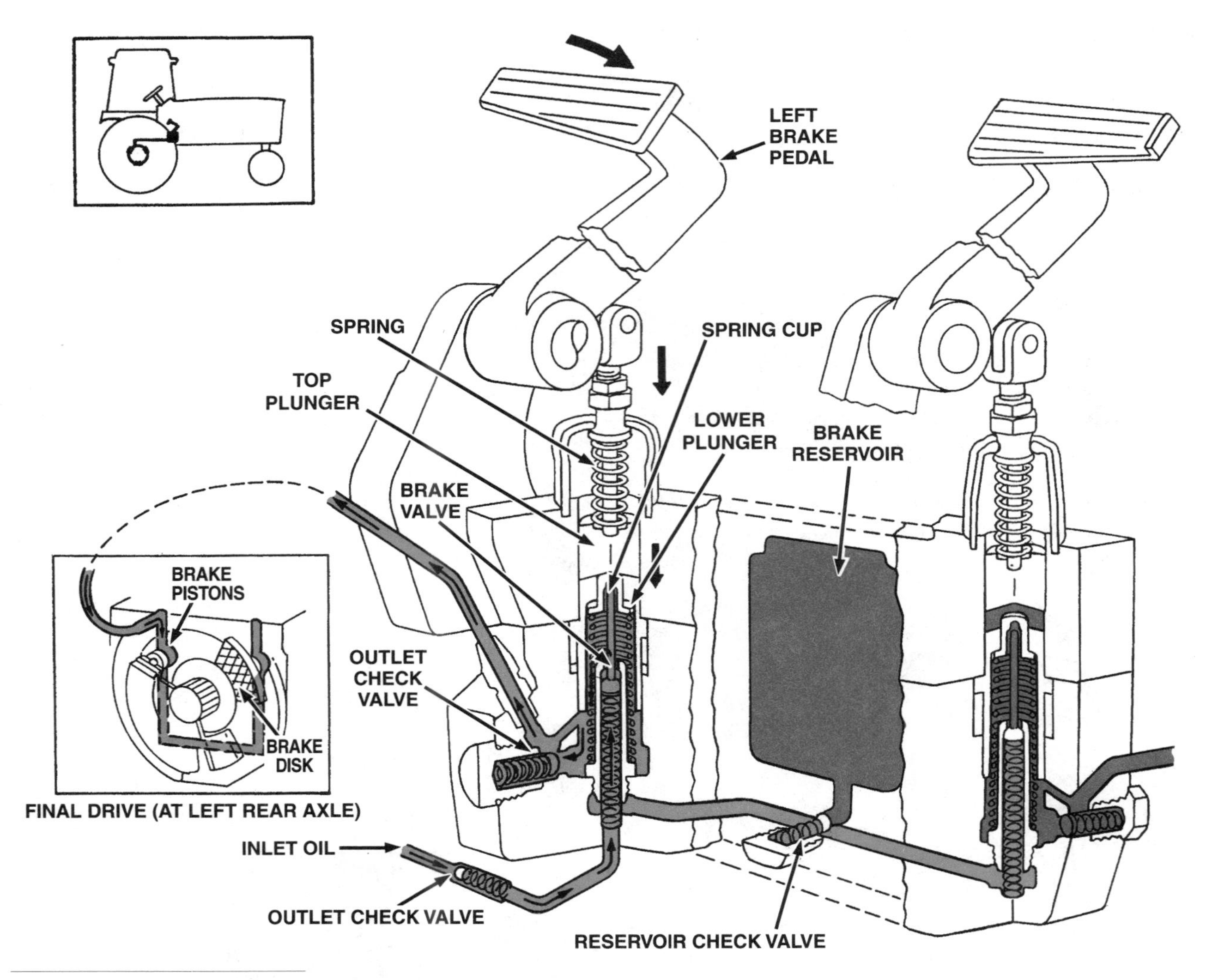

Fig. 42 - Power Brakes (Left Turn)

On most two-wheel drive tractors, brakes are located on each rear axle or wheel. For turning, the operator presses down the pedal for the left or right wheel. For stopping, he presses down on both pedals at once.

On four-wheel drive machines, a single brake mechanism controls the whole unit.

Let's discuss the operation of hydraulic and power brakes in more detail.

HYDRAULIC BRAKES

Fig. 41 shows hydraulic brakes on a typical tractor. Operation is shown during a left turn.

For a sharp left turn, the operator presses down the left brake pedal. This rotates the pedal arm against the brake piston as shown and moves it to the rear. The piston closes the inlet check valve from the reservoir, trapping oil in the cylinder. As the piston moves farther, it forces the trapped oil out of the cylinder, unseating the outlet check valve.

The oil is pushed through a pipe to the final drive at the left rear axle, where it applies force against the brake pressure plate (see inset). This presses the revolving brake disk against the side of a fixed plate, braking the left axle and wheel.

When the brake pedal is released, the force against the brake disk is relieved. The outlet check valve and retainer (inset, Fig. 41), may meter oil returning from the axle unit. Spring force pushes the piston to the front again. This opens the reservoir check valve allowing more oil to enter the cylinder as needed for the next braking.

When both brake pedals are pressed down at once, oil is sent by both brake valves to both final drives.

To assure equal oil pressure on both sides, equalizing valves under each brake piston are opened, connecting the two brake cylinders. Note in Fig. 41 that when the left pedal was

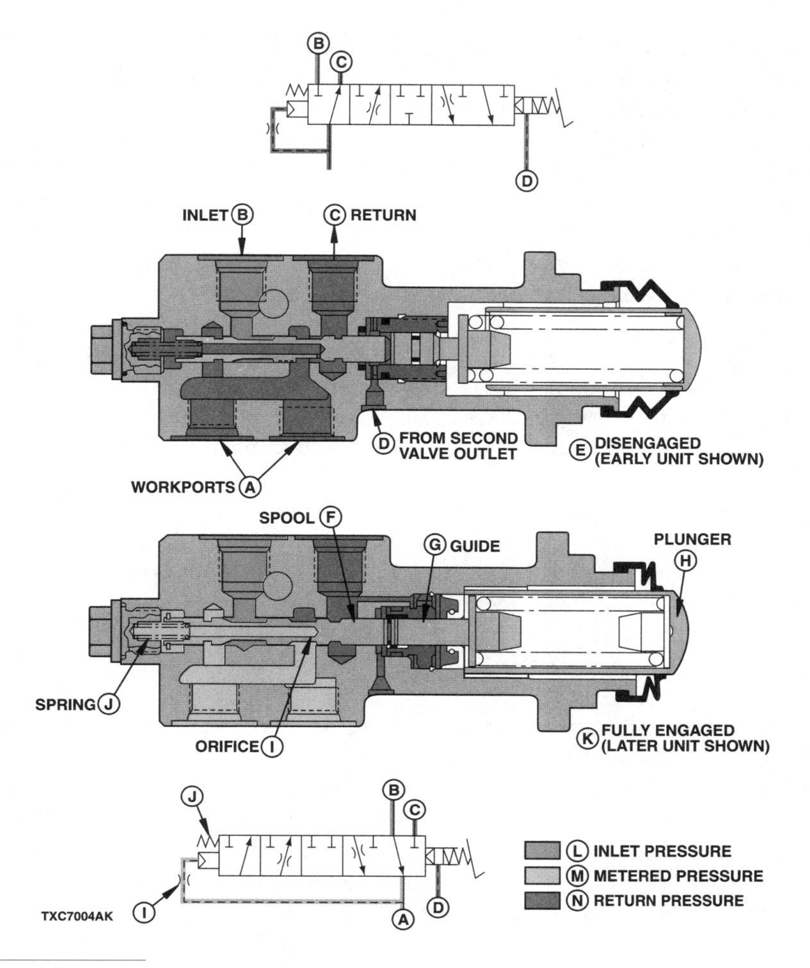

Fig. 43 - Spool Type Power Brake Valve

pushed, that pressurized oil was available at the equalizing valve for the right side. If the right valve is moved just a little bit, it will open the equalizing valve and allow pressurized oil to act on the right brake with the same force as the left. This is particularly important for higher speed roading operations.

These brakes are not affected by engine or hydraulic failure. Oil may be supplied to the brake reservoirs from the tractor hydraulic system, however, many need to be checked and filled periodically.

POWER BRAKES

"Power" brakes mean that oil from a hydraulic source does the braking of the machine. Power brake valves do have an operator "feel" built into them so the brakes can be applied smoothly. The valve is designed so that the more foot pedal force the operator applies, the more pressure will be sent to the brake pistons thus making it possible to control the amount of braking.

Fig. 42 shows poppet type power brakes with a closed-center hydraulic system. The left brake is being operated.

When the operator presses down on the left brake pedal it puts spring pressure against the top plunger. The top plunger pushes the spring cap into the lower plunger, blocking return oil flow to the reservoir.

The guide pushes a rod linkage down to open the brake valve. Inlet oil under pressure now rushes in through the open valve, forces the outlet check valve open, and flows on to the final drive at the left rear axle (see inset).

Here the oil forces the brake pistons and pressure plates to press the revolving brake disk against a fixed plate which is connected to the final drive shaft. This brakes the left axle and wheel.

When pressure builds in the brake cavity it will push the spring cap and top plunger up against the spring. When the hydraulic force equals the spring force, the valve will close. The amount of pressure built in the brakes therefore will depend on how much force is applied to the top spring by the operator. The brake pressure will be in direct proportion to the pedal force applied.

When the brake pedal is released, the brake valve is closed again by its spring, and inlet oil is shut off. This relieves pressure on the brake disk at the axle, and braking stops as some oil flows back to the brake valve area. This oil flows through the spring cap into the brake reservoir.

To insure equal braking on two wheel drive machines when both brake pedals are pressed down at once, equalizing valves (not shown) are opened, connecting the two brake valves.

In case of engine or hydraulic failure, the brakes will operate as hydraulic brakes rather than power brakes. When the pedal is depressed the top plunger makes contact with the lower plunger. The lower plunger becomes the piston for hydraulic brakes and forces oil out to operate the brakes. Inlet and reservoir check valves keep oil trapped in the circuit for continued operation.

On some large machines, the valve is also backed up with an accumulator which holds enough "charge" of pressurized oil in reserve to insure several power brake applications after power is lost.

Fig. 43 shows one type of a spool type power brake valve. It is shown in the neutral and engaged positions. When in neutral the inlet oil is trapped (closed-center system) and the line to the brakes connected to the return.

Note: Return lines on brakes are always returned to the reservoir and not to the pump to insure that pump charge pressure does not keep the brakes engaged.

When the brake pedal is depressed, the plunger is pushed against the spring. The spring in turn pushes the valve spool. As the spool moves, it first closes the return passage, then opens the high pressure passage allowing oil to go to the brake piston(s).

Oil is also directed to the end of the spool causing it to push back against the pedal spring. When pressure in the brake line can overcome the spring, the valve will close maintaining that pressure to the brakes. The pressure to the brakes is therefore determined by how hard the operator pushes on the pedal. This gives complete control of the degree or "feel" of braking.

When the pedal effort is removed, the valve returns to neutral, shutting off the high-pressure oil and allowing the oil in the brakes to return to the reservoir.

Because this valve has no hydraulic brake capability, it is backed up with an accumulator which stores enough oil under pressure for several emergency brake applications.

HYDRAULIC POWER LIFT (ROCKSHAFT) FOR REAR-MOUNTED EQUIPMENT

On modern tractors, there are many uses for rear mounted equipment. On some equipment, it is necessary to have precise control of implement depth with the use of the control lever. On others, such as plows or rear blades, it is desirable to have the equipment respond to the load or draft that the implement is putting on the tractor. In other applications, it is desirable to use a combination of the two. There are therefore three modes of operation:

1. **Depth Control**

2. **Draft Control**

3. **Draft and Depth Control**

DEPTH CONTROL

In Fig. 44, the operator has selected the depth position on the selector lever. This puts the valve linkage in line with the rockshaft cam. It will take a signal only from the cam. Any movement of the draft linkage will not move the valve link.

When the operator moves the lever to the raised position, the linkage rotates the valve lever to open the pressure valve. Oil is directed to the piston that raises the rockshaft. As the rockshaft rotates, the cam rotates also. The valve link rotates clockwise and closes the valve when the rockshaft reaches the position called for by the control lever position.

When the control lever is moved to the lower position, the valve lever rotates clockwise opening the return valve allowing oil in the rockshaft piston to return to the reservoir. As the rockshaft and cam lower, the cam follower and link will rotate the valve linkage until it closes the return valve.

In the depth position, every position of the control lever has a corresponding position of the rockshaft. This provides excellent control of rockshaft depth by the operator.

DRAFT CONTROL

Fig. 45 shows the rockshaft linkage moved to the Draft position. The follow-up linkage has been moved to the bottom of the cam follower. In this position, the valve linkage will receive a signal from the draft linkage but none from the rockshaft cam.

When the control lever is at the top, the pressure valve will be held open, holding the rockshaft in the full raised position.

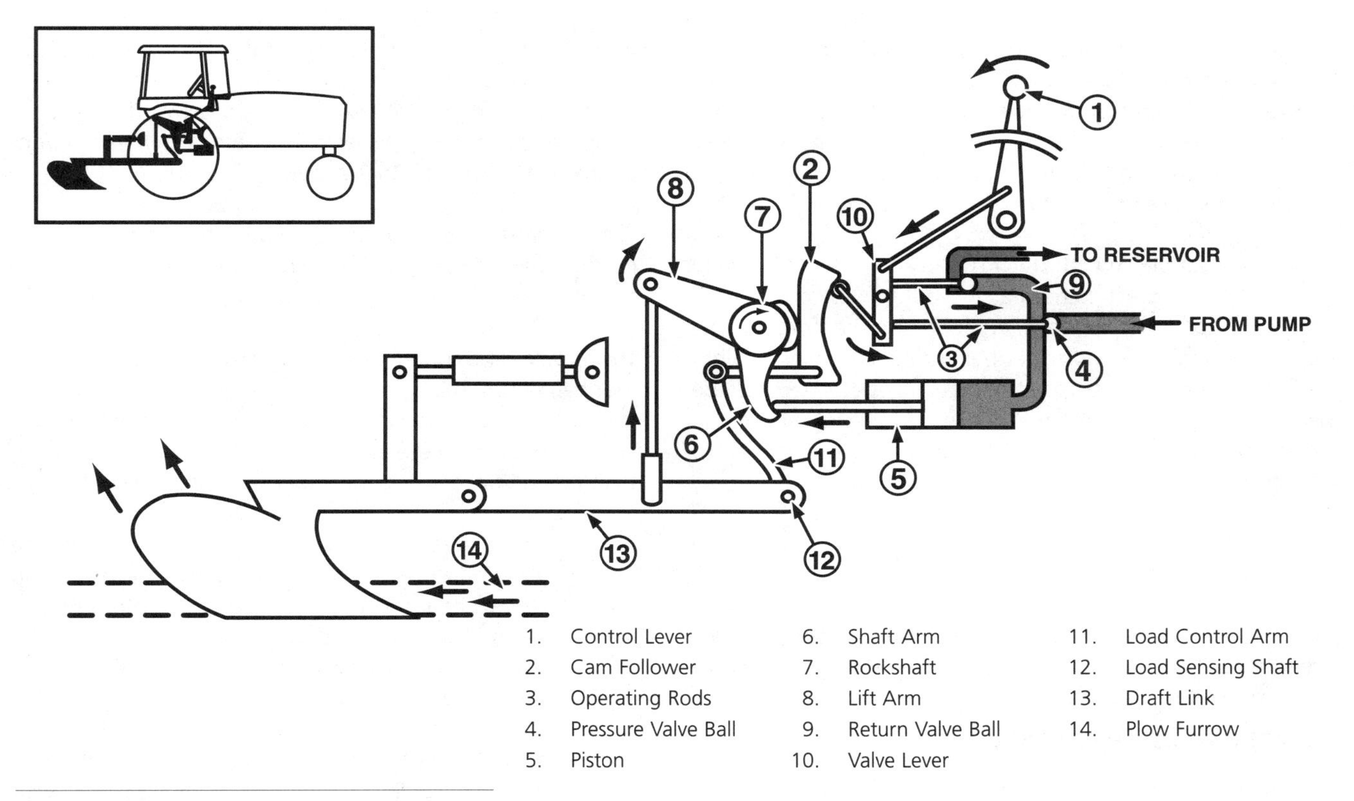

Fig. 44 - Raising an Implement in Depth Control

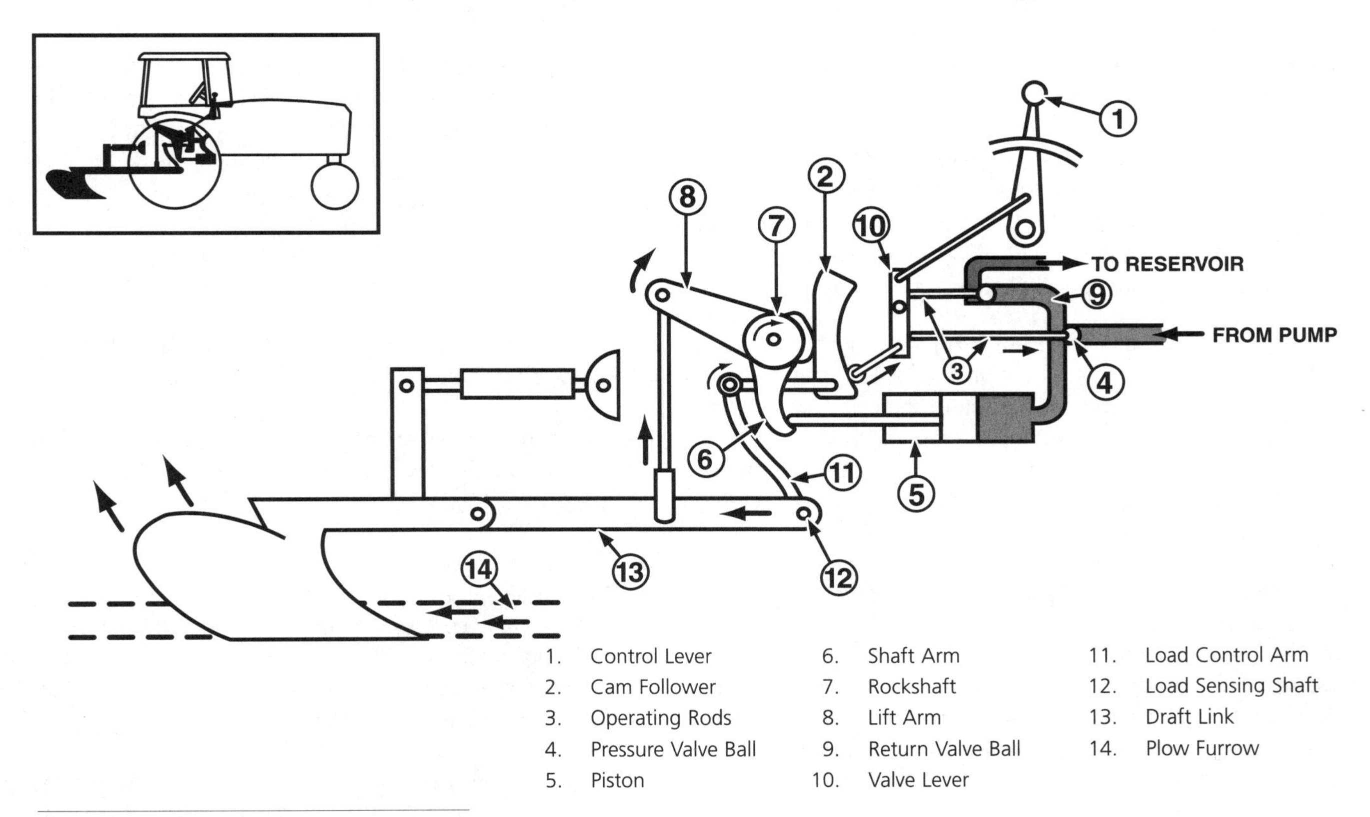

Fig. 45 - Rockshaft Operating in Draft Control

When the control lever is lowered, the valve link will rotate clockwise opening the return valve to lower the rockshaft. The rockshaft will continue to lower until the draft of the implement is enough to flex a load-sensing bar or spring and move the linkage. As the linkage is moved, it will rotate the valve shaft counterclockwise closing the return valve.

For our example, we'll attach a plow to the rockshaft in Fig. 45. If the draft increased, it would need to be raised in order to maintain an even load on the tractor. The increased draft would further flex the torsion bar. The linkage would move the valve linkage to open the pressure valve and raise the plow. When it had raised enough to reach the normal draft load, the linkage would move to close the valve.

When the draft reduced, the linkage would move to lower the plow until normal draft was achieved.

In draft control, the position of the control lever sets a certain draft load. The plow will be raised and lowered automatically to maintain that draft load.

DRAFT AND DEPTH CONTROL

The draft and depth control is a combination of the two positions. The operator moves the draft and depth selector lever so that the valve linkage will contact the cam follower in one of several intermediate positions. It is usually at mid point. In this position, the linkage receives a signal from both the rockshaft and the draft linkage.

This gives the operator good control of the implement depth and the automatic load control will move the implement up and down a little even though it does not maintain a constant draft load. This is the position most used in plowing.

HYDRAULIC SENSING ROCKSHAFT

The hydraulic load sensing system in Fig. 46 consists of a load control valve (1) and a sensing cylinder (2). The load control selector (3) is in a position at the top of the cam follower (4) to allow maximum load sensing. The rod end of the sensing cylinder piston (5) is attached to one draft arm (6). A shaft connects the two draft arms. The draft links (7) are attached to the draft arms.

As the plow enters hard ground (8), the soil resistance increases the draft load on the draft arms. The draft force is transmitted to the sensing cylinder (2) by the draft arms and pulls the sensing cylinder piston and valve (9) rearward. More oil then flows into the sensing cylinder through a variable orifice (10), causing sensing pressure on the front of the load control valve (l) to increase.

The increase in sensing pressure results in rearward movement of the load control valve (1), cam follower (4), and valve operating link (12) which causes the valve operating cam (13) to rotate clockwise. Note that an orifice (11) allows some oil to escape from the front of the load control valve. This results in a variable pressure on the valve depending on the amount of oil that enters through the variable orifice (10).

The clockwise rotation of the valve operating cam causes the pressure valve (14) to open and direct pressure oil through the throttle valve (15) to the backside of the rockshaft piston (16).

The throttle valve controls the speed of oil flow to and from the rockshaft piston. The piston moves forward and causes the rockshaft (22) to rotate, lifting the draft links (7) and raising the plow. The check ball (21) prevents return oil from the front of the rockshaft piston from entering the return valve housing (19).

When the plow passes the hard ground, the draft force on the sensing cylinder decreases. The sensing cylinder piston (5) and valve (9) move forward permitting less oil to flow through the variable orifice, decreasing sensing pressure at the front of the load control valve. The spring in the load control valve housing (17) pushes the valve forward.

The spring (18) causes the valve-operating cam to rotate counterclockwise and forces the cam follower (4) forward, along with the load control valve. The pressure valve (14) closes and, if necessary, the return valve (19) opens to lower the plow.

When the pressure and relief valves are closed, oil trapped at the rear of the rockshaft may expand if the oil temperature rises. The thermal relief valve (20) senses thermal expansion of hydraulic oil in the system and opens if the expansion is too great.

ELECTRONICALLY CONTROLLED ROCKSHAFT

Fig. 47 shows a tractor with two external cylinders that raise or lower the three-point hitch and thus the implement.

Hydraulic pump (D) transfers pressurized oil via the inlet priority valve to rockshaft valve (F), which controls the rockshaft cylinders.

The rockshaft cylinders act on draft links (H), enabling implements to be raised or lowered. The rockshaft assembly is controlled electronically.

Electronic control box (B) receives signals from the operation unit (A) and also from draft sensor (G) or position sensor (C). These values are coordinated in electronic control box (B) before being passed on to stepper motor (E). The stepper motor opens the raising or lowering valve in the rockshaft valve assembly by means of a cam.

When operating in depth control, the signal received from the operation unit (A) is coordinated with the actual value from position sensor (C).

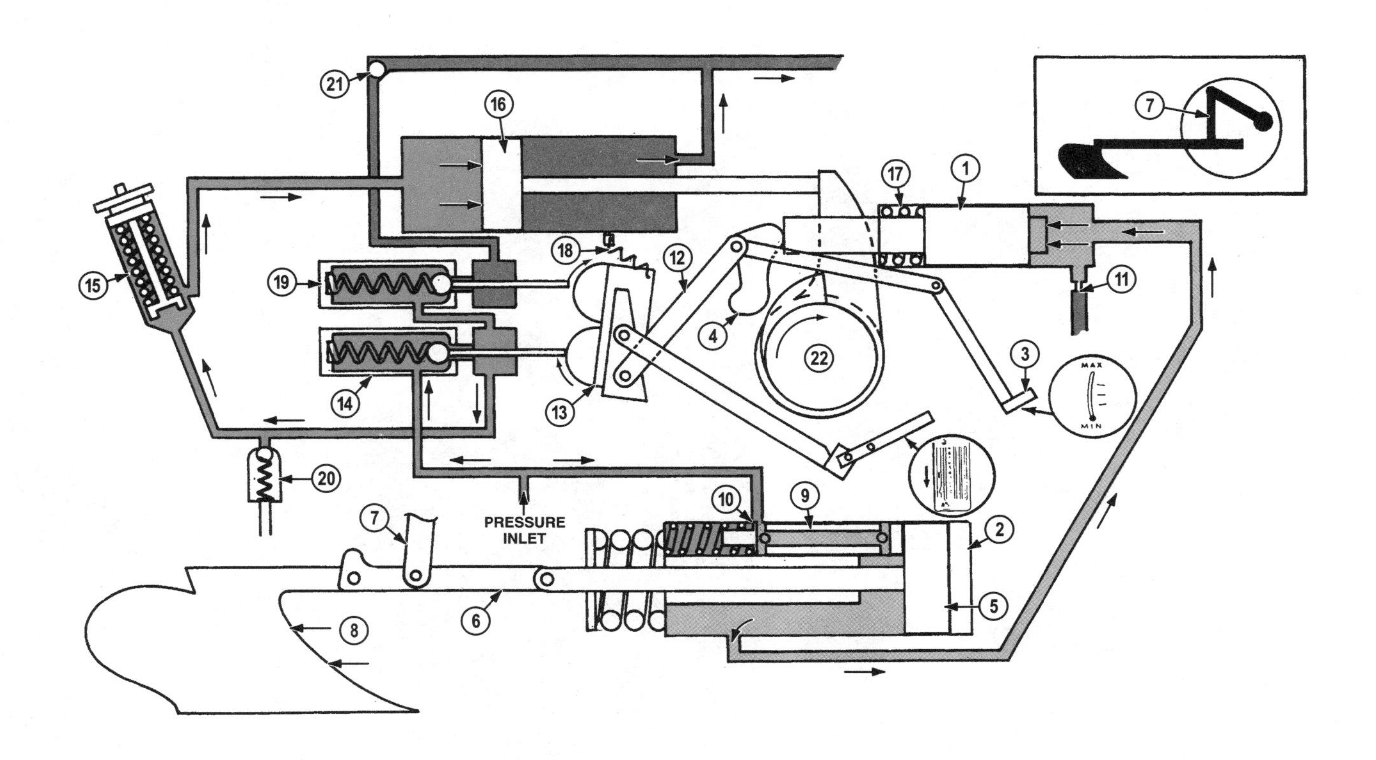

Fig. 46 - Hydraulic Sensing Rockshaft

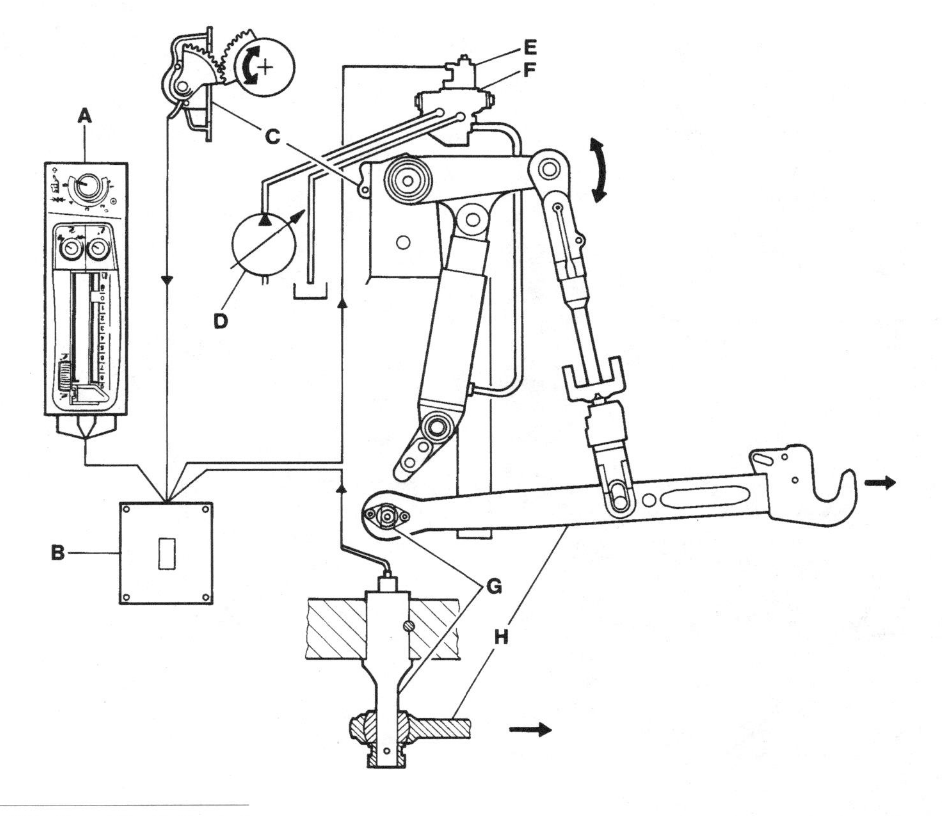

Fig. 47 Electronic Controlled Draft and Depth Control

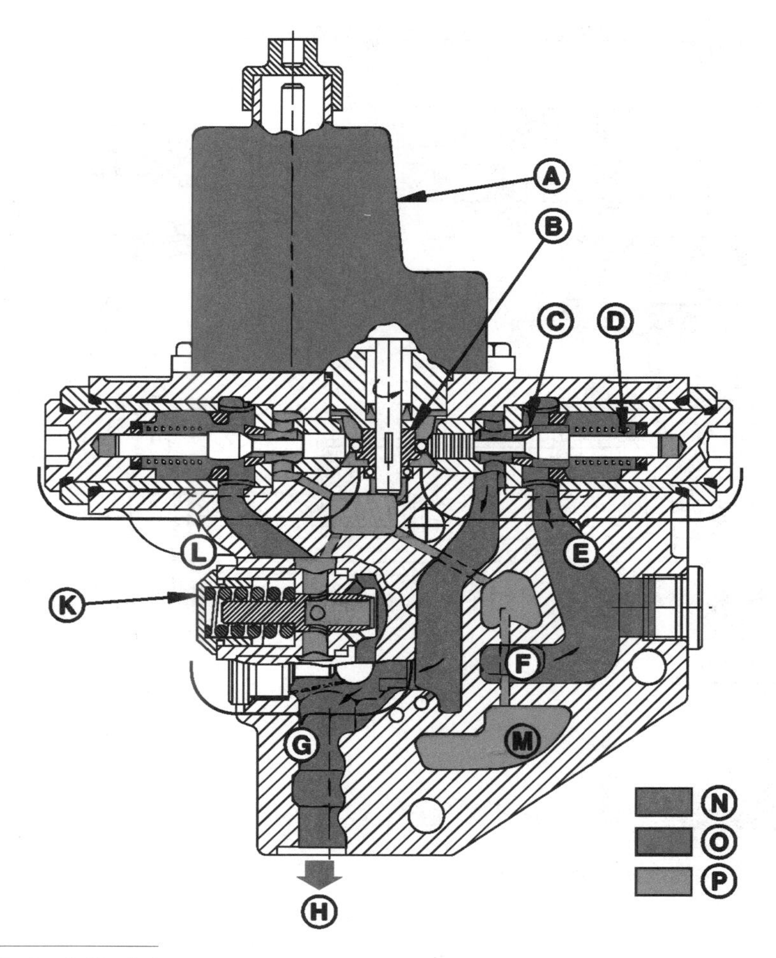

Fig. 48 - Rockshaft Valve - Electronic Control

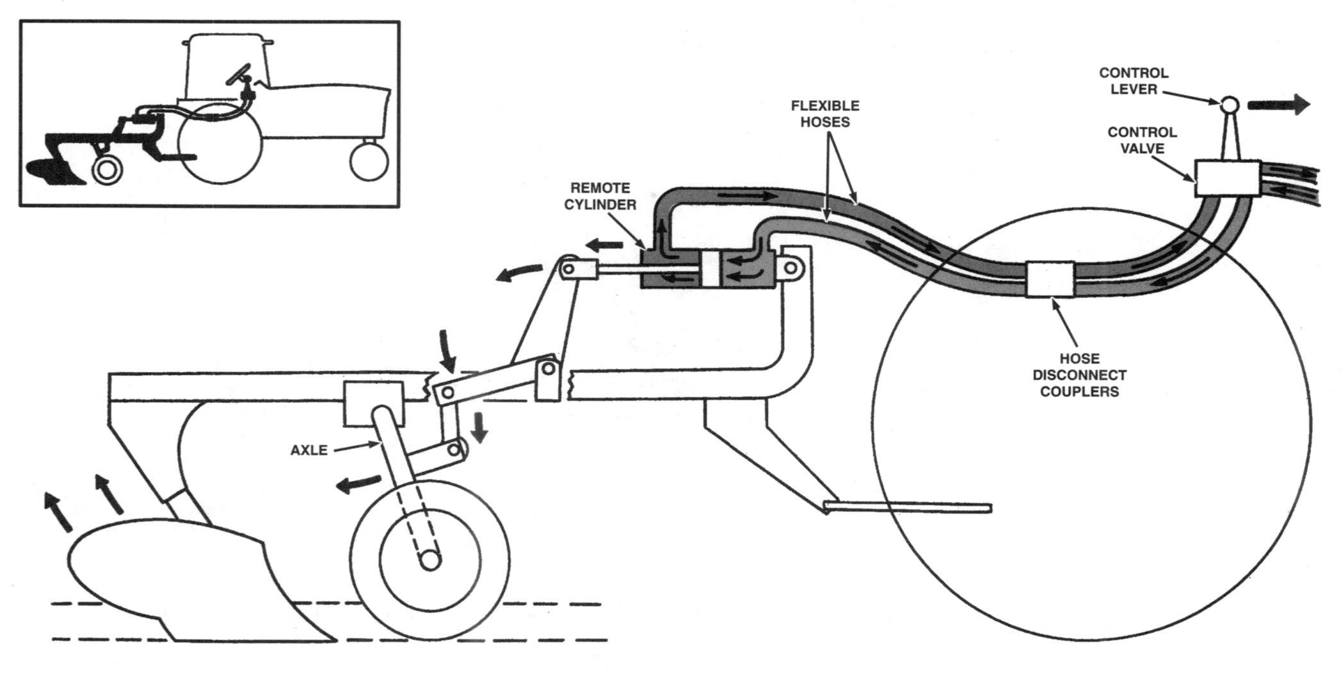

Fig. 49 - Remote Control Hydraulics

When operating in draft control, the signal from the operation unit (A) is coordinated with the actual load value at draft sensor (G).

The rockshaft valve (Fig. 48) has the raising and lowering valves (E) and (L), check valve (G), surge relief valve (K) and stepper motor (A). It is shown raising.

During the lifting process, stepper motor (A) receives a signal and rotates cam (B), which raises valve spool (D) from the seat (C) of raising valve (E). High-pressure oil (N) can now flow through check valve (G) to the two rockshaft cylinders.

During the lowering process, lowering valve (L) is opened by the cam and the oil flows out of the rockshaft cylinders by the weight of the implement.

When the rockshaft valve is in its neutral position, check valve (G) traps the hydraulic oil in the rockshaft cylinders and prevents the lift arms from lowering when the engine is shut off.

Surge relief valve (K) reduces pressure peaks caused when transporting heavy implements.

Pilot-pressure oil (O) gives a signal to a load sensing hydraulic pump.

REMOTE CONTROL OF EQUIPMENT

Tractors may operate equipment that is not mounted, but is pulled or pushed. To control this equipment with hydraulics, a remote actuator such as a cylinder or a motor is needed. It is separate from the tractor and connected by flexible hoses.

Let's take the case of a plow again, this time, one that is pulled behind the tractor (Fig. 49). The operator wants to raise it so he moves the control lever to the front as shown. This actuates the control valve that sends pressure oil to the front of the remote cylinder. As this oil pushes the piston to the rear, the cylinder rod extends. Oil from the other side of the cylinder is forced out and returns through the valve to the reservoir. As the cylinder extends it pivots linkage to raise the plow.

Note- See Chapter 5 for details of Valves and Chapter 6 for cylinders

DIAGNOSIS AND TESTING OF HYDRAULIC SYSTEMS

In the final chapter of this manual, we will once again return to complete hydraulic systems for diagnosis and testing. But because the most important tool in troubleshooting is knowledge of the system, we must look at the various working parts in more detail.

The following chapters will do that. We will then use our knowledge of "how the system works" to find out "why the system fails" and "how to remedy" these failures. Chapter 15 is titled "Diagnosis and Testing of Hydraulic Systems."

HYDRAULIC FACTS

Here are some key facts that will help you understand hydraulics:

1. There are two basic types of hydraulics:

   a) Hydrodynamics is the use of fluids at high speeds that impact another member to supply power (kinetic energy). It is the weight and speed of the oil that is harnessed.

      Example: a torque converter.

   b) Hydrostatics is where fluids are forced through tubes at relatively low speeds but at higher pressures to transfer power (hydraulic energy) from the power source (pump) to the work (cylinder or motor). Example: most hydraulic systems. It is the only type covered in this manual.

2. Hydraulic power is generated from mechanical power. Example: A hydraulic pump driven by an engine crankshaft.

3. Output is achieved by converting hydraulic power back to mechanical energy. Example: A cylinder that raises a heavy plow.

4. Hydraulic energy is neither created nor destroyed, only converted to another form. All energy put into a hydraulic system must come out either as work (gain) or as heat (loss).

5. Flow through an orifice or restriction causes a pressure drop.

6. When a moving liquid is restricted, heat is created and there is a loss of potential energy (pressure) for doing work.

   Example: A tube or hose that is too small or is restricted. Orifices and relief valves are also restrictions but they are purposely designed into systems.

7. Oil takes the course of least resistance.

8. Oil is pushed into a pump by atmospheric pressure, not drawn into it.

9. A pump does not pump pressure; it creates flow. Pressure is caused by resistance to flow.

10. Two hydraulic systems may produce the same power output, one at high pressure and low flow, the other at low pressure and high flow.

11. A basic hydraulic system must include four components: a reservoir to store the oil, a pump to push the oil through the system, valves to control oil pressure and flow, and a cylinder (or motor) to convert the fluid movement into work.

12. Compare the two major hydraulic systems:

    Open-Center System = Valves used have an open passage when in the neutral position to allow oil to pass from the valve inlet to the outlet and back to the reservoir.

    Closed-Center System = Valves used have all paths for the oil blocked when in the neutral position. There is no oil flow through the valve. This requires that the system be able to supply oil only when needed.

TEST YOURSELF

QUESTIONS

1. (True or False) "A pump creates pressure?"

2. Oil flow through an orifice causes pressure beyond the orifice to:

    a. rise

    b. drop

3. How do you determine the force exerted by a piston?

4. What four components are needed to complete a very basic hydraulic system?

5. (True or False) "Hydraulic pumps convert hydraulic power to mechanical power?"

6. (Fill in the blank.) "Speed is a factor in determining __________."

    a. displacement

    b. work

    c. power

7. (Fill in the blank.) "To determine if a system is an open-center or closed-center system you must look at __________."

    a. the pump

    b. the control valve

    c. both the pump and valve

8. Describe the difference in the control valve during neutral in an open-center system, as compared to a closed-center system.

9. Is fluid pushed into a pump or drawn into it?

10. In what system do both the flow and pressure vary?

SAFETY RULES FOR HYDRAULICS

INTRODUCTION

In the remainder of this book we will explain the high forces and pressures that can be generated in the hydraulic system. These can cause serious injuries and even death, if reasonable care is not exercised. The following information is designed to alert you to the potential hazards, and to keep you safe if you follow the instructions.

Fig. 1 — Always Exercise Safety

HYDRAULICS SYSTEMS

Hydraulics systems store energy. Hydraulic systems must confine fluid under high pressure often higher than 2,000 pounds per square inch.

A lot of energy may be stored in a hydraulic system, and because there is often no visible motion, operators do not recognize it as a potential hazard. Carelessly servicing, adjusting, or replacing parts can result in serious injury. Fluid under pressure attempts to escape (Fig. 2). In doing so it can do helpful work, or it can be harmful.

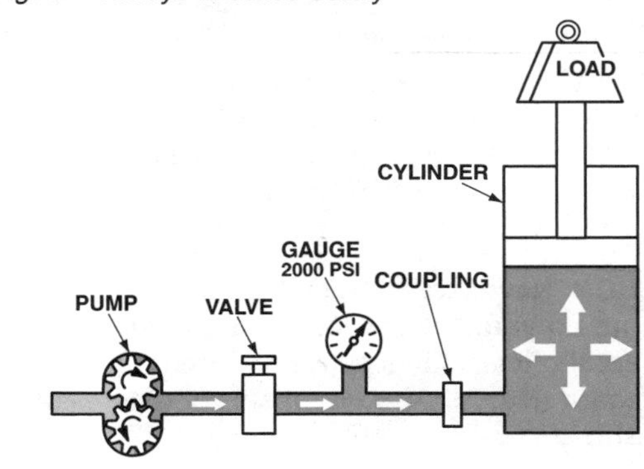

Fig. 2 — Hydraulic Fluid Under Pressure Attempts to Escape Or Move To A Point Of Lower Pressure

 **Caution: Never service or adjust systems under pressure.**

Adjusting and removing components when hydraulic fluid is under pressure can be hazardous (Fig. 3). Imagine attempting to remove a faucet from your kitchen sink without relieving the water pressure. You'd get a face full of water! It is much more dangerous with hydraulic systems. Instead of just getting wet from water at 40 psi, you would be seriously injured by oil under 2,000 psi, or more. You could be injured by hot, high pressure spray of fluid and by the part you are removing when it is thrown at you (Fig. 3).

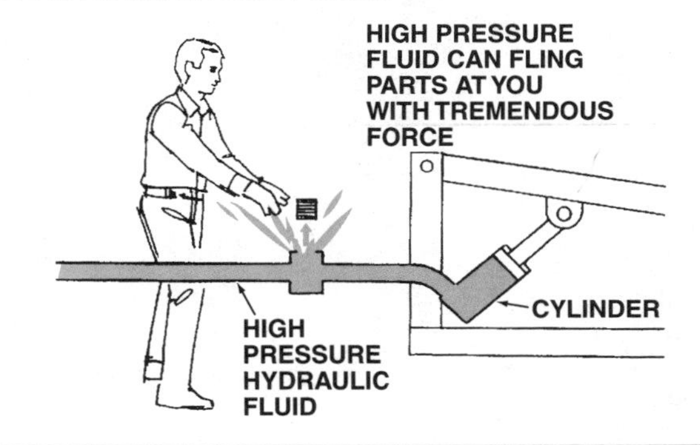

Fig. 3 — Always Relieve Hydraulic Pressure Before Adjusting Hydraulic Fittings. You Could Be Injured By A Hot, High, Pressure Spray Of Hydraulic Fluid Or By A Part Flung At You

AVOID HIGH-PRESSURE FLUIDS

Escaping fluid under pressure can penetrate the skin causing serious injury.

Avoid the hazard by relieving pressure before disconnecting hydraulic or other lines. Tighten all connections before applying pressure.

Search for leaks with a piece of cardboard. Protect hands and body from high pressure fluids.

If an accident occurs, see a doctor immediately. Any fluid injected into the skin must be surgically removed within a few hours or gangrene may result. Doctors unfamiliar with this type of injury should consult with medical source to obtain pertinent information.

Fig. 4 — Avoid High-Pressure Fluids

Pinhole Leaks Can Be Dangerous

If fluids, under pressure, escapes through an extremely small opening, it comes out as a fine stream (Fig. 5). The stream is called a pinhole leak. Pinhole leaks in hydraulic systems are hard to see and they can be very dangerous. High-pressure streams of oil from a pinhole leak can penetrate human flesh. Hydraulic systems often have pressure over 2,000 psi. That's higher than the pressure in hydraulic syringes used to give injections. Injuries caused by fluids (oil) injected into human flesh can be very serious. Consult a physician immediately, if you believe that you have been injured by a pinhole stream.

CAUTION: Never try to detect the pinhole leak by running your hand over the area where you suspect the leak. Always use a piece of cardboard (Fig. 5). Also, wear safety glasses or a face shield.

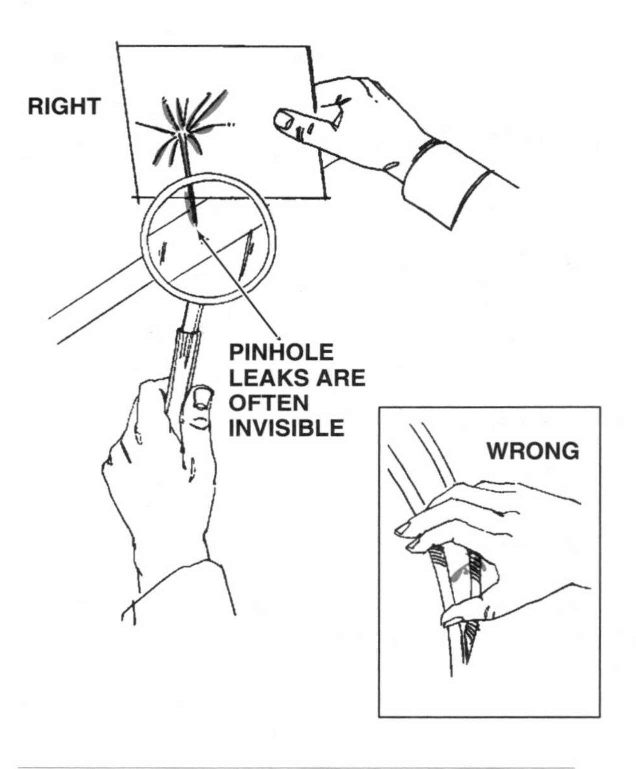

Fig. 5 — The Jet Stream Or Mist From A Pinhole Leak In A Hydraulic System Can Penetrate Your Skin — Don't touch it!

AVOID HAZARD OF STORED ENERGY FROM HYDRAULIC ACCUMULATOR

Some hydraulic systems have accumulators to store energy. They may also be used to absorb shock loads and to maintain a constant pressure in the system. Recognize that the accumulators and the entire hydraulic system may have energy stored in it, if the pressure has not been relieved. Even though the pump may be stopped, or an implement has been disconnected from the tractor, energy is stored in the accumulator unless the pressure was relieved before shutdown. The nitrogen is under pressure, so the hydraulic fluid is also under pressure (Fig. 6).

 **CAUTION: Observe these basic safety rules for hydraulic accumulators:**

1. *Recognize accumulators as sources of stored energy.*
2. *Relieve all hydraulic system pressure before adjusting or servicing any part of an accumulator system.*
3. *Relieve all hydraulic pressure before leaving a machine unattended for the safety of others as well as for you.*
4. *Make sure pneumatic accumulators are properly charged with the proper inert gas (Dry Nitrogen). A pneumatic accumulator without gas is a potential "bomb".*
5. *Read and follow manufacturers instructions for servicing accumulators.*

Generally, manufacturers recommend only authorized dealers service gas charged accumulators. Read and follow the manufacturer's instructions thoroughly.

AVOID HAZARD OF TRAPPED OIL AND THERMAL EXPANSION

Another hazard with trapped oil (Fig. 7) is heat. Heat from the sun can expand oil and increase pressure. The pressure can blow seals and move parts of an implement or machine.

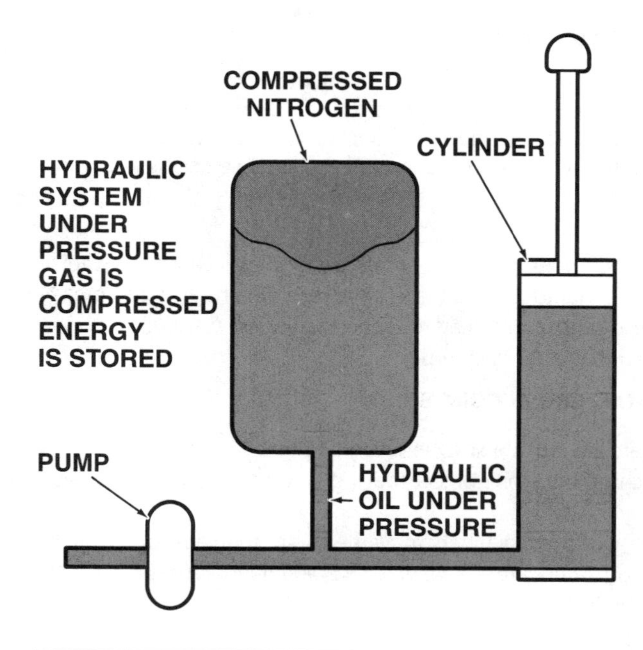

Fig. 6 — Hydraulic Accumulators Store Energy

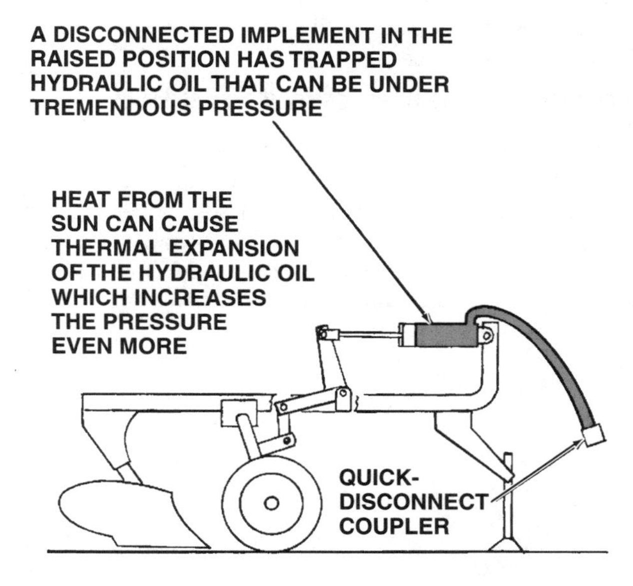

Fig. 7 — Thermal Expansion Causes An Increase In Pressure

CONNECT LINES CORRECTLY

The movement of hydraulic components should correspond to the movement of the controls. If a control lever is placed in the "raise" position, the function should raise; or if the steering wheel is turned to the left the wheels should turn to the left.

Wrong hook-up of lines or hoses (Fig. 8) will cause the reverse of the intended action. This may result in an "element of surprise" and could lead to serious injuries. Carefully test the machines after each repair.

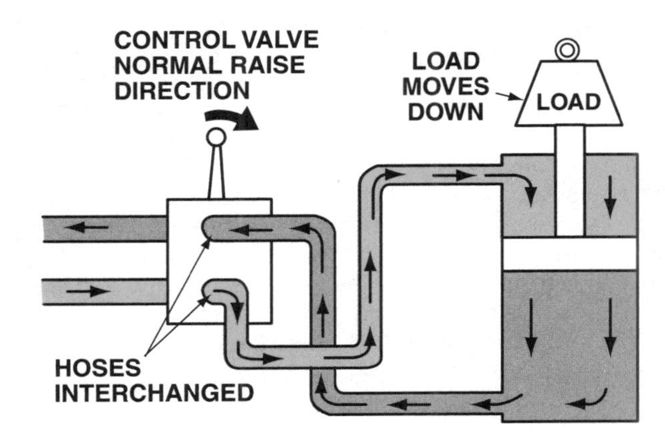

Fig. 8 — Avoid Incorrect Hose Connection

AVOID CRUSH POINTS

There are crush points between two objects that move toward each other or one object moving toward a stationary object.

Many of the machinery movements are triggered by hydraulic action. There are, for instance, several dangerous crush points on hydraulic hitches.

Another dangerous crush point is between the tires or frame parts on tractors with articulate steering (Fig. 9). There can be immediate movement upon start-up of the tractor when the steering wheel isn't even moved. Hydrostatic steering systems are very sensitive. NEVER stand between the tires during start-up or any time the engines are running. NEVER allow anyone else to stand there either.

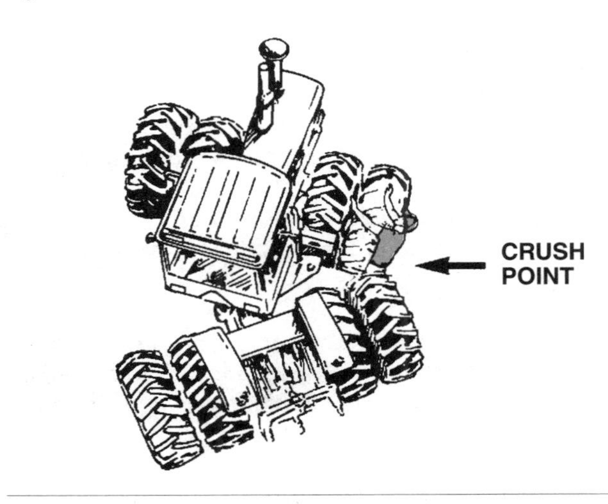

Fig. 9 — Four-Wheel Drive Tractors With Articulated Steering Can Create A Crush Point

AVOID HEATING NEAR PRESSURIZED FUEL LINES

Flammable spray can be generated by heating near pressurized fluid lines (Fig. 10), resulting in severe burns to yourself and bystanders. Do not heat by welding, soldering, or using a torch near pressurized fluid lines or other flammable materials. Pressurized lines can be accidentally cut when heat goes beyond the immediate flame area.

Fig. 10 — Four-Wheel Drive Tractors With Articulated Steering Can Create A Crush Point

GENERAL SAFETY INSTRUCTIONS

On the previous pages, you have been given safety information that is specific to the hydraulic system. But when working on the hydraulic system, it is often necessary to remove and replace non-hydraulic components. Therefore, total care has to be exercised in all service activities.

The following safety instructions, if followed, are designed to keep you safe. Make it a practice to work safely. It's the right thing to do.

Fig. 11 — Always Exercise Safety

RECOGNIZE SAFETY INFORMATION

This is the safety-alert symbol. When you see this symbol on your machine, in this manual, or the machine's manual, be alert to the potential for personal injury.

Follow the recommended precautions and safe operating practices.

Fig. 12 — Safety Alert Symbol

UNDERSTANDING SIGNAL WORDS

A signal word — DANGER, WARNING, or CAUTION — is used with the safety alert symbols. DANGER identifies the most serious hazards.

DANGER or WARNING safety signs are located near specific hazards. General precautions are listed on CAUTION safety signs. CAUTION also calls attention to safety messages in this manual, and the machine's manual.

⚠WARNING

⚠CAUTION

Fig. 13 — Signal Words

FOLLOWING SAFETY INSTRUCTIONS

Carefully read all safety messages in this manual and on the machine's safety signs. No two machines are exactly alike, so don't take any chances. Become familiar with the safety instructions and follow them. Also insist that those working with you follow the instructions. It will keep you and them safe.

Fig. 14 — Follow Safety Instructions

PRACTICE SAFE MAINTENANCE

Understand the service procedure before doing the work. Keep the are clean and dry.

Never lubricate or service the machine while it is moving. Keep hands, feet, and clothing from power driven parts. Disengage all power and operate the controls to relieve pressure. Lower the equipment to the ground. Stop engine. Remove key. Allow the machine to cool.

Securely support any machine elements that must be raised for service work.

Keep all parts in good condition and properly installed. Fix damage immediately. Replace worn or broken parts. Remove any buildup of grease, oil, or debris.

Disconnect the battery ground cable (-) before making adjustments on the electrical systems or welding on the machine.

Fig. 15 — Practice Safe Maintenance

WORK IN CLEAN AREA

Before starting a job:

- **Clean the work area and the machine.**
- **Make sure you have all the necessary tools to do your job.**
- **Have the right parts on hand.**
- **Read all the instructions thoroughly: do not attempt shortcuts.**

Fig. 16 — Work In Clean Area

PREPARE BEFORE EMERGENCIES

Be prepared if a fire starts.

Keep a first aid kit and fire extinguisher handy.

Keep emergency numbers for doctors, ambulance service, hospital, and fire department near your telephone.

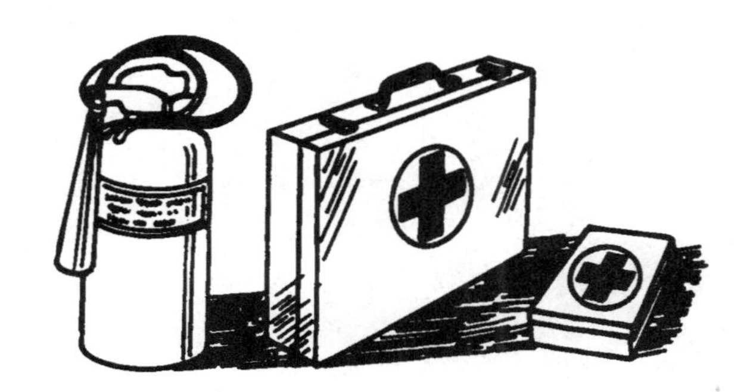

Fig. 17 — Prepare For Emergencies

PARK MACHINE SAFELY

Before working on the machine:

- **Lower all equipment to the ground.**
- **Stop the engine and remove the key.**
- **Disconnect the battery ground strap.**
- **Hang a "DO NOT OPERATE" tag in operator station.**

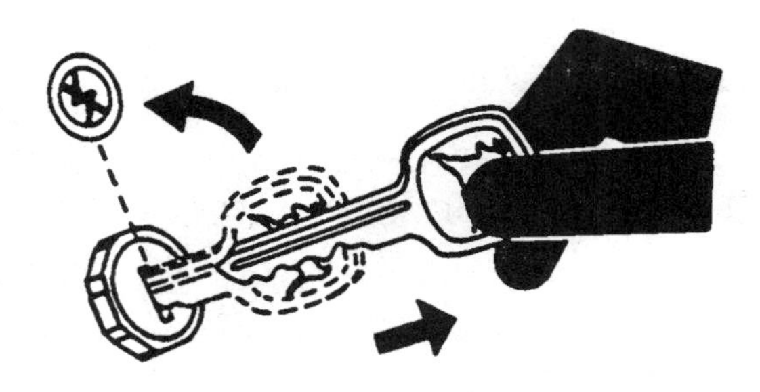

Fig. 18 — Park Machine Safely

SERVICE MACHINES SAFELY

Tie long hair behind your head. Do not wear a necktie, scarf or loose clothing, or necklace when you work near machine tools or moving parts. If these items were to get caught, severe injury could result.

Remove rings and other jewelry to prevent electrical shorts and entanglement in moving parts.

Fig. 19— Service Machine Safely

USE PROPER TOOLS

Use tools appropriate to the work. Makeshift tools and procedures can create safety hazards.

Use power tools only to loosen threaded parts and fasteners.

For loosening and tightening hardware, use the correct size tools. DO NOT use U.S. measurement tools on metric fasteners. Avoid bodily injury caused by slipping wrenches.

Fig. 20 — Use Proper Tools

HANDLE FLUIDS SAFELY—AVOID FIRES

When you work around fuel, do not smoke or work near heaters or other fire hazards.

Store flammable fluids away from fire hazards. Do not incinerate or puncture pressurized containers.

Fig. 21 — Avoid Fires

PREVENT MACHINE RUNAWAY

Avoid possible injury or death from machinery runaway.

Do not start the engine by shortening across the starter terminals. The machine will start in gear if normal circuitry is bypassed.

NEVER start the engine while standing on the ground. Start the engine only from the operator's seat, with the transmission in neutral or park.

Fig. 22 — Prevent Machine Runaways

DISPOSE OF FLUIDS PROPERLY

Improperly disposing of fluids can harm the environment and ecology. Before draining any fluids, find out the proper way to dispose of waste from your local environmental agency.

Use proper containers when draining fluids. Do not use food or beverage containers that may mislead someone into drinking from them.

DO NOT pour oil into the ground, down a drain, or into a stream, pond, or lake. Observe relevant environmental protection regulations when disposing of oil, fuel, coolant, brake fluid, filters, batteries, and other harmful waste.

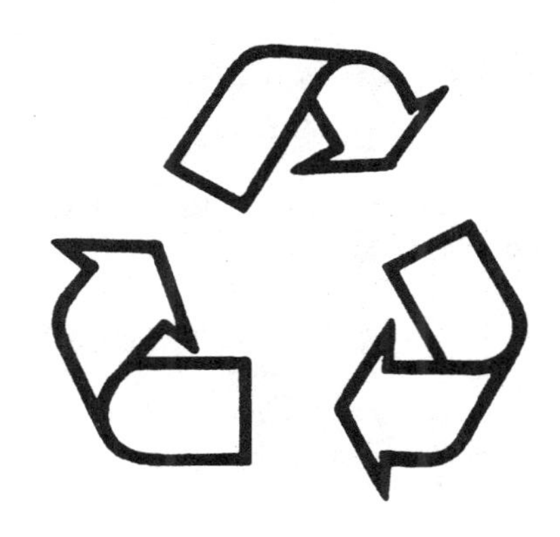

Fig. 23 — Dispose of Fluids Properly

TEST YOURSELF

QUESTIONS

1. What does the safety alert symbol indicate?

2. What are the three safety signal words?

3. What must be used to locate a high pressure leak?

4. Hydraulic accumulators store ____________________.

5. What is a crush point?

6. What four things must be done before any service is performed?

HYDRAULIC SYMBOLS

READING HYDRAULIC SYMBOL DRAWINGS

Symbolic drawings and diagrams are the most popular way of representing hydraulic components and systems. These symbols are easier to draw, read, and standardize than the other types of engineering drawings. In order to be literate in the maintenance and manufacturing process, you must understand the rules that make symbolic drawings such a unique language.

In this chapter we will look at the guidelines that you can use to interpret and understand symbolic drawings. We will also look at how components are interconnected to make a working hydraulic system.

As opposed to pictorial drawings, symbols do not show the exact shape of the component that they represent. Symbols are two dimensional. They are lines drawn on paper. They have no additional explanation (such as color coding or words) (Fig. 1).

Instead symbols are a visual short-hand method of communication. They rely on figures, such as squares and circles, and marks, such as arrows, to represent hydraulic components. Unlike cutaway drawings, symbols do not show the parts of the hydraulic component. Symbols and symbolic diagrams do show:

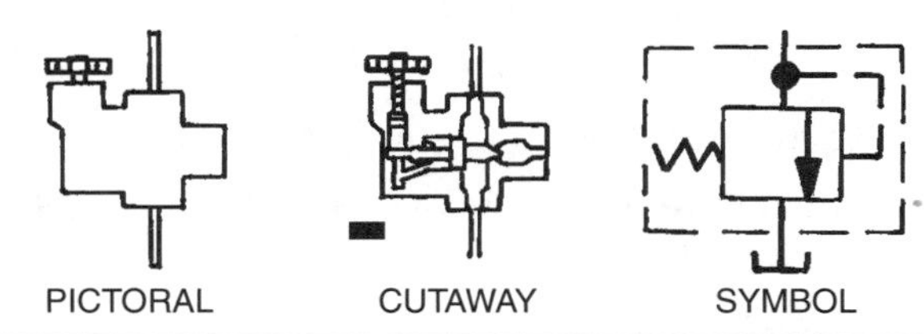

Fig. 1 — Symbols Do Not Show the Visual Detail of a Pictorial or Cutaway Drawing.

How components are interconnected in a system (Fig. 2).

- **Flow paths of hydraulic fluid.**
- **The general way a component works.**
- **The number of ports or connections on the component.**

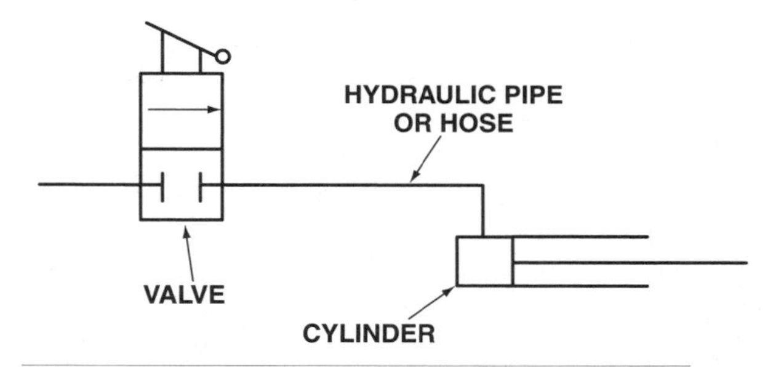

Fig. 2 — Symbolic Diagrams Show How Components are Interconnected.

Some Basic guidelines that you can use in this chapter to help interpret symbols are:

- **Symbols and diagrams do not show the internal conditions of the hydraulic system. This includes fluid temperature and pressure.**
- **Each symbol is drawn to represent the neutral or normal position of the component before it is actuated. A normally closed value will be shown in the closed position (Fig. 3).**
- **Symbols can be rotated or put into a location that is not their normal position. This does not alter their meaning. They must be shown correctly connected if they are drawn in a hydraulic system diagram.**

- **Symbols can be drawn in any size without altering their meaning.**
- **All line width should be read as the same thickness. Unlike orthographic drawing that relies on line width to convey an idea, symbols are straight forward and do not need to be interpreted.**

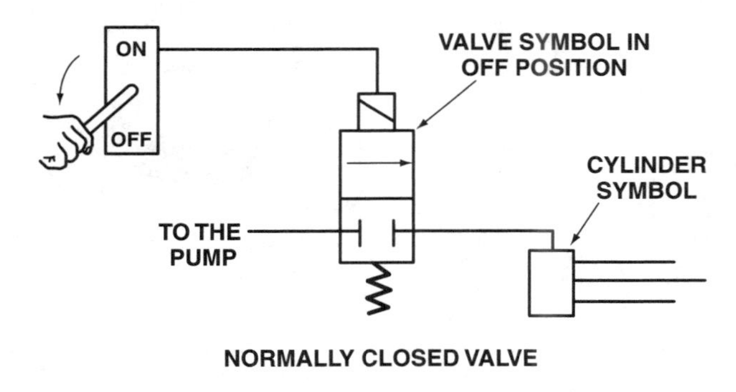

Fig. 3 — The normal or Nonactuated State of the Component is Represented by the Symbol.

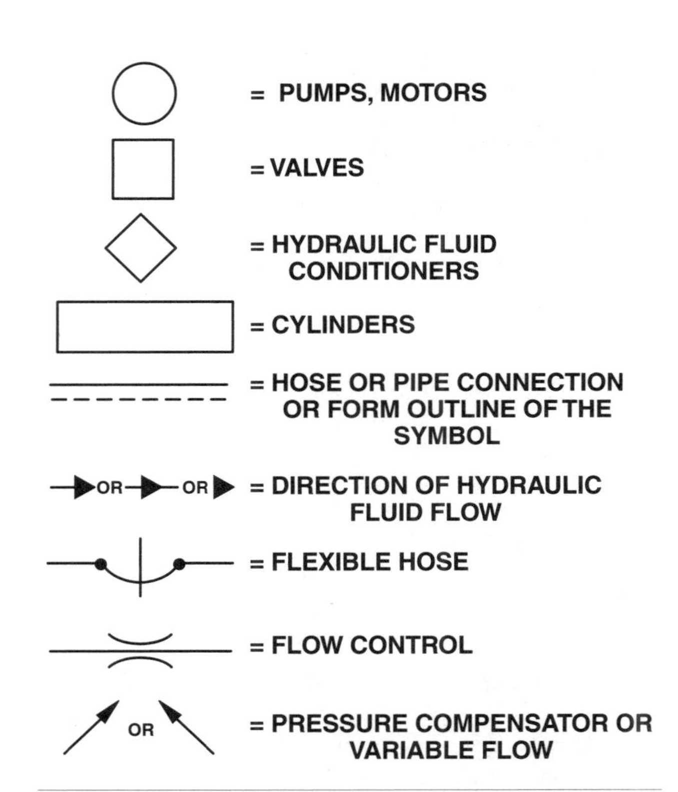

Fig. 4 — There are Four Basic Shapes, a Broken or Unbroken Line, and Five Basic Marks That are Used to Make Hydraulic Symbols.

CHARACTERISTICS OF SYMBOLS

The graphical symbols used as examples in this chapter are in accordance with ISO 1219-1 (Fluid Power Systems and Components —Graphical Symbols and Circuit Diagrams — Part 1: Graphical Symbols).

The symbols that we look at are **based** on this ISO standard. Any **color**, words, or lines outside of the symbol are not part of the symbol. These external marks have been added to help clarify some particular point.

SHAPES

The basic shapes used to make hydraulic symbols are:

- **Circles, semicircles**
- **Squares**
- **Diamonds**
- **Rectangles**

Marks

A symbol is made by using one of the four basic shapes and adding the appropriate marks. These marks include:

- **Lines**
- **Arrows, arrowheads**
- **Arcs**

Circles, Semicircles

When you see a circle or a semicircle with marks inside of it, you are looking at a pump or motor symbol. The circle represents circular rotation such as the internal rotating parts of a motor (Fig. 4).

Squares

Squares are also called **envelopes**. They are used to represent one position or hydraulic fluid path through a valve. Two envelopes together make a two **position** valve. Three together make a three position valve. Four together make a four position valve.

Diamonds

Diamonds represent a part that helps condition the hydraulic fluid in a system. These include fluid coolers, and filters.

Rectangles

Rectangles represent hydraulic cylinders.

Lines

There are four basic types of lines (Fig. 5).

- **The solid line represents the route that hydraulic fluid flows through the system.**
- **The dash shows a pilot line that connects from a pressurized part of the system back to a part that helps control the system. This helps keep the system stable.**

- **Center lines are used to enclose assemblies.**
- **The double line is used to show a mechanical connection (shaft, rod, lever, ect.).**

MAIN WORKING LINE (SOLID)

PILOT LINE FOR CONTROL (DASH)

ENCLOSURE OUTLINE (CENTER)

MECHANICAL CONNECTION (DOUBLE)

Fig 5 — There Are Four Symbols For Lines.

Lines (pipes or hose) that cross in a hydraulic system but do not connect, are represented as uninterrupted lines. They are also shown as arcs looped over one another (Fig. 6). Lines that show hose or tubing that connect are joined by a bullet, or are shown as perpendicular joining lines (Fig. 7).

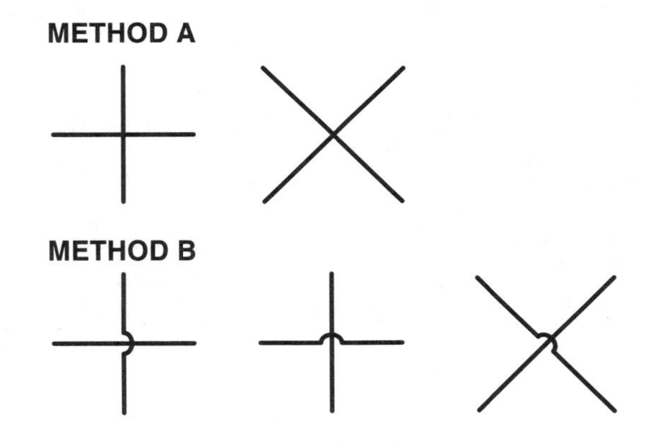

Fig. 6 — Uninterrupted Lines Represent Hose or Pipe that Cross Over Each Other But Do Not Connect.

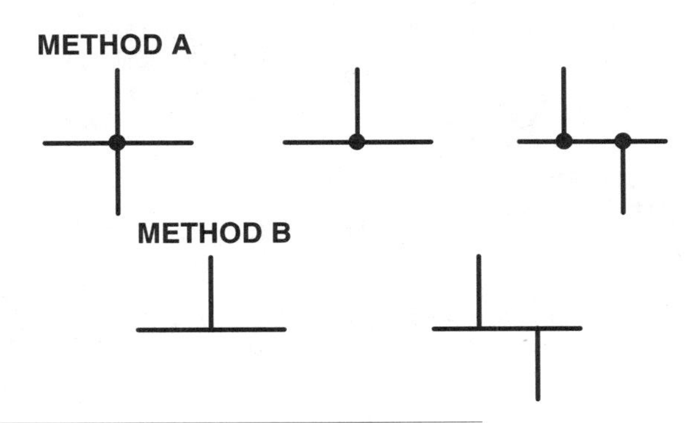

Fig. 7 — Connected Pipes or Hose Symbols

Arrows

Arrows show the direction of the flow of hydraulic fluid. They also show direction of movement in pumps and motors. An arrow inside of a circle or through two arcs can represent an adjustment point for the amount of hydraulic fluid flow or pressure (Fig. 8).

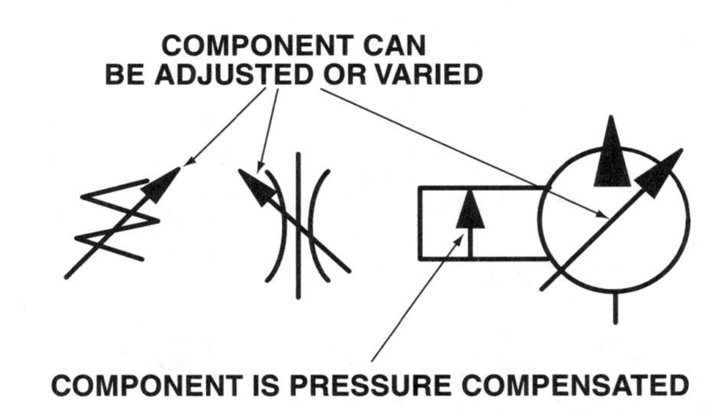

Fig. 8 — Arrows Can Represent Adjustment Points.

Arcs

Arcs show a point of adjustment when they are used together. See Fig. 4 "a flow control." They are also used to show a flexible hose line in a hydraulic system.

HYDRAULIC COMPONENTS AND SYMBOLS

PUMPS

The basic symbol for hydraulic pumps is a circle or semicircle (Fig. 9). The circle or semicircle alone isn't enough. It doesn't tell you anything at all about the pump. For example it doesn't tell you how many ports it has, whether it has fixed or variable displacement; whether or not it is unidirectional or bidirectional (sometimes referred to as over-center): or if it is pressure compensated.

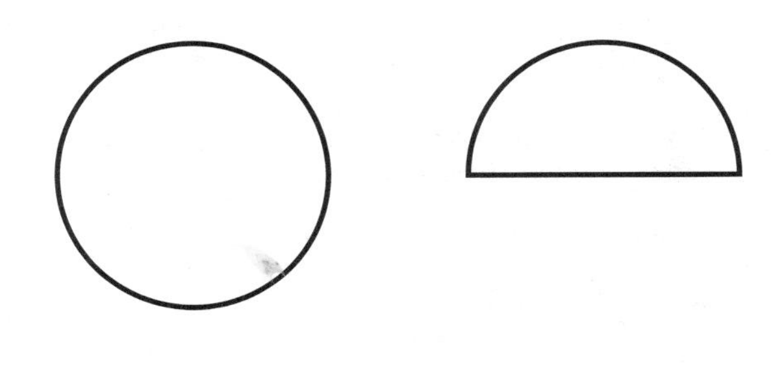

Fig. 9 — Circles and Semicircles Represent Pumps and Motors.

Even the cutaway and pictorial views shown in Fig. 10 do not tell us much more than the ISO symbol. We **can** tell from the cutaway and pictorial views that the pump shown is an external gear pump. We can also see how the pump is constructed and how the gears, shafts, wear plates, and housing fit together.

When a hydraulic system is shown, the ISO symbol tells us all we need to know. It represents the function of the pump within the circuit and its relationship to the other components that make up the circuit.

Now we will start adding some arrows, lines, arrowheads, and rectangles to it and see how we can use a circle to represent many different types of pumps.

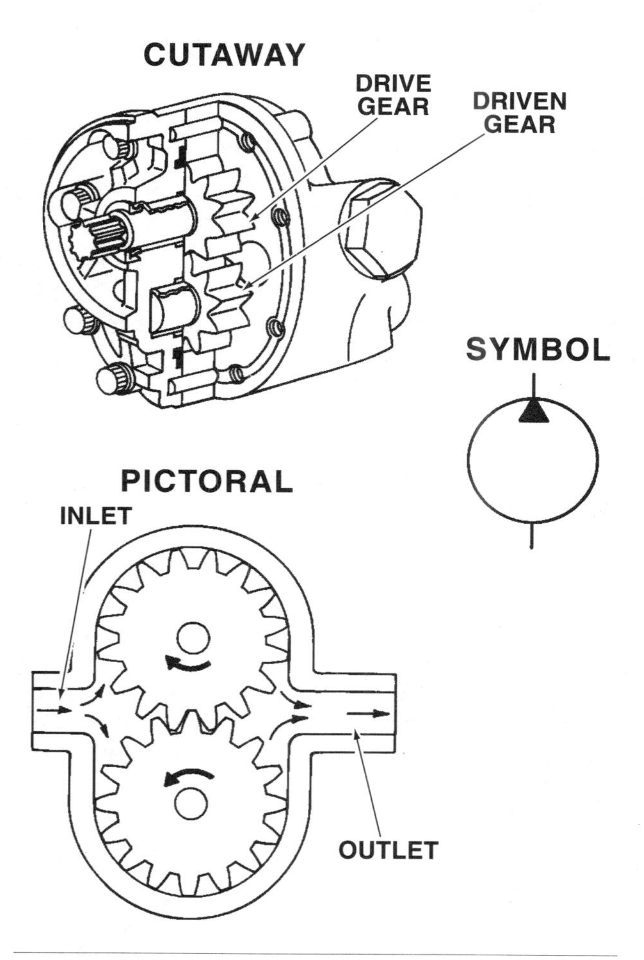

Fig. 10 — Symbols Often Tell You As Much Or More Information Than Cutaway or Pictorial Drawings.

Fixed Displacement Pumps

When we look at Fig. 11, we start to learn a little more. View 1 symbolizes a unidirectional pump with two parts. View 2 is a bidirectional pump (hydraulic oil can flow in either direction) as indicated by the two triangles. Both pumps shown in these two views are fixed displacement pumps.

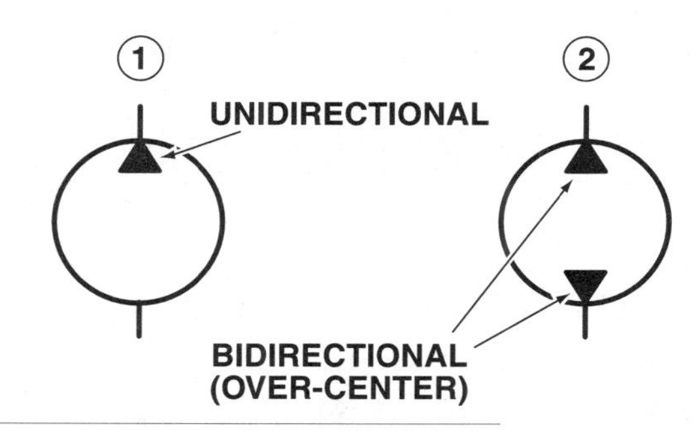

Fig. 11 — Arrows Show Oil Flow in Pumps

Now, let's add some arrows and see how they affect our symbol.

Variable Displacement Pumps

View 1 of Fig. 12 is a unidirectional variable displacement pump as indicated by the arrows going through the circle symbol. In other words, the output of the pump can be adjusted or varied. Also, a shaft and its direction of rotation is shown. When indicating direction, the arrow is assumed to be on the near side of the shaft.

View 2 of Fig. 12 is an even more sophisticated pump. It is a bidirectional, variable displacement, pressure compensated pump with a reversing shaft and a drain line back to the hydraulic reservoir of the machine. Notice the arrowhead at the bottom of the circle.

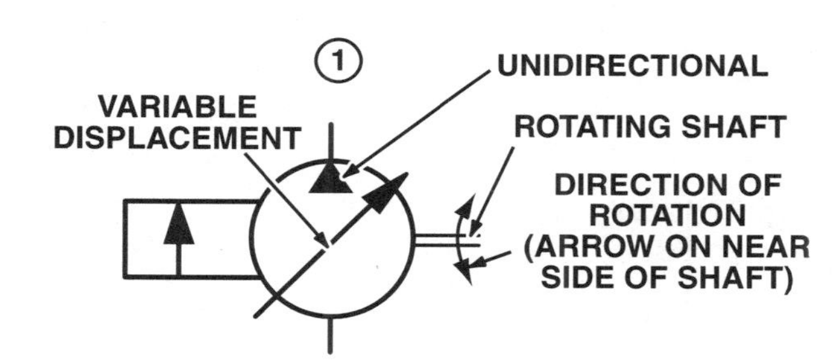

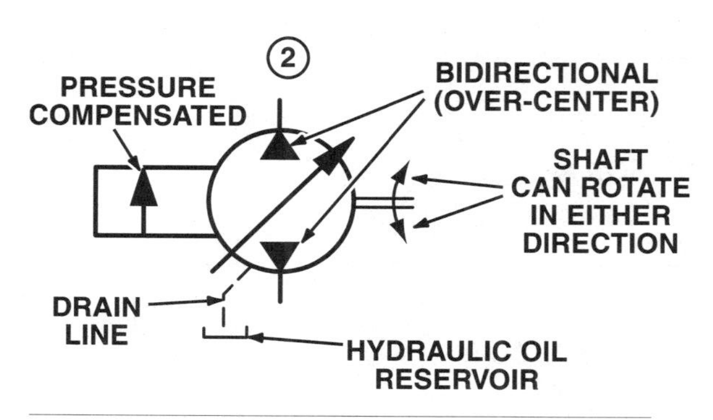

Fig. 12 — Complex Pump Symbols Are Made From Basic Symbols.

Summary

The circle is the basic symbol for pumps. The type of pump the circle is representing will determine the need for additional symbols. The addition of arrows tell if the pump is pressure compensated, if it is a variable displacement type, and in what direction the shaft is rotating. Arrowheads indicate whether it is unidirectional or bidirectional.

MOTORS

The basic symbol for a hydraulic motor is the same as for a pump. Motors and pumps are similar in design (see Chapter 5).

Like pumps, motors can be either unidirectional or bidirectional. They can be fixed or variable displacement, and be pressure compensated. These characteristics can all be shown symbolically.

It is not enough to show a circle with the parts added. You would have no way of knowing whether the symbol represents a pump or a motor. It is represented by simply inverting the arrowheads within the circle symbol. This also indicates whether the pump or motor is unidirectional or bidirectional. Since all pumps and motors are either unidirectional or bidirectional, arrowheads indicate the type of pump or motor.

In Fig. 13 we see a bidirectional, variable displacement, pressure compensated pump and motor with a reversing shaft and drain line. View 1 is the pump; view 2 is the motor.

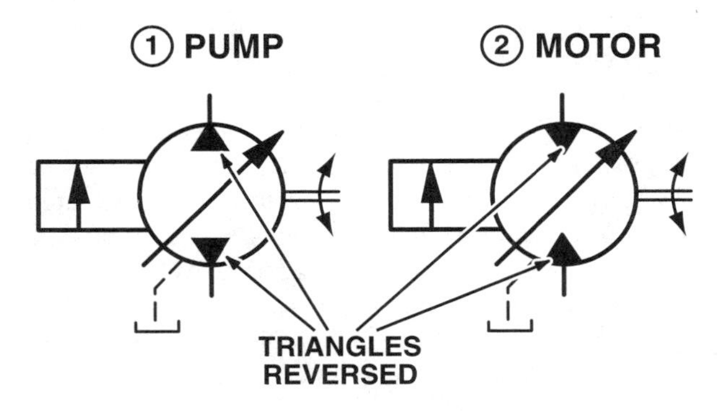

Fig. 13 — The Position of Arrowheads Differentiates Pumps from Motors.

Both Pump and Motor

In some hydraulic systems, a component will operate as a pump part of the time and as a motor part of the time, depending on what the system was designed to do.

How do we read this on our fluid power diagrams? Again, it's simply the way the arrowheads are placed within the circle symbol.

Let's look at some examples.

In Fig. 14 we see the symbol for a part that is both pump and motor. View 1 indicates that the component operates in one direction as a pump and in the other direction as a motor. View 2 shows a component that operates in only one direction but either as a pump or motor. Note the placement of the arrowheads.

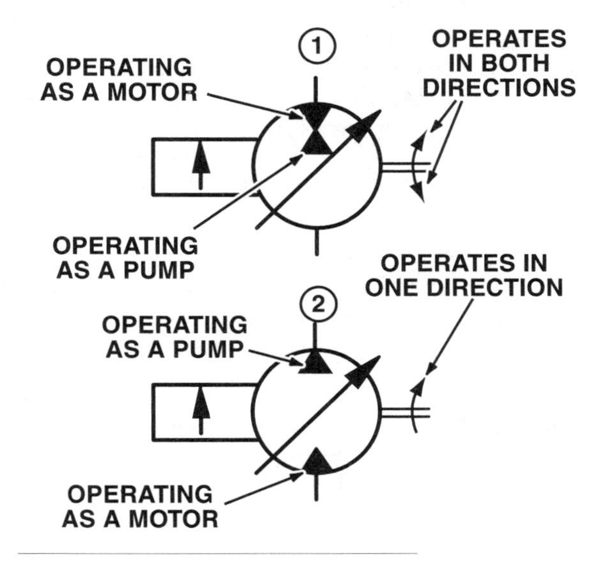

Fig. 14 — Pumps Can Operate as Motors.

The component shown in Fig. 15 can operate as a pump or motor in either direction. Note the four arrowheads. Two indicate a bidirectional pump and two indicate a bidirectional motor.

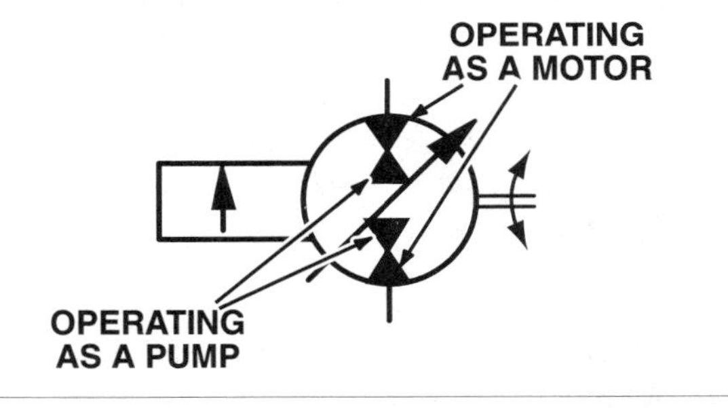

Fig. 15 — A Bidirectional Pump That Can Operate as a Motor Has Two Sets of Arrowheads.

Summary

The circle is the basic symbol for a motor and a pump. Arrows and arrowheads indicate the same things for a motor that they do for a pump. The key to whether a circle is representing a motor or a pump is the way that the arrowheads are placed. This also indicates a unidirectional or a bidirectional working of the motor or pump. In a pump symbol, the arrowheads point toward the outer circumference of the circle; in a motor symbol, they point toward the center of the circle.

VALVES

There are three major types of valves. They are the directional control, pressure control, and volume control valve. Three different types of ANSI symbols are used to represent them.

The square is the basic ISO symbol used to represent one of the most common directional valves the spool valve (Fig. 16). Two or more squares or envelopes indicate a valve having as many distinct positions as there are squares. Before we go further in our discussion of spool valves, let us review some rules in the correct use of ISO symbols for spool valves.

Correct usage of ISO Symbols in Directional Control Spool Valves

ISO symbols are internationally used and recognized. Because of this, certain rules must be followed when using them in fluid power diagrams to represent valves.

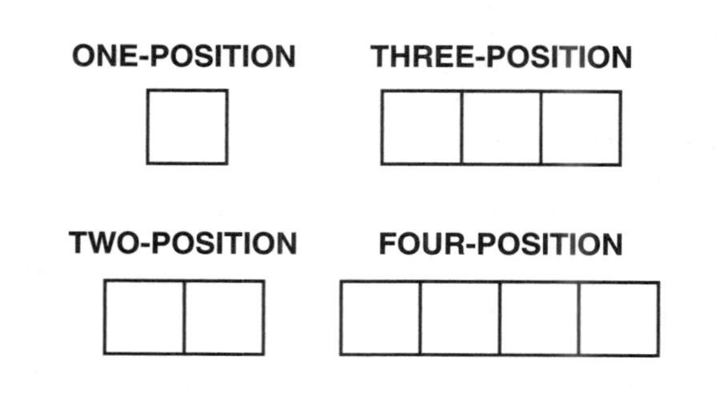

Fig. 16 — Squares Symbolize One Possibility of Oil Flow in a Valve.

1. Since the valves can only be in one valving position at a time, all connections from the valve to the rest of the circuit must be made from one block or envelope only. See Fig. 17 "hydraulic line to rest circuit."

2. All connections to the external circuit from ISO symbol should be made from the square which shows flow when the valve is in its nonactuated or normal state. In Fig. 17 we see a 2-way normally closed spool valve with manual actuating lever and return spring. In its nonactuated or normal state, the valve is closed. Therefore, it is shown connected to the rest of the circuit from the square that indicates no flow or closed position.

3. All valve actuators — hand lever, foot pedal and solenoid — should always be visualized as "pushing" the entire assembly of porting squares or envelopes in a lateral direction. Therefore, external circuit connections should be made to the porting block or square farthest from the actuator. This situation is shown in Fig. 17. If you were to "push" on the symbol representing the manual actuating lever, the two squares would slide to the right causing the arrow that indicates flow to align with the hydraulic lines leading to the rest of the circuit.

4. Arrows inside the squares or envelopes show direction of flow when that particular square is moved into working position. Look at Fig. 17 again. If the lever is pushed, valve slides the left square or envelope into alignment with the hydraulic lines. The arrow indicates the direction of flow. In some circuits, fluid may be free to flow in either direction. In these cases a double-headed arrow is used.

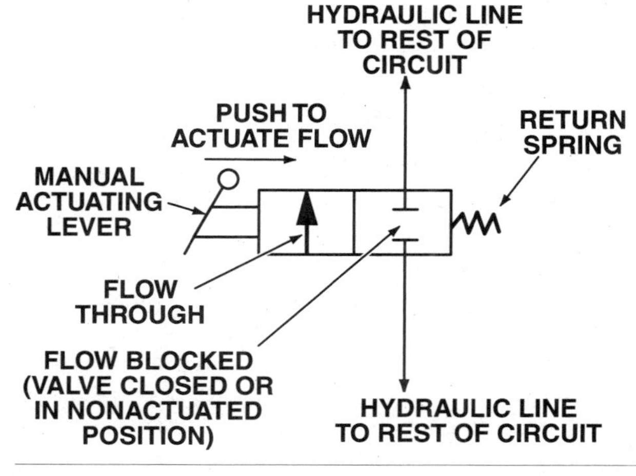

Fig. 17 — The Valve Must Slide To the Right in Order to Open the Hydraulic Circuit.

Two-Way Valves

In Fig. 18 we see a typical spool-type, two-way, normally closed valve both pictorially and by using ISO symbols. View 1 is a picture drawing of the valve in its normally closed or nonactuated state. The valve passage is blocked to oil flow. Graphically, the nonactuated valve is shown by a simple square (view 2) that indicates a blocked flow. The actuated valve is shown pictorially in view 3. Note that the ports are open to full flow. Graphically, the actuated valve is again shown by a square indicating full flow (view 4). When the two squares are put together (view 5), both flow functions of this 2-way, normally closed valve, are shown.

To complete the ISO symbol, the primary actuator (a manual lever in this case) is placed on one end and the return actuator (a spring) is placed on the other end (view 5).

Because the valve shown in Fig. 18 is normally closed the operator must hold it in the actuated or open state by keeping the handle firmly depressed. When the handle is released, spring tension returns the valve to its normally closed state.

In Fig. 19 we see a 2-way, normally open spool valve. It operates in reverse of the normally closed valve in that the spring returns it to the open position when it is nonactuated. Note that in the

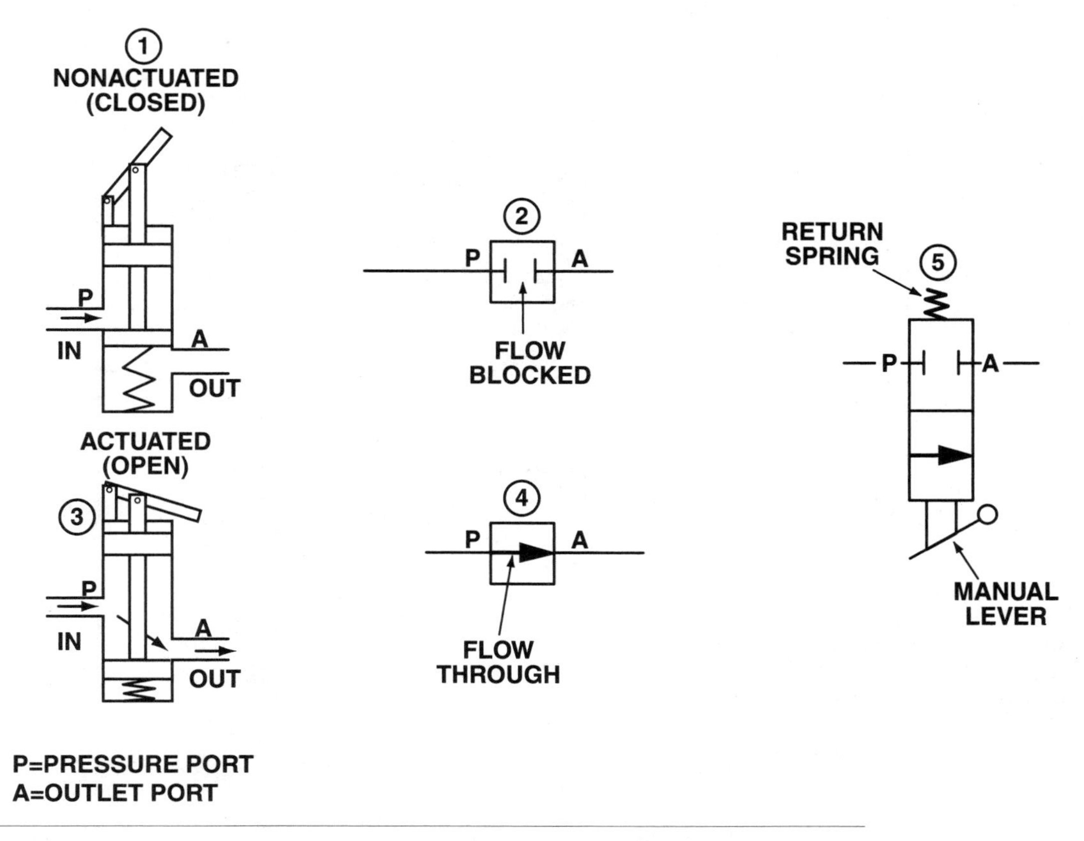

Fig. 18 — The Operator Must Keep the Valve Lever Depressed to Keep the HYdraulic Oil Through the Valve.

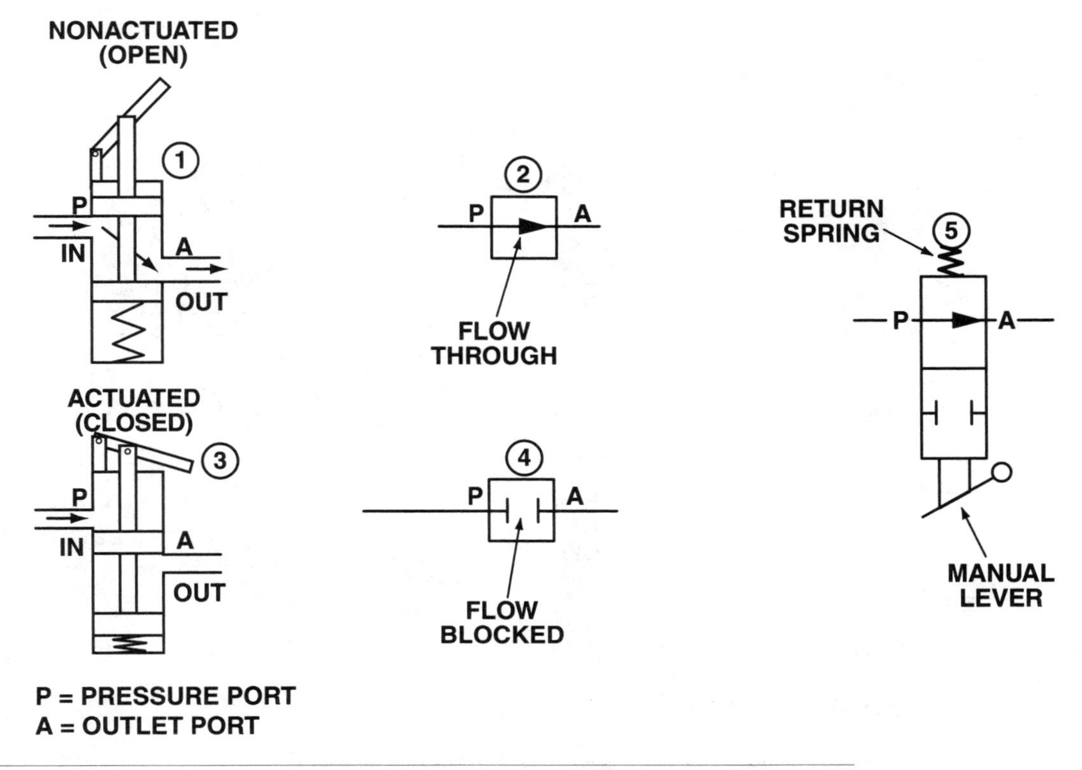

Fig. 19 — The Valve Lever Must Be Depressed In Order to Keep the Hydraulic Oil Blocked at the Valve.

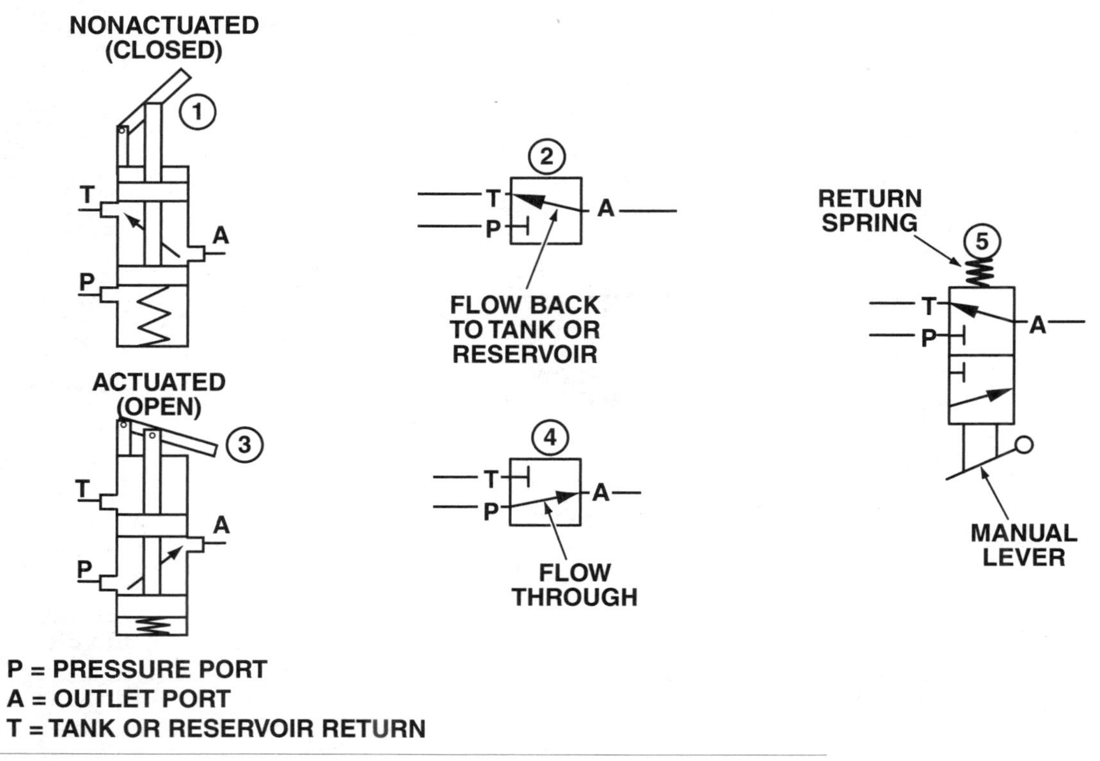

Fig. 20 — In Normal Positions This Valve is Closed and Oil Flows Back to the Tank or Reservoir.

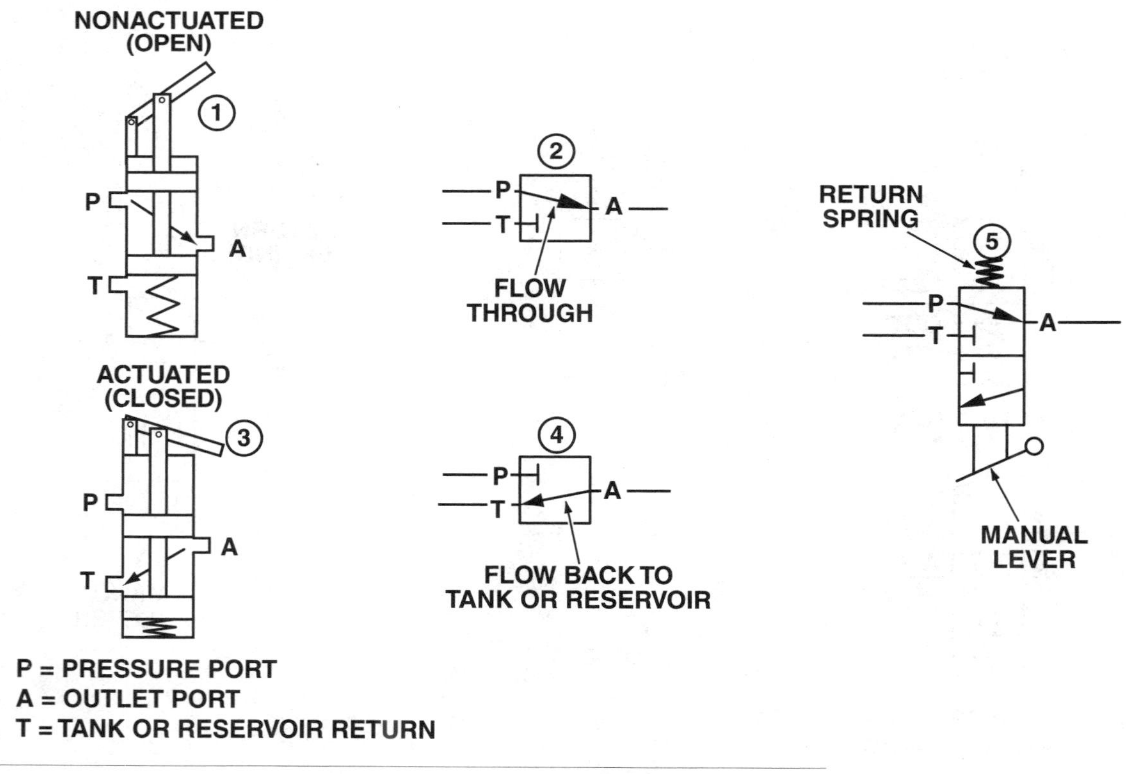

Fig. 21 — In Normal Position, This Valve is Open and Pressurized Oil Flows Through the Valve.

graphic illustration of this valve (view 5), the square indicating flow block is shown on the lever end of the symbol (remember rule 3).

Now, let's look at 3-way valves. They have three possible directions that the hydraulic fluid can flow.

Three-Way Valves

View 1 of Fig. 20 is a pictorial representation of a 3-way, normally closed valve in the nonactuated or closed state. Note that the pressure port is blocked. The flow of hydraulic fluid is only the return to the tank or reservoir. View 2 is the ISO symbol showing the same thing. Note the reversing arrow and the blocked pressure point.

View 3 of Fig. 20 shows the valve in the actuated or open position. Note that the pressure port is open to full flow to the outlet port. The return port to the tank or reservoir is blocked. View 4 is the ISO symbol. Note the arrow indicating flow from the pressure port to the outlet port. View 5 shows both squares or envelopes together with a manual lever and return spring added.

In Fig. 21 we see a 3-way, normally open spool valve. It operates in the reverse of the normally closed valve. They behave the same as the normally open and normally closed 2-way valves do.

Three-Position Directional Control Valve

Although the valve shown in Fig. 22 looks very similar to the ones shown in Fig. 20 and 21, it is quite different. It has no normal position — it remains in whatever position the operator places it. It does not have an exhaust or reservoir return port, but it has two outlet ports.

A valve of this type is called a three-position, closed center, directional control valve. It permits the operation of two systems from one pump. When the handle is in the up position (view 1, Fig. 22), hydraulic pressure enters pressure port P and exits through outlet port A to one system (a cylinder, for example). Pressure is closed off to outlet port B. When the stem or handle is in the down position (view 3), pressure exits through port B to operate a second system (such as another cylinder). Port A is then closed to pressure. When the stem is centered (view 2), both outlet ports are closed with neither receiving pressurized hydraulic fluid.

That is why the valve is called a three-position, closed center, directional control valve. It can be placed in any one of three positions; pressure can be directed to either one of two systems; and when the stem is centered, the valve is closed. Notice that it takes three squares for the complete symbol of this valve - a square for each of the three positions.

Now let us look at a three-position, 4-way valve and see how we represent it with an ISO symbol.

Three Position, 4-Way Valves

In Fig. 23 we see a three-position, 4-way valve with a closed center both pictorially and graphically. When the spool is shifted left, pressure enters through port P and is directed out port B to a cylinder. This allows port A to exhaust or drain back to the tank or reservoir through port T. When the spool is shifted right, the opposite happens — pressure from port P is routed to port A and port B exhausts to the tank (port T). When the spool is centered, all four ports are closed. Pressure is still present at port P but it can't go anywhere. The ISO symbols show this 4-way flow by using arrows to connect the ports.

The ISO symbol in Fig. 24 is the same valve shown in Fig. 23. The actuators and centering mechanisms have been added in this case solenoids are the actuators and springs are the centering devices.

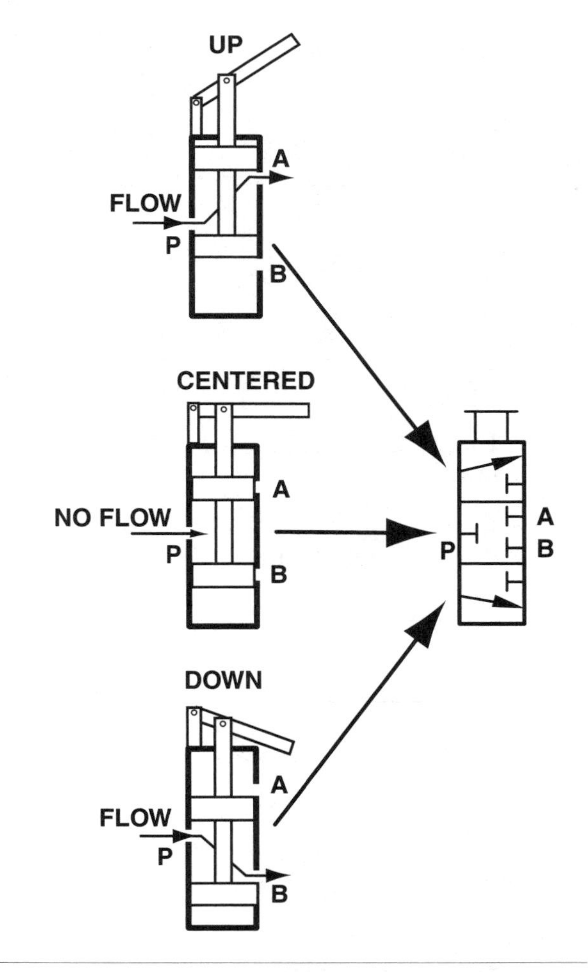

Fig. 22 — Two Systems Can Be Operated at the Same Time with This Three-Position, Closed Center Valve.

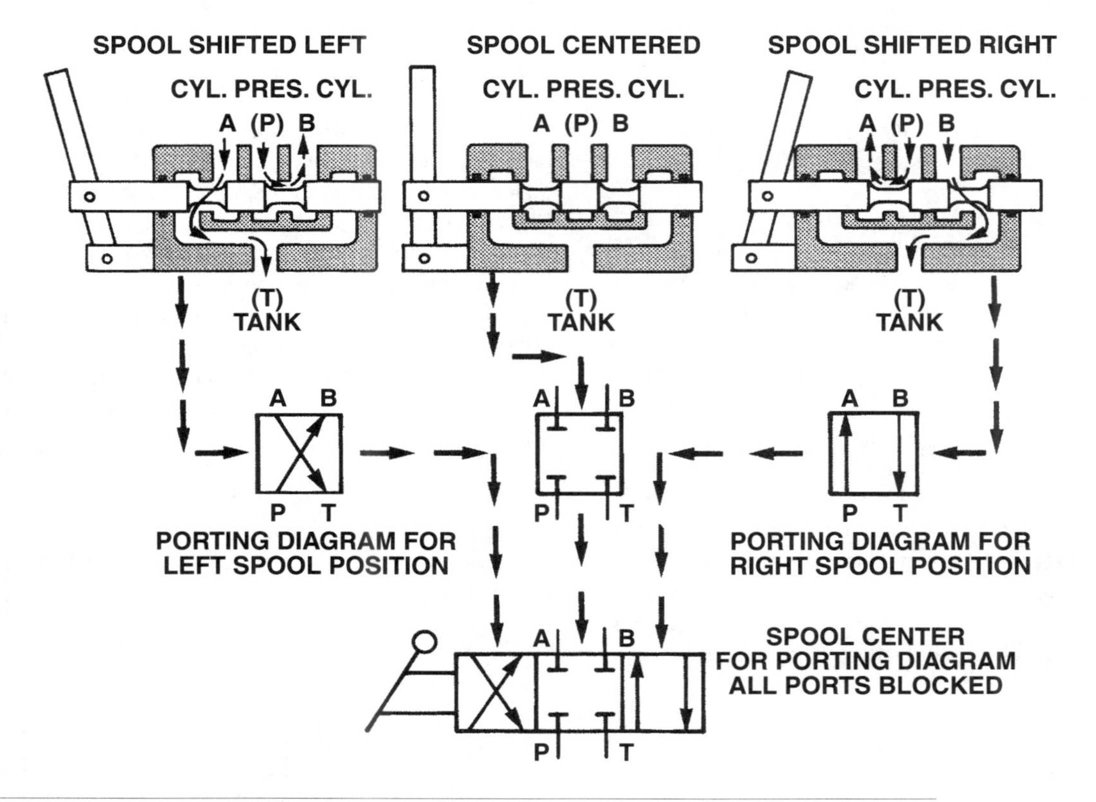

Fig. 23 — Two Systems Can Be Operated at the same Time with This Three-Position, Closed Center Valve.

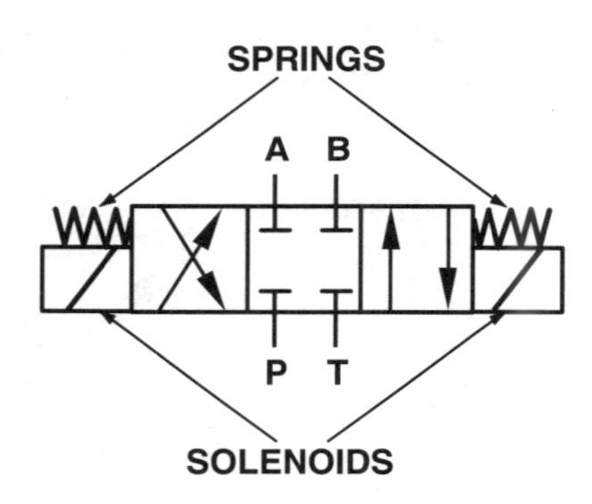

Fig. 24 — Electronic Solenoids Can Be Used to Control Valves.

Flow Control Valves

A second type of control valves are needle valves and check valves.

Not all valves use the square as their basic symbol. Needle valves and check valves are types of valves that are represented by different symbols.

A flow control valve is shown in Fig. 25. It consists of a needle valve and check valve. Flow is controlled in one direction only. It is toward the right. Fluid flowing toward the left can pass in free flow through the internal check valve.

The ISO symbol of this valve (Fig. 25) has an arrow marking the direction of controlled flow and is an integral part of the symbol. An arrow (not a part of the symbol) shows free flow direction back through the check valve. Note also that the pressure port and free flow port are marked.

The valve in Fig. 25 is actually two commonly used hydraulic fluid flow restrictors. First, the adjustable needle valve is shown with the two arcs and the arrow through them. (This use of the arrow symbol shows an adjustment point whether it is used alone or with another symbol).

The second is the directional flow restrictor. The ball inside the arrowhead outline symbolizes that hydraulic fluid is free to flow from right to left in the valve (Fig. 25). Fluid cannot flow from left to right. This also is a common symbol that can be used alone. Remember, the two arrows shown outside of this valve are for clarification purposes only and are not part of the symbol.

SUMMARY

The main point to remember in constructing an ISO symbol for a spool valve is that all positions, flow directions, and method of actuation are shown correctly.

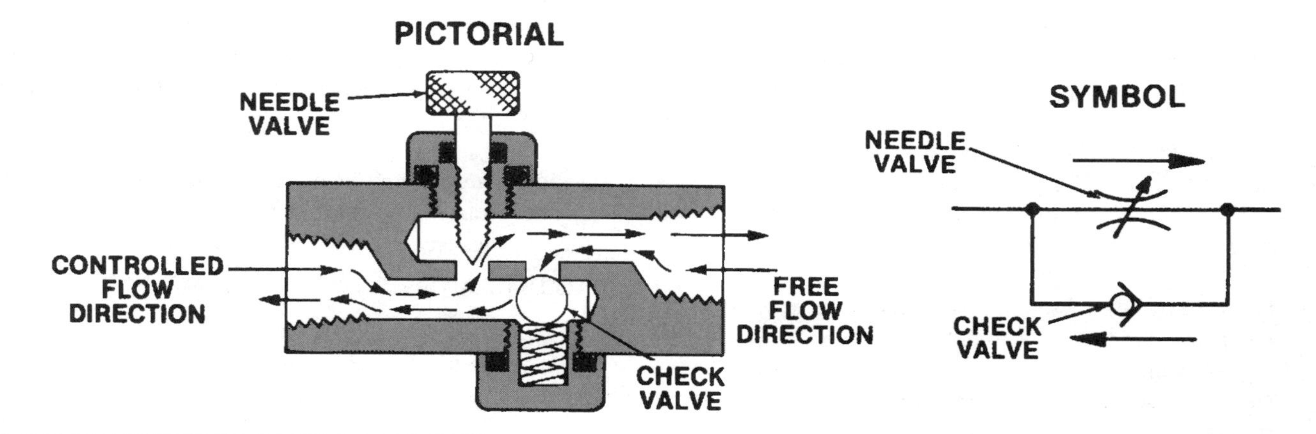

Fig. 25 — Needle Valves Also Are Fluid Control Valves

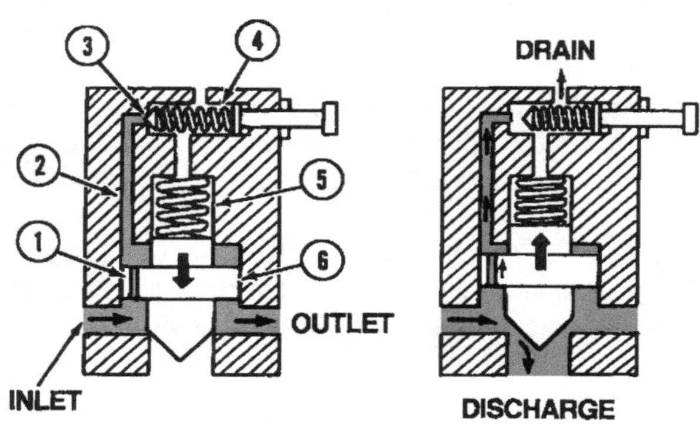

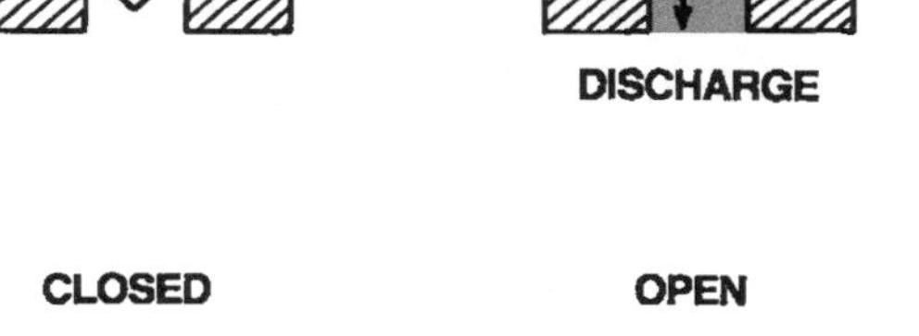

Fig. 26 — A Pressure Relief Valve Controls Hydraulic Pressure Internally in the System.

Remember:

- **A four-position valve must be represented by four squares (or envelopes).**
- **A three-position valve must be represented by three squares.**
- **A two-position valve must be represented by two squares.**
- **One arrow must be used for each direction of flow. The ISO symbol for a 4-way valve would have four arrows, each indicating a direction of flow.**
- **The method by which the valve is actuated must be shown. This includes a manual lever, solenoid, pilot, or other controls.**
- **Flow control valves also restrict the direction of fluid flow, or the speed of fluid flow.**

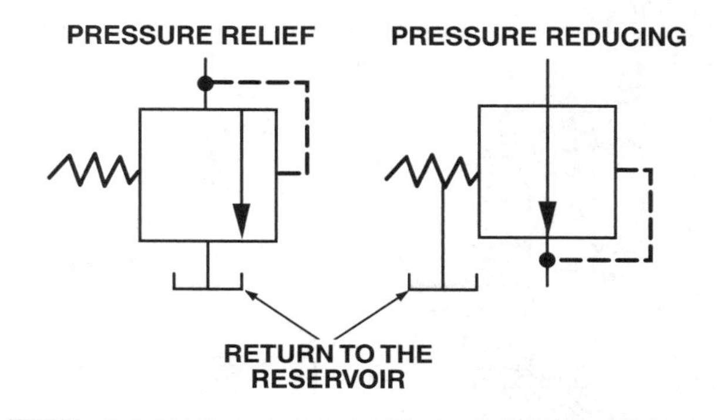

Fig. 27 — A pressure Control Valve Either Relieves A System With Too Much Pressure or Helps Reduce the Pressure.

Pressure Control Valves

The third common type of valve is the pressure control valve. This type of valve can be a pressure **relief** or **pressure** regulating type.

The pressure relief valve helps control the hydraulic system pressure by opening if the system pressure gets too high. Hydraulic fluid drains back to the reservoir until the system pressure reaches the desired setting. Pressure relief valves are usually **adjustable** and can be pilot operated by back pressure in the system (Fig. 26).

The symbol for a common type of pressure relief valve is in Fig. 27.

The pressure reducing valve is usually a preset valve that is intended to reduce hydraulic pressure in a certain part of the hydraulic system. In Fig. 27 we can see the symbolic representation. In Fig. 28 we can see a pictorial view of a common pressure reducing valve.

Again we can see how the basic arrow and square (envelope) can be used to represent a hydraulic valve.

CYLINDERS

The ISO symbol for a hydraulic cylinder resembles a cylinder. In Fig. 29 we see both a pictorial representation and the ISO symbol for a typical double-acting hydraulic cylinder. (As you will recall from Chapter4, a double-acting cylinder provides force in both directions of travel. Pressure is applied both the extend and then to retract it).

Clearly evident in the ISO symbol are the piston, rod, and cylinder housing. The ISO symbol does not indicate the piston seals, the type of clevis, or the internal configuration of the cylinder housing. This isn't necessary for a hydraulic circuit diagram.

In Fig. 30 we see a diagram that shows how two cylinders can be operated at the same time. In this figure the cylinders are labeled double-acting, but in an actual circuit diagram, nothing would be labeled. How would we know, then, that the cylinders are double-acting instead of single-acting? Note that

FROM MAIN CIRCUIT

TO SECONDARY CIRCUIT

VALVE OPEN, NOT OPERATING

VALVE PARTLY CLOSES TO REDUCE PRESSURE

Fig. 28 — A Common Pressure Reducing Valve-Pictorial View.

both ports on the cylinders can be connected to the pressure ports on the valves when the valves are actuated. (Remember rule 3 earlier in this chapter for the correct usage of ISO symbols.)

If the cylinders were single-acting, one of the ports on each cylinder would be a vent (not connected to the valves). Also, 4-way valves would not be necessary.

ACCUMULATORS

As we will learn in Chapter 8, accumulators can be used in hydraulic systems to dampen pressure spikes and vibrations, to gradually build pressure, and to store oil under pressure for use later.

One of the applications for accumulator use is the closed-center system with fixed displacement pump and an accumulator shown in Fig. 29, Chapter 1. In the discussion of that system, we saw how a small pump could be used in a system that had short, infrequent work cycles. This reduces component size and costs and provides a closed-center system.

In Fig. 32, we see a comparison of this system to a conventional open-center hydraulic system. The pictorial view of the open-center system is shown in Fig. 19, Chapter 1.

Remember that a true hydraulic symbolic drawing does not label the parts of the system.

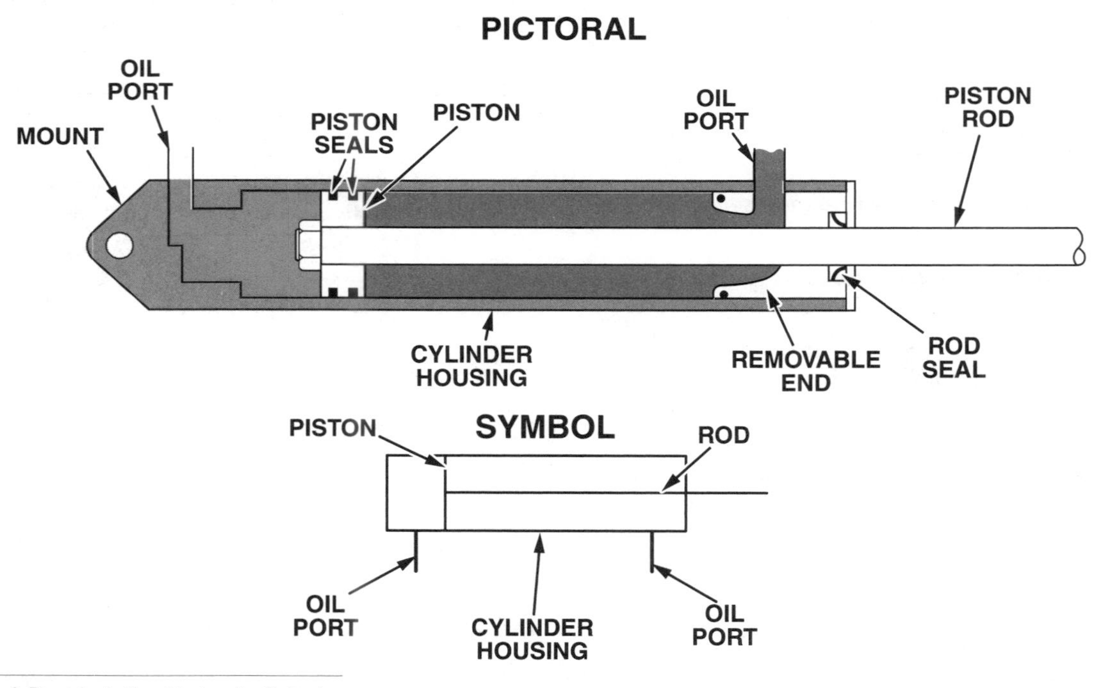

Fig. 29 — A Double-Acting Hydraulic Cylinder.

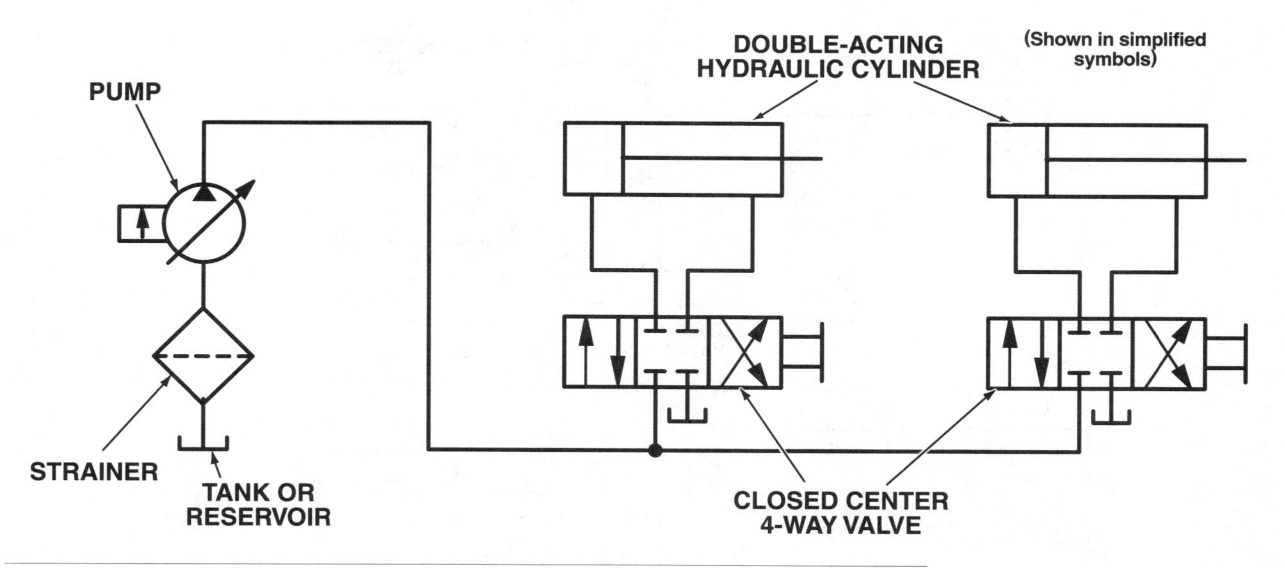

Fig. 30 — Two Double-Acting Cylinders Can Be Operated at the Same Time in the Same System.

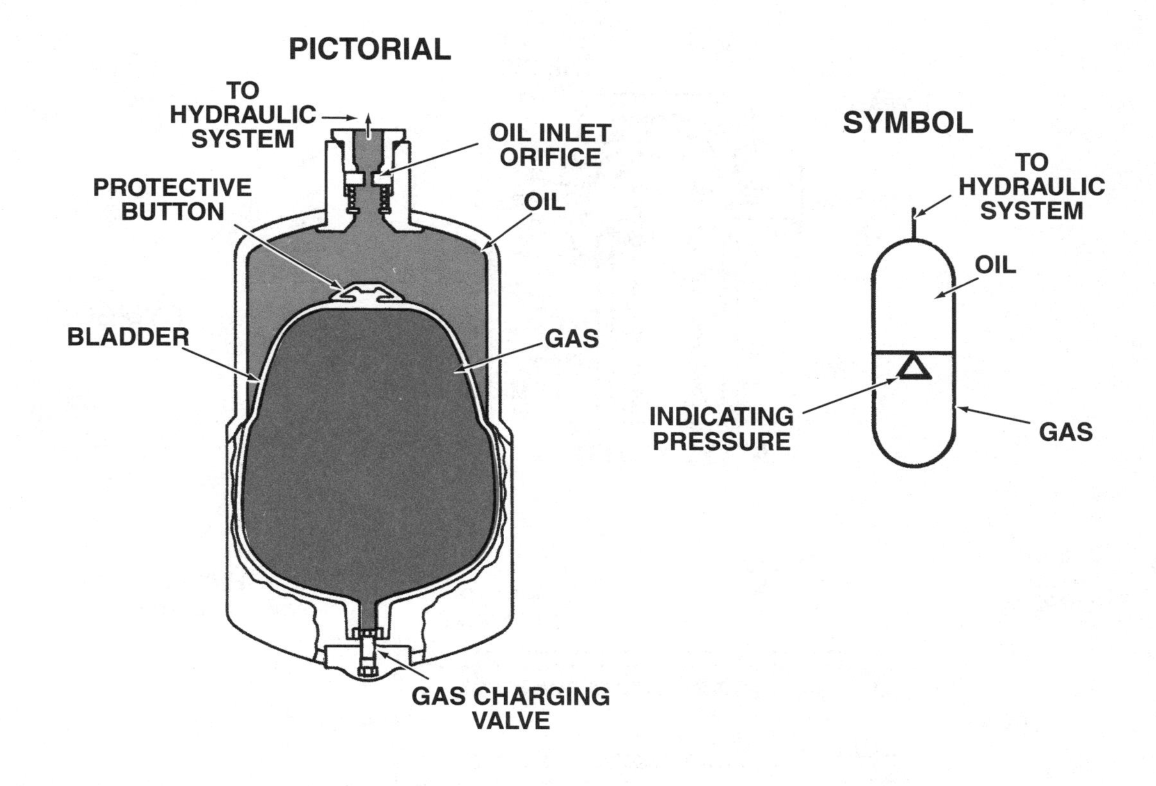

Fig. 31 — Accumulators Store Pressurized Hydraulic Fluid.

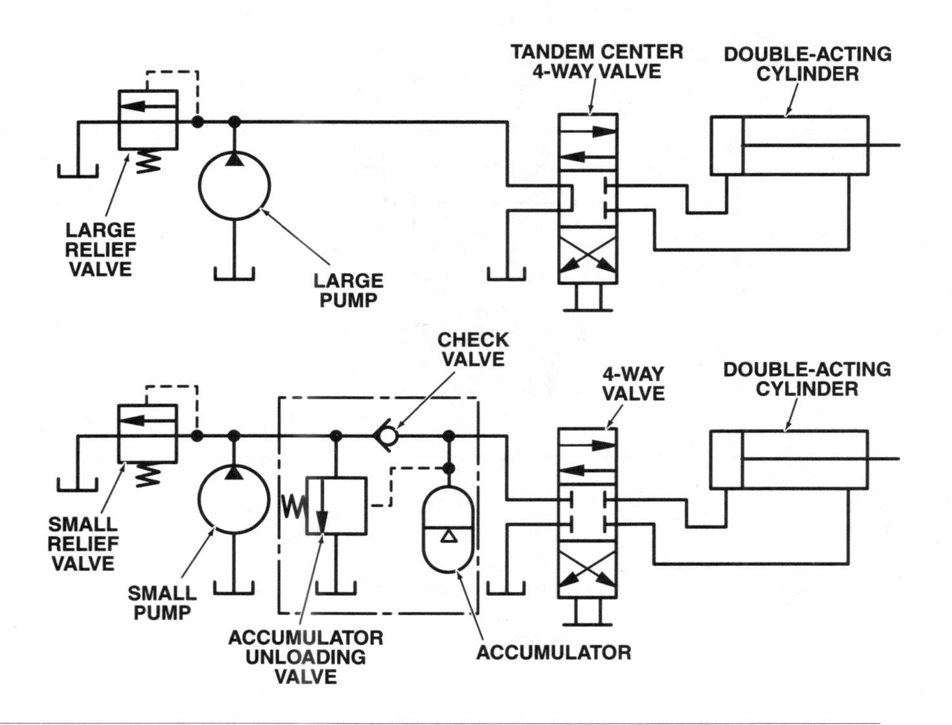

Fig. 32 — An Accumulator May be Economical if it Can Reduce the Size of the Pump and the Power Supply.

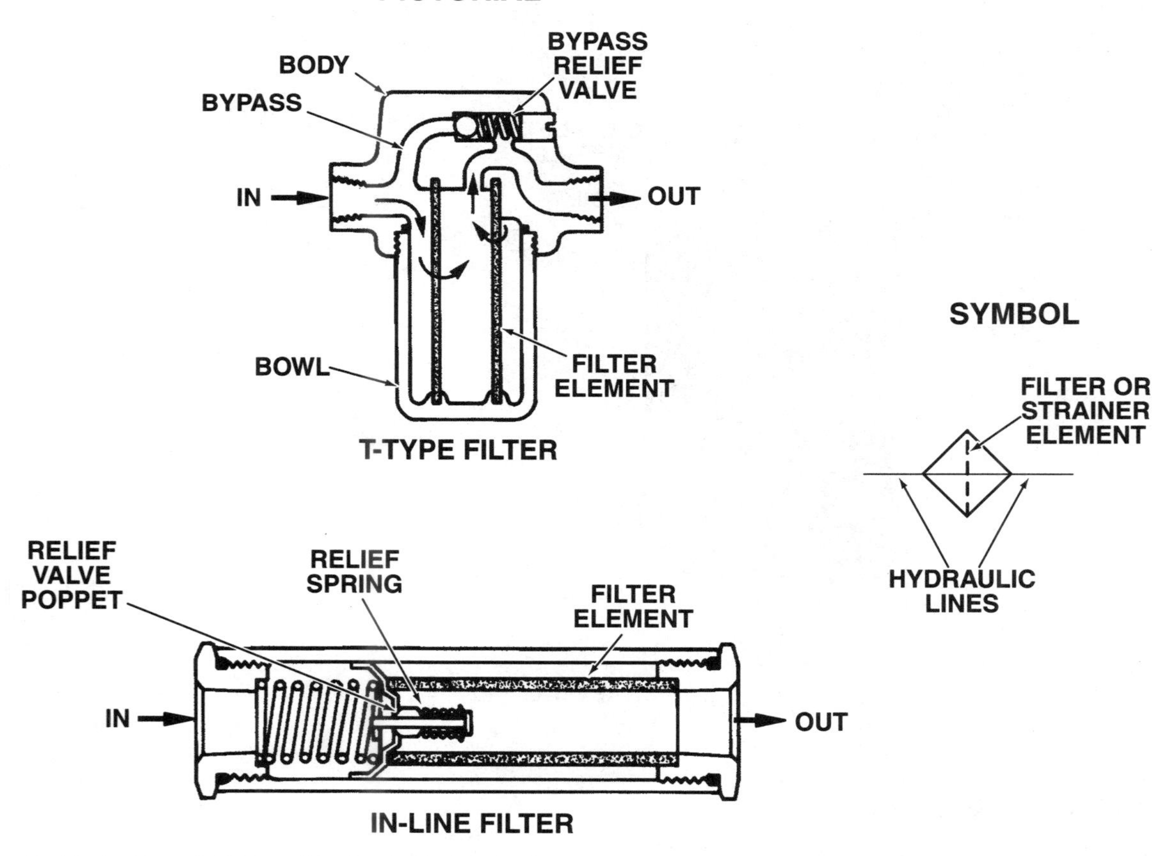

Fig. 33 — Oil Conditioners are Symbolized by Diamond Shapes.

FILTERS

Filters, strainers, and other hydraulic fluid conditioners, all use the same symbol. In Fig. 33 we see a pictorial cross section view of a T-type and in-line filter along with the ISO symbol. The symbol is a diamond. The system hydraulic lines are connected to two of the corners. The filter or strainer element is represented by a dotted line connecting the other two corners.

Fig. 34 shows two circuits both using a suction strainer in the same location. The top diagram shows a micronic filter in the high-pressure pump line. The lower diagram shows the filter placed in the reservoir return line from the system relief valve.

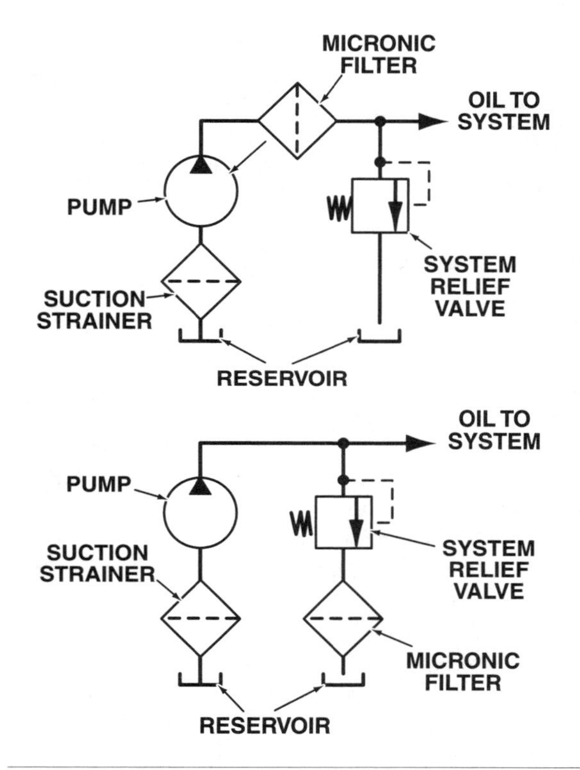

Fig. 34 — Filters Can be placed in Many Different Parts of the Hydraulic System.

OIL COOLERS

The two types of oil coolers (sometimes referred to as heat exchangers) that are widely used are air-to-oil and water-to-oil. Both are represented by the same ISO symbol.

In Fig. 35 we see a water-to-oil cooler along with the approved symbol.

Oil coolers will not stand high pressures. Therefore, they are installed in a low-pressure part of the system. This typical location is the reservoir return line. It is shown in Fig. 36.

HYDRAULIC RESERVOIRS

Hydraulic reservoirs are represented by the simplest ISO symbol (Fig. 37). You have been seeing this symbol on just about all the circuit diagrams in this chapter.

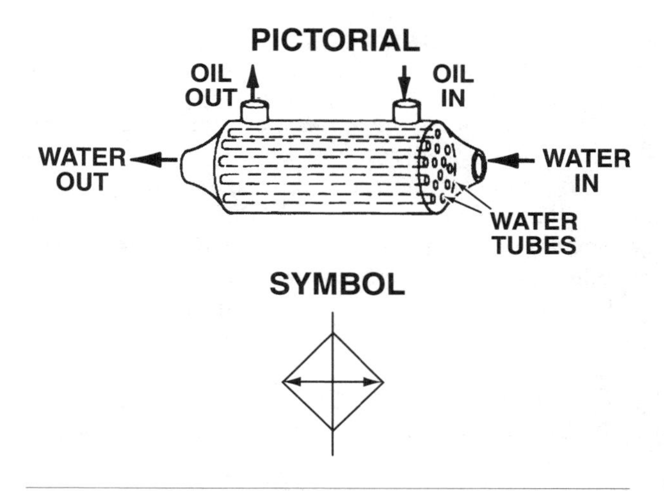

Fig. 35 — Water-to-Oil Coolers Transfer Heat from the Oil to the Water to Help Protect the Oil Quality.

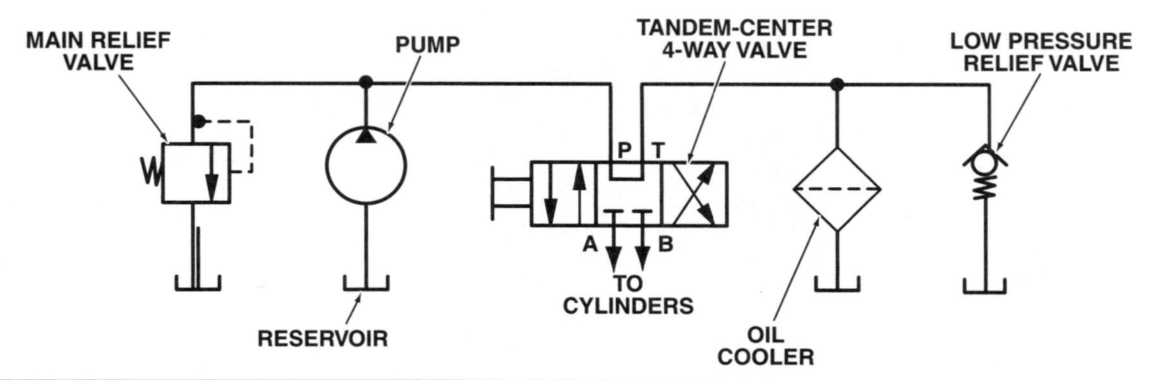

Fig. 36 — Oil Coolers are installed in a Low Pressure Area of the Hydraulic System.

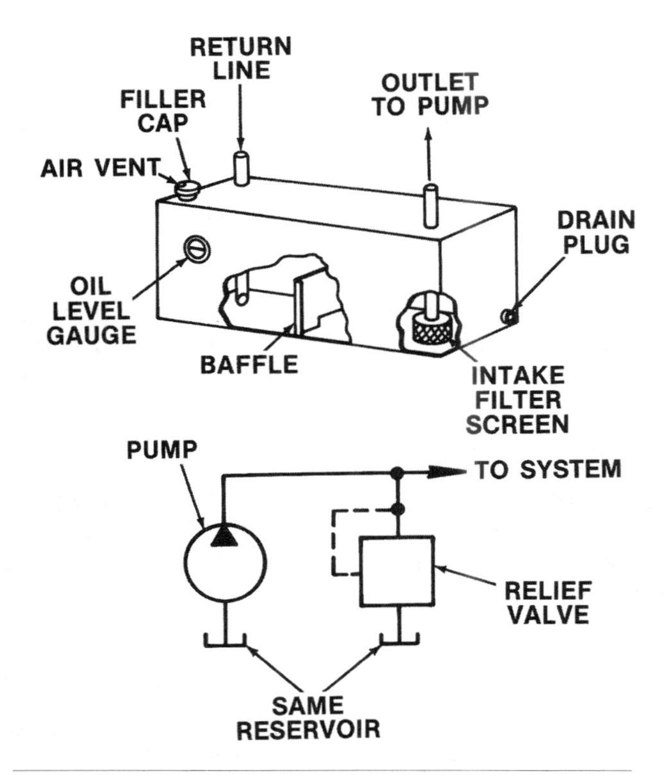

Fig. 37 — There Usually is Only One Hydraulic Fluid Reservoir in the Hydraulic System Even though Many Symbols May Be Used.

There is one main difference in its use and the other ISO symbols. Any given circuit diagram Fig. 36, for example may use the symbol several times, but in each instance, it represents the **same** reservoir. The symbol is used four times in Fig. 36 to represent different returns to the one reservoir. This prevents cluttering the diagram with so many return lines.

If you see two filter symbols (Fig. 34), then the system has two filters. The same applies to pumps, motors, cylinders, and valves.

The reason the symbol can be so simple is that all reservoirs are about the same. They may consist of a tank, gauge, baffle, air vent, intake screen, drain plug, and outlet and return lines as shown in Fig. 37.

SUMMARY

A hydraulic circuit diagram made with ISO symbols has certain advantages over a pictorial diagram.

- **They are internationally understood**
- **They simplify design, fabrication, analysis, and service of hydraulic circuits**
- **They emphasize component function**
- **They show connections and flow paths**
- **They are easier to draw than the pictorial and cutaway drawings**

To emphasis these points, let us look at Fig. 38 and Fig. 39. They are diagrams of the same circuit. One is pictorial, and the other is an ISO symbol diagram. Note how much simpler and easier to draw Fig. 39 is over Fig. 38. This is why the large majority of system diagrams that you will see will be drawn with symbols.

You must be knowledgeable about ISO symbols.

To help your learn ISO symbols, study Fig. 40. It contains all the basic symbols that you need to help interpret hydraulic circuit diagrams.

All symbolic diagrams are drawn with the basic forms — the circle, square, diamond, and rectangle. They are further defined with basic marks including arrows, lines, arrowheads, and arcs. With these basic concepts you will be able to decode even the most complex symbolic diagrams.

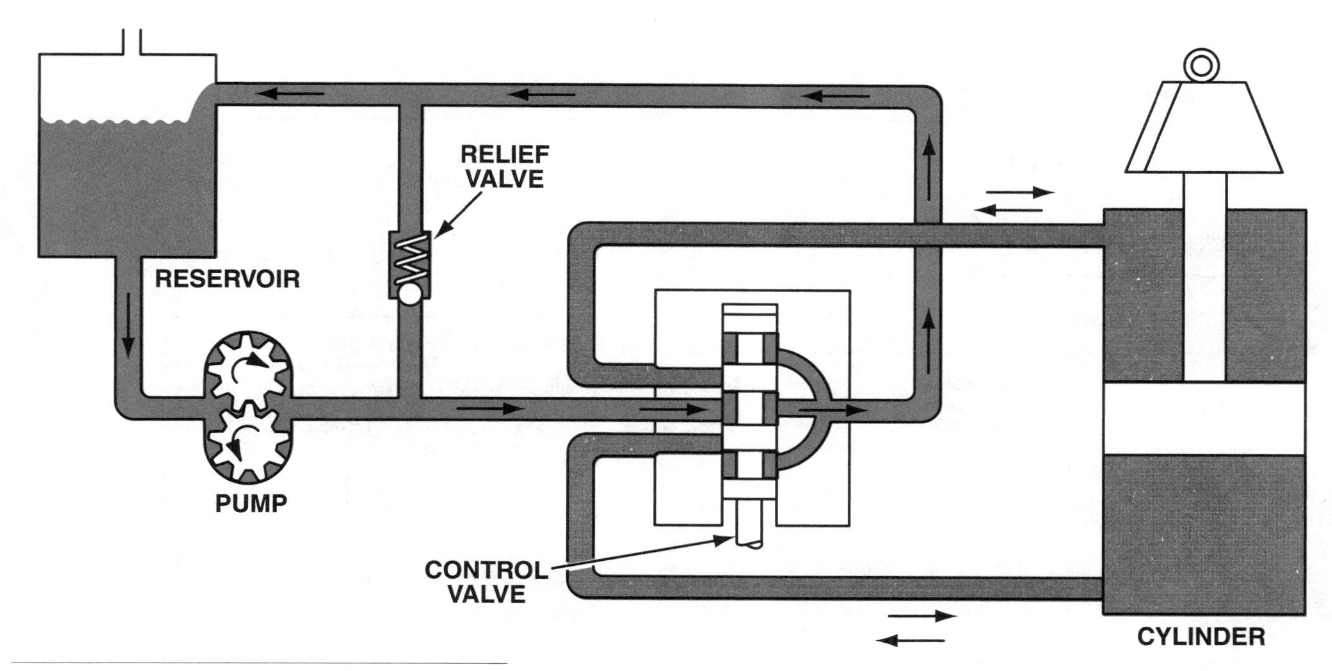

Fig. 38— A Pictorial Diagram of a Hydraulic System.

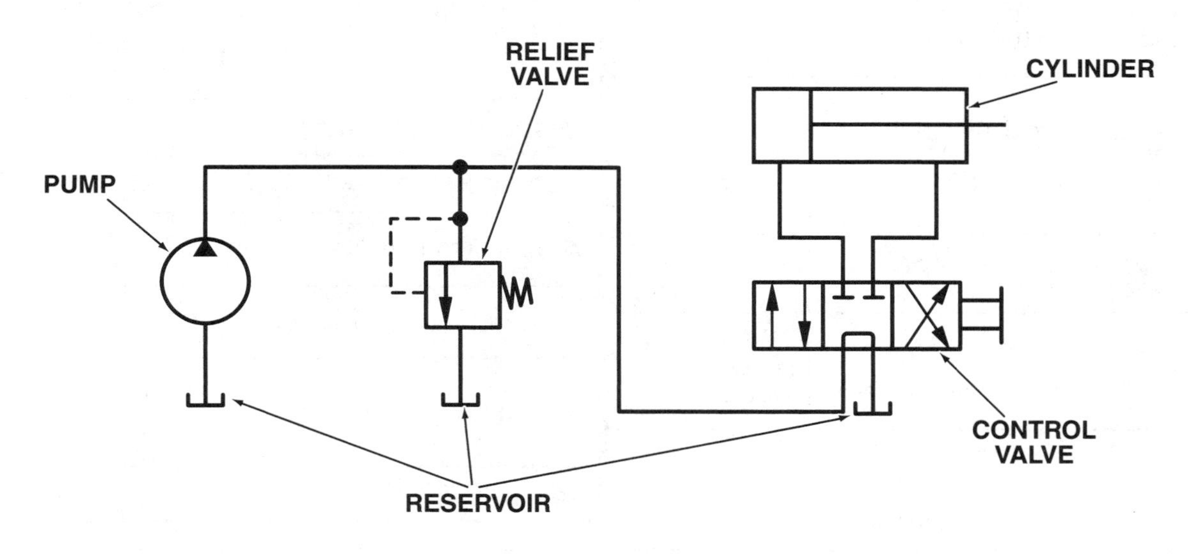

Fig. 39 — A Symbolic Diagram of a Hydraulic System.

Lines

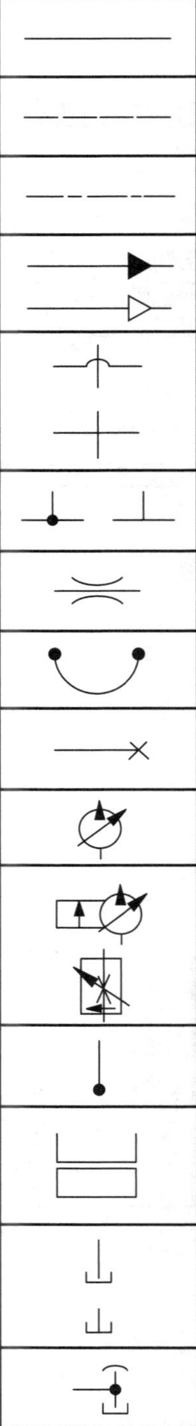

LINE, WORKING (MAIN)	
LINE, PILOT (FOR CONTROL)	
LINE, ENCLOSURE OUTLINE	
FLOW, DIRECTION OF — HYDRAULIC / PNEUMATIC	
LINES CROSSING	
LINES JOINING	
LINE WITH FIXED RESTRICTION	
LINE, FLEXIBLE	
STATION, TESTING, MEASUREMENT OR POWER TAKE-OFF	
VARIABLE COMPONENT (RUN ARROW THROUGH SYMBOL AT 45°)	
PRESSURE COMPENSATED UNITS (ARROW PARALLEL TO SHORT SIDE OF SYMBOL)	
TEMPERATURE CAUSE OR EFFECT	
RESERVOIR — VENTED / PRESSURIZED	
LINE, TO RESERVOIR — ABOVE FLUID LEVEL / BELOW FLUID LEVEL	
VENTED MANIFOLD	

Pumps

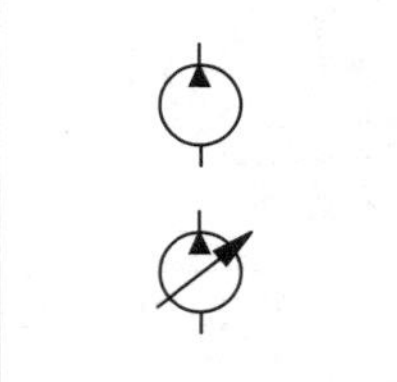

HYDRAULIC PUMP — FIXED DISPLACEMENT / VARIABLE DISPLACEMENT	

Motors and Cylinders

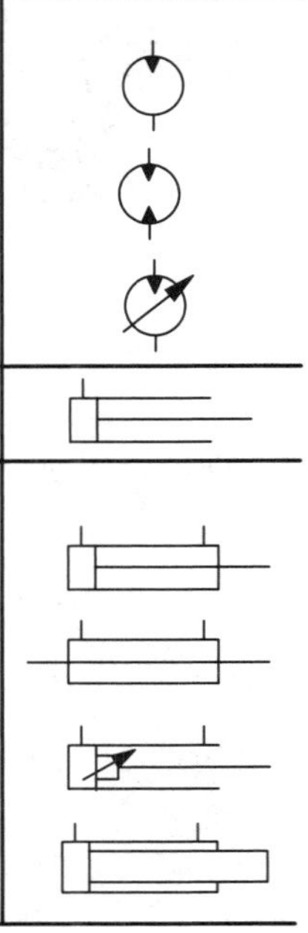

HYDRAULIC MOTOR — FIXED DISPLACEMENT ONE DIRECTION / FIXED DISPLACEMENT REVERSIBLE / VARIABLE DISPLACEMENT	
★CYLINDER, DOUBLE ACTING — SINGLE END ROD / DOUBLE END ROD / ADJUSTABLE CUSHION ADVANCE ONLY / DIFFERENTIAL PISTON	

Miscellaneous Units

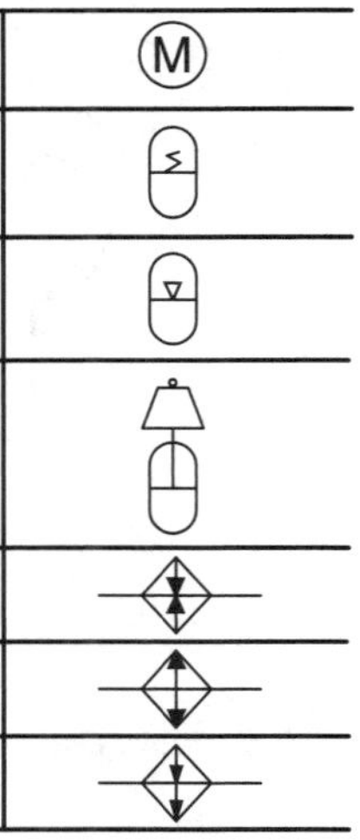

ELECTRIC MOTOR	M
ACCUMULATOR, SPRING LOADED	
ACCUMULATOR, GAS CHARGED	
ACCUMULATOR, WEIGHTED	
HEATER	
COOLER	
TEMPERATURE CONTROLLER	

★ Cylinder symbol shown in simplified version

Miscellaneous Units (cont.)

FILTER, STRAINER	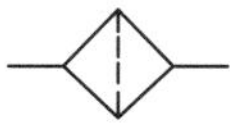
PRESSURE SWITCH	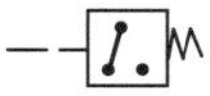
PRESSURE INDICATOR	
TEMPERATURE INDICATOR	
COMPONENT ENCLOSURE	
DIRECTION OF SHAFT ROTATION (ASSUME ARROW ON NEAR SIDE OF SHAFT)	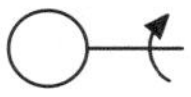

Methods of Operation

SPRING	
MANUAL	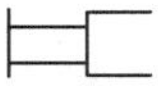
PUSH BUTTON	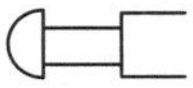
PUSH-PULL LEVER	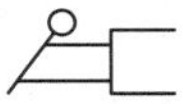
PEDAL OR TREADLE	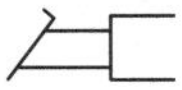
MECHANICAL	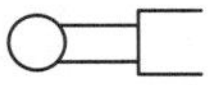
DETENT	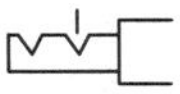
PRESSURE COMPENSATED	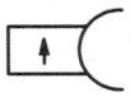
SOLENOID, SINGLE WINDING	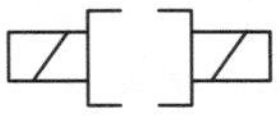
SERVO MOTOR	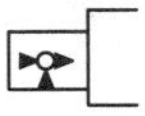

PILOT PRESSURE REMOTE SUPPLY INTERNAL SUPPLY	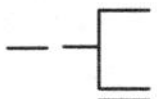 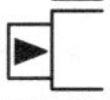

Valves

CHECK	
ON-OFF (MANUAL SHUT-OFF)	
PRESSURE RELIEF	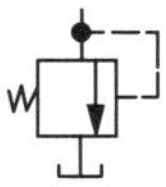
PRESSURE REDUCING TO CONSTANT PRESSURE	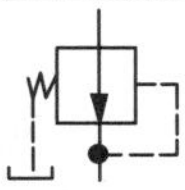
PRESSURE REDUCING CONSTANT AMOUNT	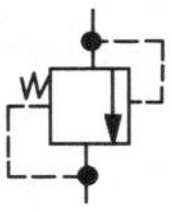
SEQUENCING	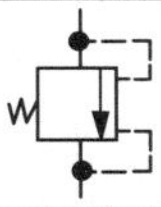
FLOW CONTROL ADJUSTABLE-NONCOMPENSATED	
FLOW CONTROL, ADJUSTABLE (TEMPERATURE AND PRESSURE COMPENSATED)	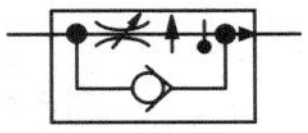
TWO POSITION TWO WAY	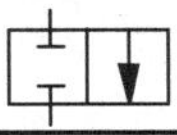
TWO POSITION THREE WAY	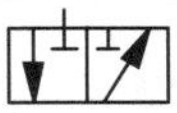
TWO POSITION FOUR WAY	
THREE POSITION FOUR WAY-OPEN CENTER	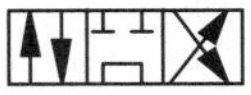
THREE POSITION FOUR WAY-CLOSED CENTER	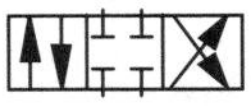
TWO POSITION IN TRANSITION	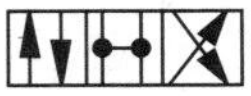
VALVES CAPABLE OF INFINITE POSITIONING (HORIZONTAL BARS INDICATE INFINITE POSITIONING ABILITY)	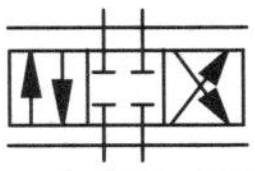

TEST YOURSELF

QUESTIONS

1. What are three advantages in using ISO symbols?

2. The circle is used to represent both pumps and motors. How do you tell which is which on a circuit diagram?

3. What does the large arrow running through the circle symbol at a 45 degree angle tell you?

4. (True or false) "A 4-way valve is always represented by four squares."

5. "A three-position valve is represented by ________ squares."

6. In an ISO symbol diagram, connections to the rest of the circuit from a three-position valve (three squares) are made from how many of the squares? Why?

7. "Connections from a valve to the rest of the circuit are made from the square that indicates the ____________________ state (position) of the valve." (actuated, nonactuated)

8. (True or false) "Filters and strainers are represented by the same symbol."

9. What hydraulic component may be represented several times on a diagram even though the actual system will have only one?

10. What do the initials ISO stand for?

HYDRAULIC PUMPS

INTRODUCTION

The pump is the heart of the hydraulic system. It creates the flow of fluid that supplies the whole circuit.

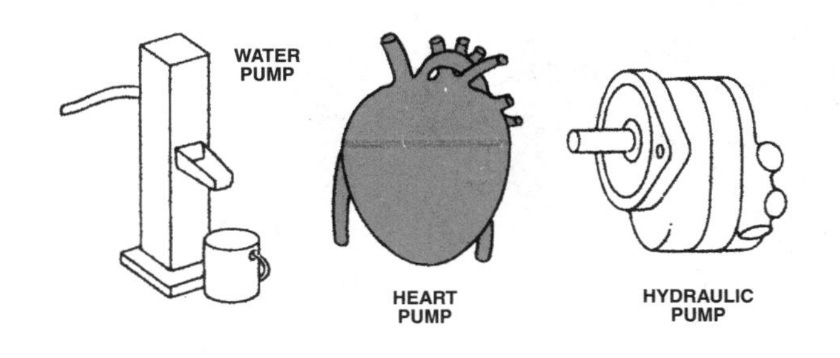

Fig. 1 — Three Kinds of Pumps

The human heart is a pump (Fig. 1). So was the old water pump once found on the farm. Somewhere in between, engineers have devised many kinds of hydraulic pumps, which do more than the old water pump, but only strive for the perfection of the human heart pump.

All pumps create flow, however, they are divided into two categories.

NON-POSITIVE DISPLACEMENT AND POSITIVE DISPLACEMENT PUMPS

Fig. 2 shows a centrifugal engine water pump that is a non-positive displacement pump. The output flow of the pump will vary or even stop depending on restriction to flow.

POSITIVE DISPLACEMENT

Fig. 2 also shows a piston pump that is a positive displacement pump. With each stroke of the pump piston, it will deliver an exact amount of fluid and will continue to do so regardless of the restriction to flow. Other types of positive displacement pumps are the gear and vane pumps. Each revolution of the pump results in a specific outlet flow.

In this chapter, we will discuss only the positive displacement pump that is the heart of the hydraulic systems. Although they are positive displacement pumps, they fall into two categories:

FIXED DISPLACEMENT pumps move the same volume of oil with every cycle. This volume is changed only when the speed of the pump is changed.

Volume can be affected by the pressure in the system, but this is due to an increase in internal leakage in the pump. This occurs when pressure rises.

VARIABLE DISPLACEMENT pumps can vary the volume of oil they move with each cycle, even at the same speed. These pumps have an internal mechanism, which varies the displacement and thus the output. They are used in applications where certain pressure or flow conditions must be maintained.

***Note:** "Displacement" is the volume of oil moved or displaced during each revolution or stroke of a pump. It is expressed in cubic inches or cubic centimeters per cycle or revolution.*

A word about pressure. Remember: A hydraulic pump does not create pressure; it creates flow. Pressure is caused by resistance to flow.

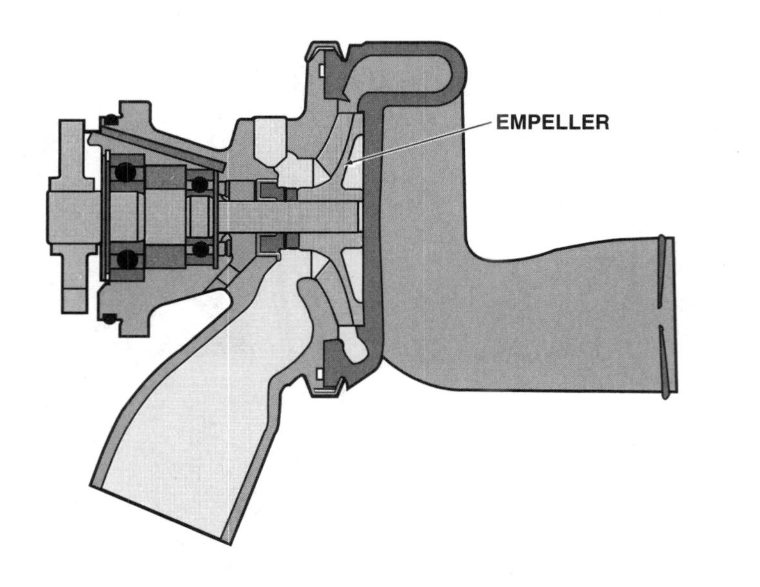

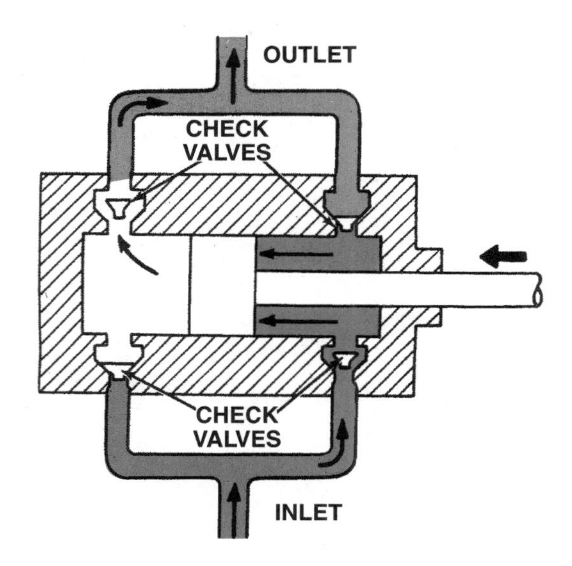

Fig. 2 — Nonpositive and Positive Displacement Pumps

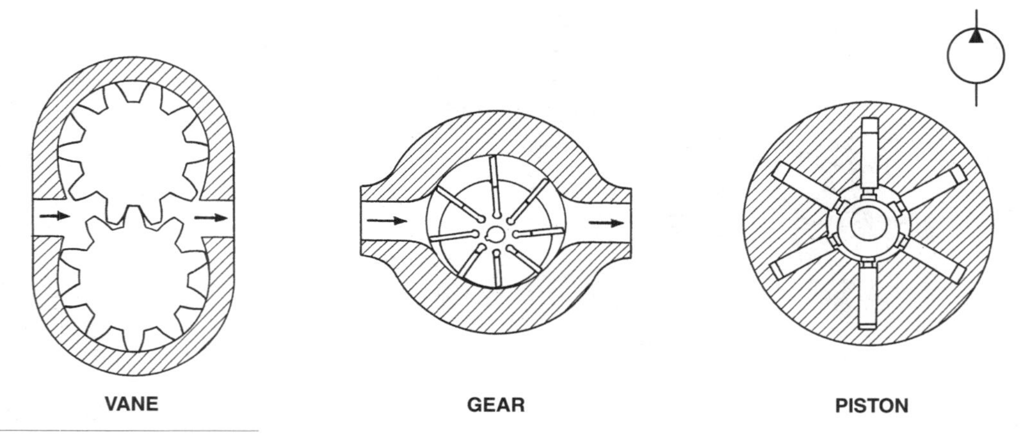

Fig. 3 — Three Types of Hydraulic Pumps

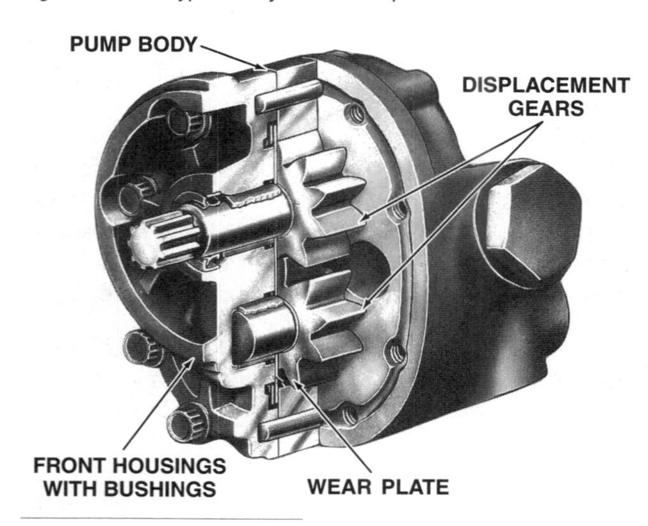

Fig. 4 — External Gear Pump

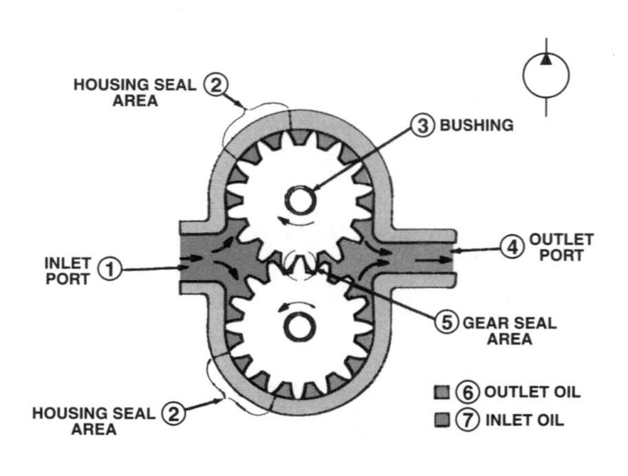

Fig. 5 — External Gear Pump in Operation

TYPES OF HYDRAULIC PUMPS

Now let's take an "inside" look at hydraulic pumps.

Most pumps used on today's machines are of three basic designs (Fig. 3):

- **Gear Pumps**
- **Vane Pumps**
- **Piston Pumps**

We will show how each type of pump operates and how it is used. A hydraulic system may use one of these pumps, or it may use two or more in combination.

All three designs work on the rotary principle: a rotating unit inside the pump moves the fluid. A rotary-positive-displacement pump meets the need for a compact, high volume component. This is needed in a mobile system where space is limited.

GEAR PUMPS

Gear pumps are the "pack horses" of hydraulic systems. They are widely used because they are simple and economical. While not capable of a variable displacement, they can produce the volume needed in most systems requiring fixed displacement pumps.

Two basic types of gear pumps are used:

- **External Gear Pumps**
- **Internal Gear Pumps**

Let's see how they work.

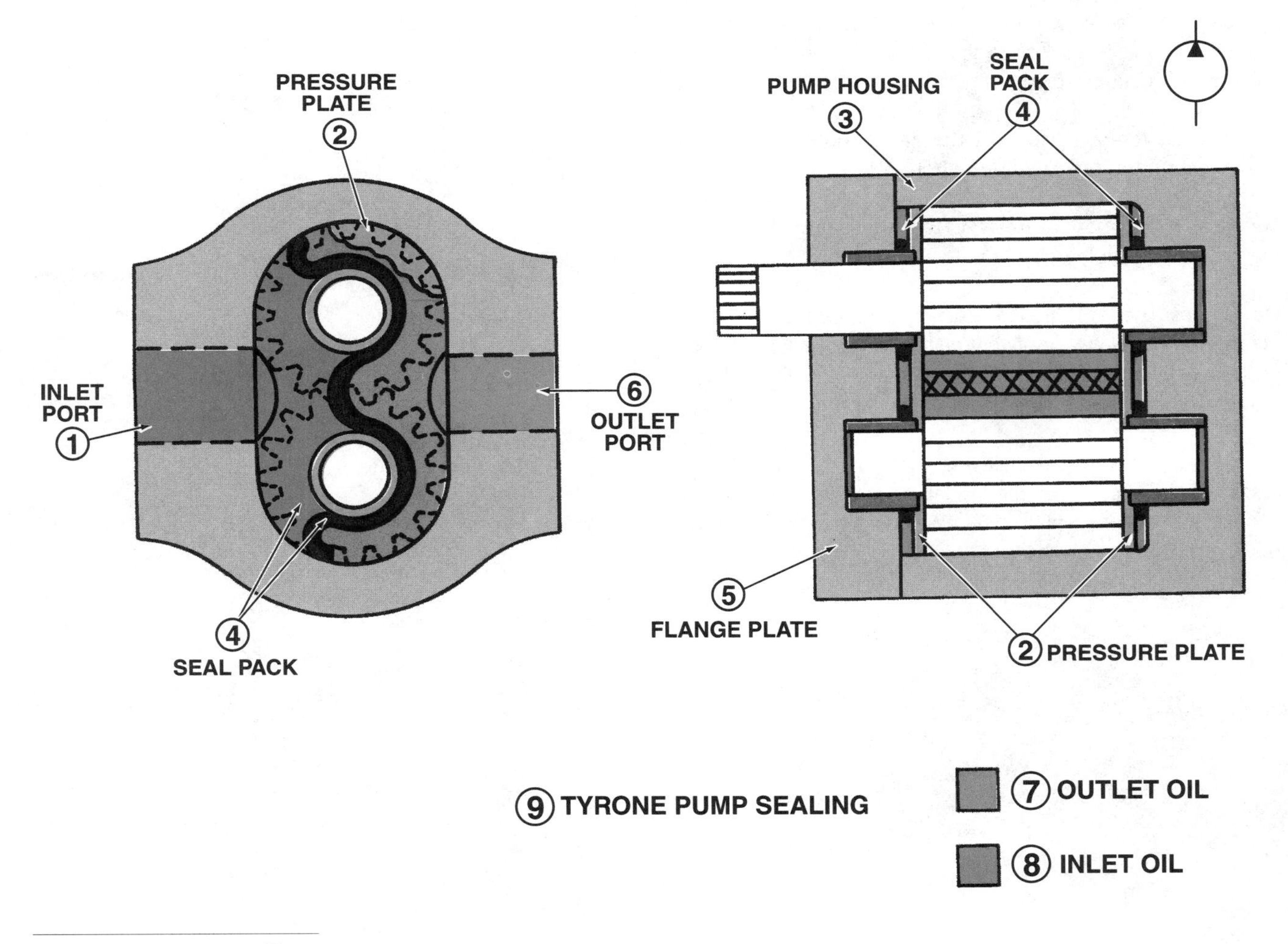

Fig. 6 — Pump Pressure Plates

EXTERNAL GEAR PUMPS

External gear pumps have two gears in mesh, closely fitted inside a housing (Fig. 4). The drive shaft drives one gear, which in turn drives the other gear. Shaft bushings and machined surfaces or wear plates seal the working gears.

Operation is quite simple (Fig. 5). As the gears rotate and come out of mesh, inlet oil fills the cavity between the gears teeth. The oil is then carried in the tooth cavity to the outlet chamber. As the gear teeth mesh again oil is forced out the outlet port and sent through the system.

Oil is pushed out in a continuous, even flow as the gears rotate.

Oil from the reservoir to replace that drawn out by the turning gears is fed into the inlet side by gravity and atmospheric pressure.

Efficiency of a gear pump is dependent on how well it prevents the high-pressure oil from leaking back to the inlet side.

The pressure in the outlet area of the pump forces the gear teeth against the center housing in the HOUSING SEAL AREA. (See Fig. 5) This is one of the seal areas required to maintain pressure in the outlet port.

Bushings on the shaft support the gears and prevent them from wearing too deeply into the housing.

The gear tooth mesh forms the second sealing point. The machined gears do a good job of sealing in this area.

On pumps used in higher-pressure applications, pressure plates, one on each side of the gear, are used. The plates are pushed against the sides of the gears to minimize the leakage of oil across the face of the gears.

Fig. 6 shows one of several ways to load the plates. In all cases, pressure oil from the outlet port is fed to an area beneath the pressure plate. This gives a force that is in proportion to the pressure being pumped.

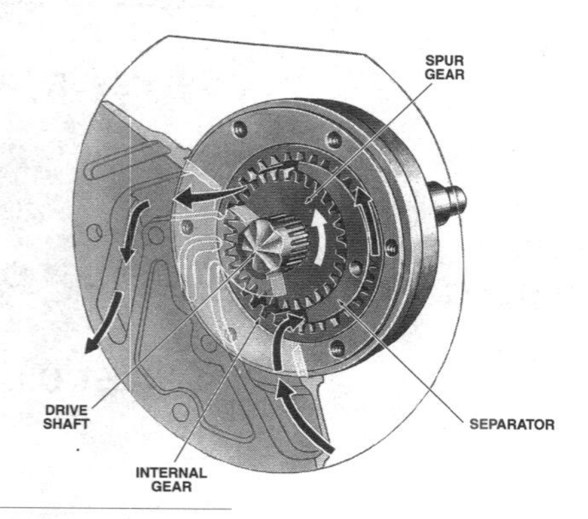

Fig. 7 — Internal Gear Pump

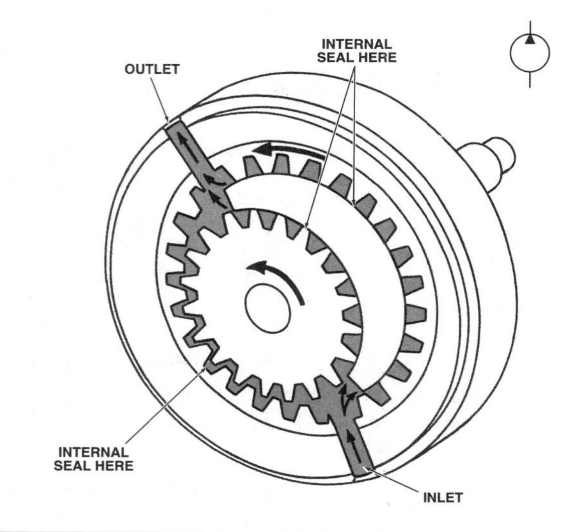

Fig. 8 — Internal Gear Pump in Operation

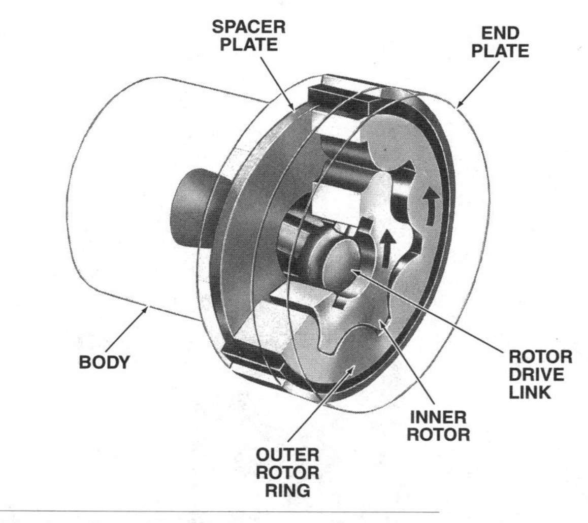

Fig. 9 — Rotor Version of Internal Gear Pump

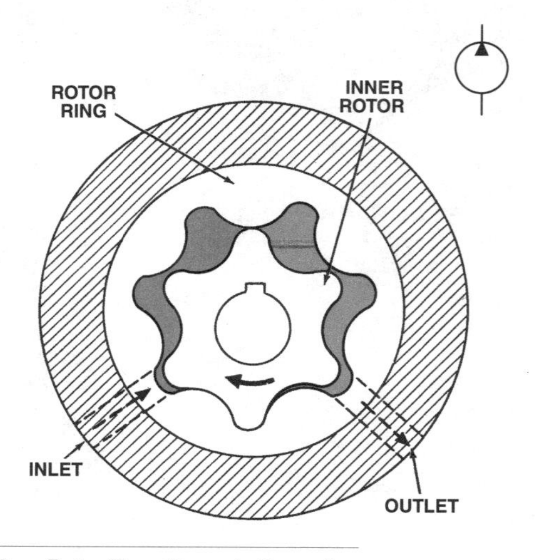

Fig. 10 — Rotor-Type Pump in Operation

INTERNAL GEAR PUMPS

The internal gear pump (Fig. 7) also uses two gears, but now a spur gear is mounted inside a larger gear. The spur gear is in mesh with one side of the larger gear, and both gears are divided on the other side by a crescent-shaped separator. For that reason, this pump is also called a crescent pump. The drive shaft turns the spur gear, which drives the larger gear.

Operation is basically the same as for the external gear pump. The major difference is that both gears turn in the same direction (Fig. 8).

As the gear teeth come out of mesh, oil is drawn into the inlet. Oil trapped between the teeth and the separator is carried around to the outlet chamber. As the gears mesh, oil is forced out the outlet.

This type of pump has a low efficiency and is used on low-pressure systems such as transmissions, charge pumps, etc.

ROTOR VERSION OF INTERNAL GEAR PUMP

The rotor pump, also called a Gerotor pump (Fig. 9), is a variation of the internal gear pump. An inner and outer rotor turns inside a housing. The rotor has rounded lobes instead of teeth. No separator is used.

In operation (Fig. 10), the inner rotor is driven inside the rotor ring. The inner rotor has one less lobe than the ring. One lobe will always be in full engagement with the outer ring. The lobe shape insures that one will always have tip contact.

As the lobes part at the inlet area, oil is drawn in to fill the void. Oil in the lobe cavity is carried to the outlet side. As the lobes again go into mesh, oil is forced out the outlet.

The lobe tips at the mid-point create the seal to prevent pressure oil from returning to the inlet.

This pump also has a low efficiency and is used in low-pressure systems.

In the valve section we'll see this unit used as a motor (follow-up) in steering valves.

VANE PUMPS

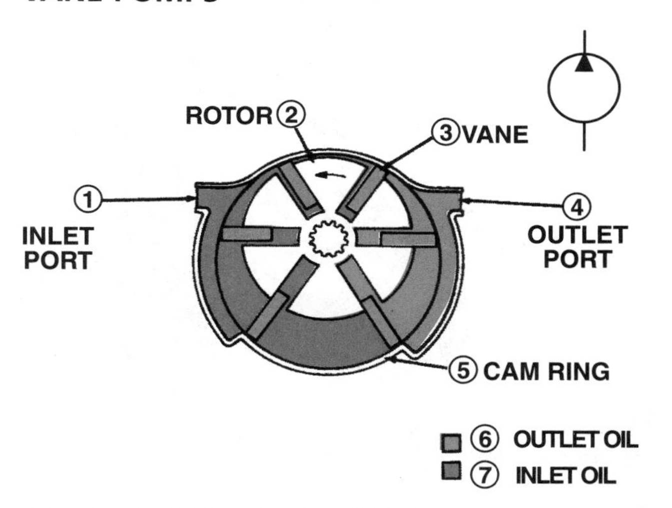

Fig. 11 — Simple Vane Pump in Operation

In a vane-type pump (Fig. 11), the shaft drives a slotted rotor. The rotor is fitted with blades that can move in and out of the slots. This assembly is offset mounted inside a circular ring. End caps seal the ends of the vanes, rotor and cam.

As the shaft and rotor are turned, the vanes are forced against the outer ring by centrifugal force.

As the assembly rotates, the cavity formed by the cam, rotor and vanes will get increasingly larger on the inlet side of the pump. Oil from the inlet will fill this void. Oil is carried to the outlet side where the cavity gets increasingly smaller. This forces oil out the outlet passage.

It is rare but this pump can be made into a variable displacement pump by moving the rotor up or down in the cam ring. When the rotor is centered in the cam, there is no change in the cavities volume as they move from the inlet to the pressure sides. Therefore, there is no output flow.

As the rotor is moved up in the cam, the cavities and therefore output would become larger. Maximum output would occur when the rotor was very near the cam as shown in Fig. 11.

The pump in Fig. 11 is called an **Unbalanced Vane Pump** because pressure oil on the outlet side of the pump pushes the rotating assembly toward the inlet side with a great force. This results in very high shaft bearing loads.

Most vane pumps used in hydraulics today are **Balanced-Vane Pumps**.

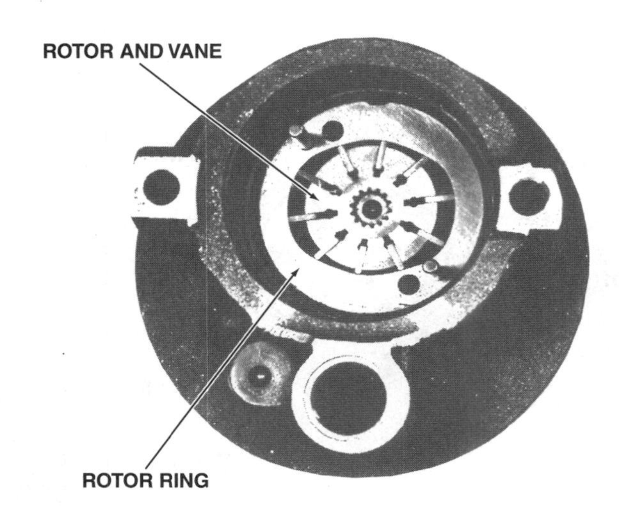

Fig. 12 — Balanced Vane Pump

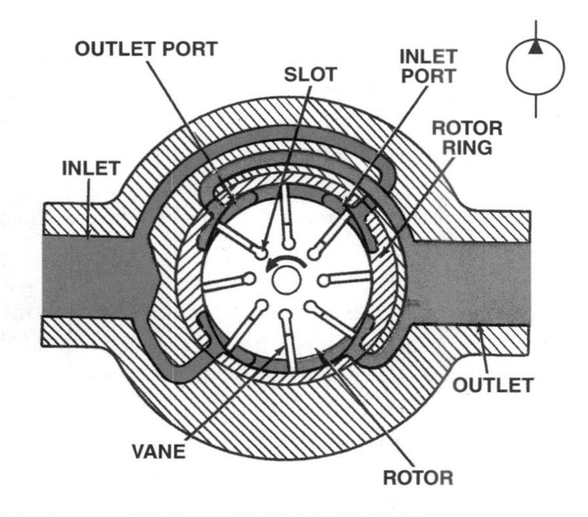

Fig. 13 — Balanced Vane Pump in Operation

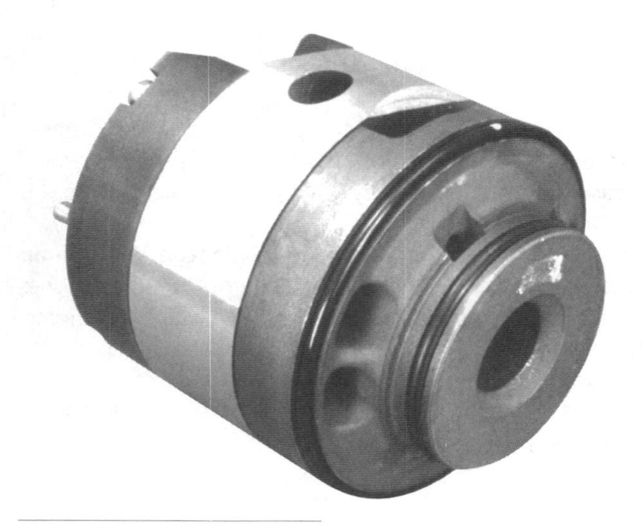

Fig . 14 — Vane Pump Cartridge

The balanced-vane pump, shown in Fig. 12, has a double cam ring. It has two inlet passages directly across the cam from each other. There are also two outlet passages. Operation (Fig. 13) is the same as the unbalanced pump but each vane cavity goes through a complete pumping cycle every 180 degrees of rotation.

This places the two equally sized pressure areas directly opposite (balanced) to minimize bearing load.

The pump center section consists of the rotor, vanes and cam. The two endcaps seal the ends of the vanes and rotor. This assembly, called a "cartridge", is placed in the housing that contains the inlet and outlet ports and passages.

The efficiency of the pump is determined by the amount of oil that leaks from the outlet cavity back to the inlet. The leakage points are the vane tips, and the sides of the vanes and rotor.

To seal the ends of the vanes and rotor, the assembly is hydraulically pressed together. The shaft end cap, as shown in Fig. 14, has two O-rings to seal the outlet passage. This outlet cavity acts as a hydraulic piston to press the assembly together, as shown in Fig. 15. This prevents leakage past the sides of the vanes and rotor.

To prevent pressure oil in the cavities from pushing the vanes down and leaking at the vane tips, passages to the bottom of the vane insure that the pressure is the same on the top and bottom of the vanes. This is done by drilling a passage in the rotor, as shown in Fig. 16, or in the vane, as shown in Fig. 17.

A small force to hold the vane against the cam at all times is provided by a small piston in the vane (Fig. 16) or the rotor (Fig. 17). Passages in the end cap and rotor direct outlet pressurized oil to the small piston.

PISTON PUMPS

Piston pumps are among the most sophisticated of all pumps. They are capable of high pressures, high volumes, high speeds, and variable displacement. For this reason, they are becoming more popular on modern hydraulic systems.

The piston pumps are more complex and more expensive than the gear and vane pumps so are not used on simpler systems.

Most types of piston pumps can be designed for either fixed or variable displacement.

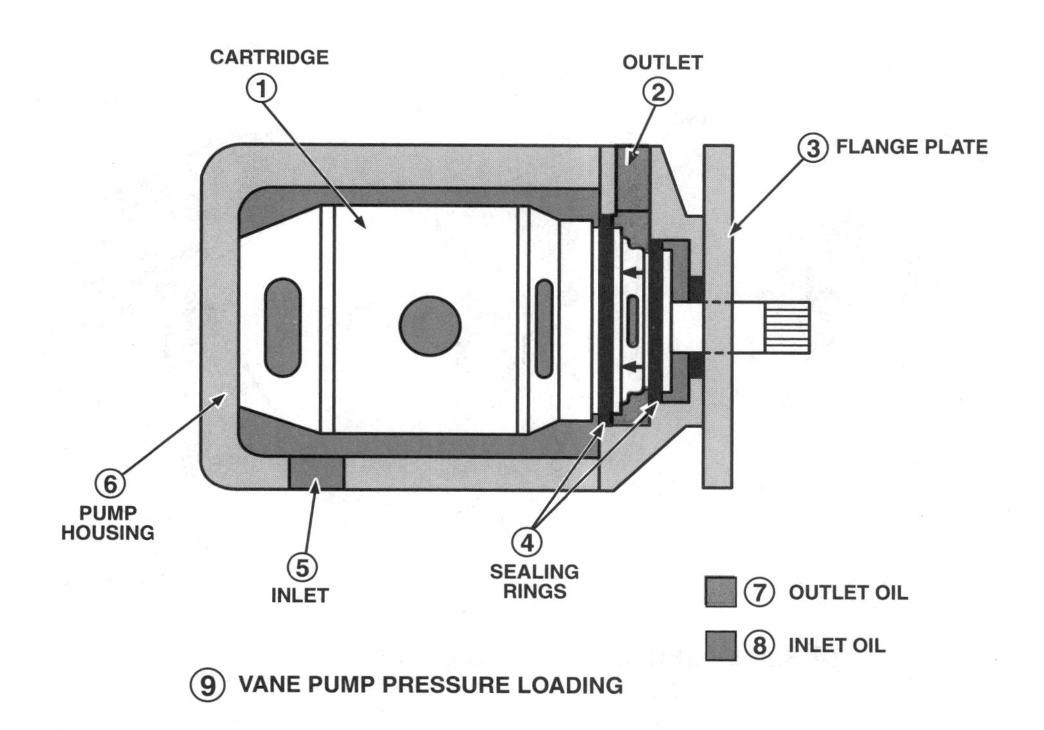

Fig. 15 — Vane Pump Pressure Loading

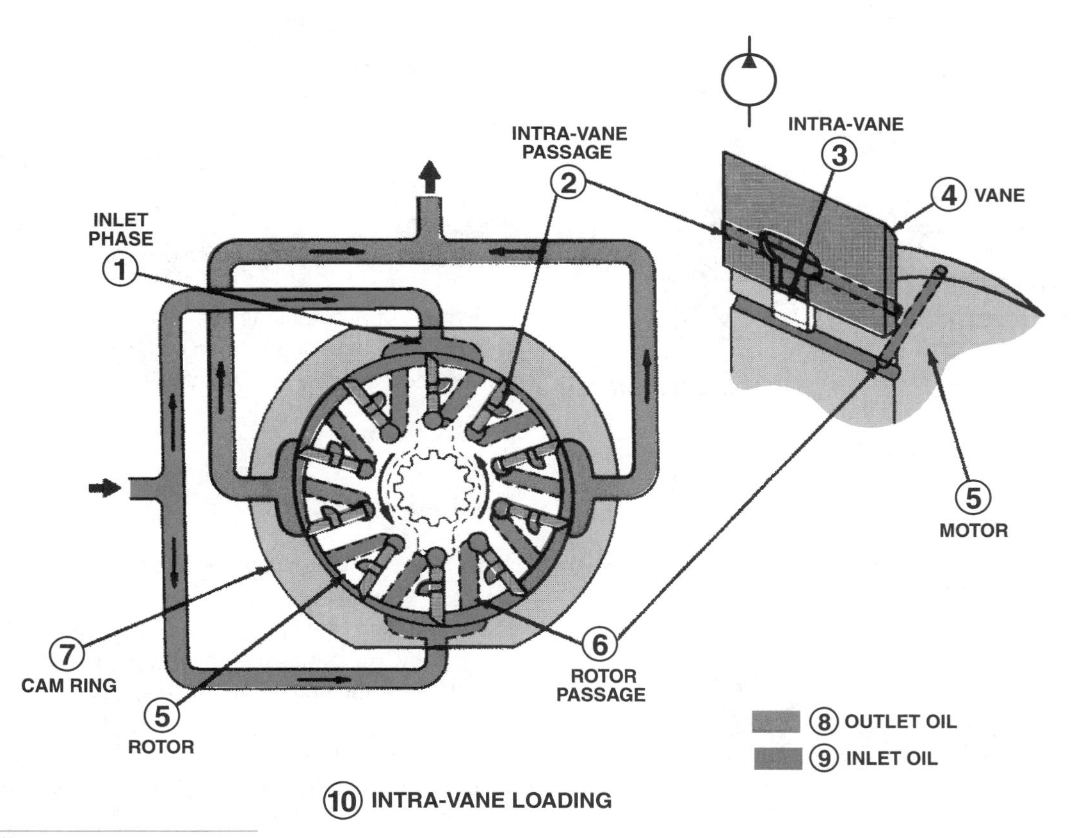

Fig. 16 — Vane Loading Through Rotor

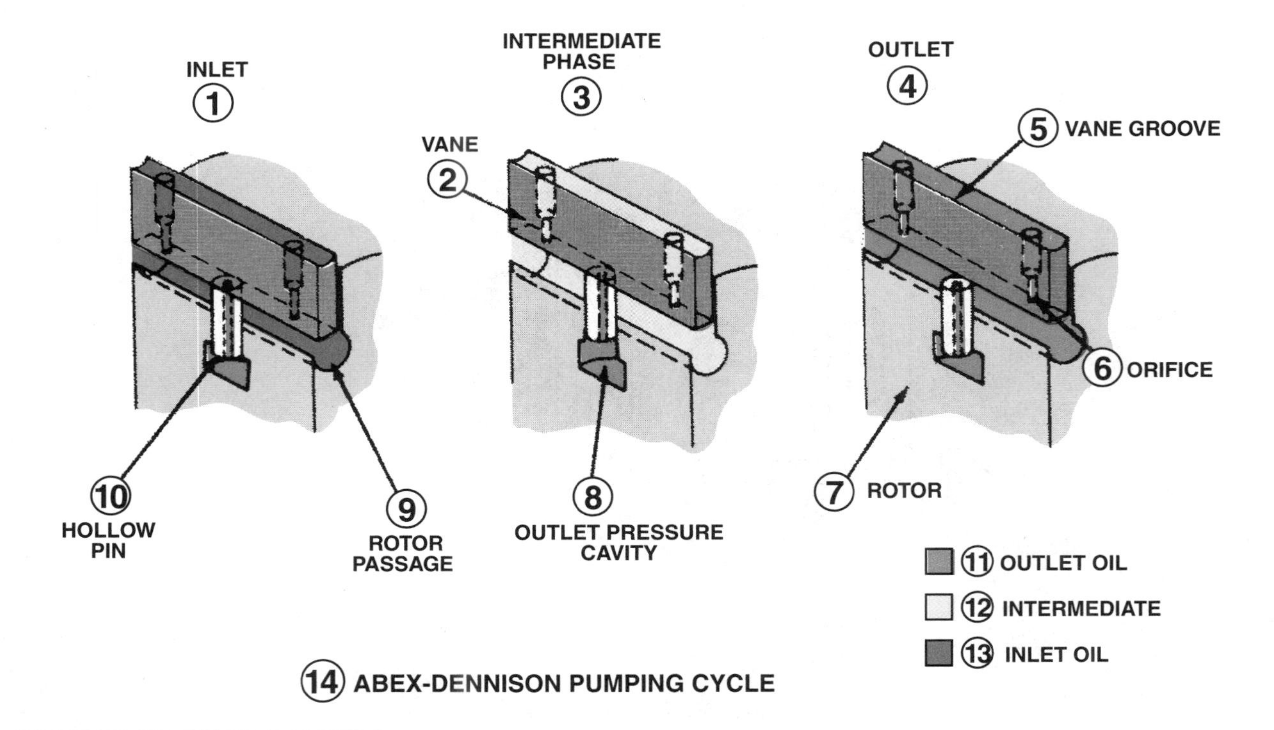

Fig. 17 — Vane Loading Through Vane

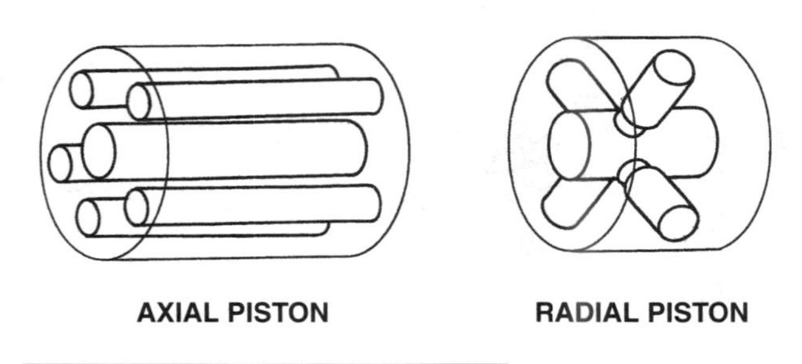

Fig. 18 — Axial and Radial Piston Pumps

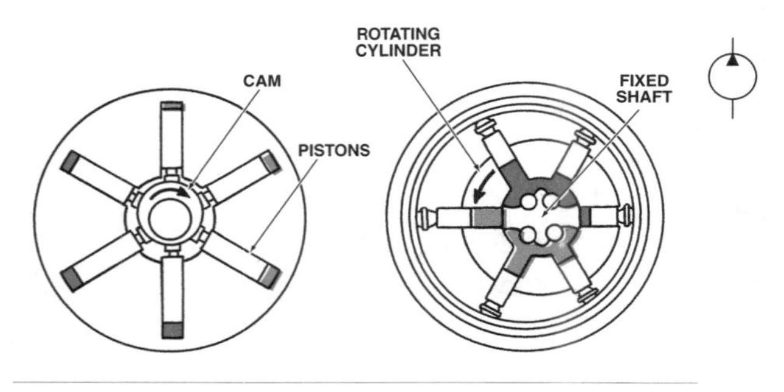

Fig. 19 — Two Operating Principles of Radial Piston Pumps

Most piston pumps are included in two types:

- **Axial Piston Pumps**
- **Radial Piston Pumps**

RADIAL piston means that the pistons are set perpendicular to the pump's centerline like the sun's rays. See Fig. 18.

AXIAL piston means that the pistons are mounted in lines parallel with the pump's "axis" (a line down the center). See Fig. 18.

Both styles of piston pumps operate using pistons that pump oil by moving back and forth in cylinder bores. (Another term for this movement is "reciprocate.")

RADIAL PISTON PUMPS

Radial piston pumps are designed to operate in two ways (Fig. 19).

In the "rotating cam" pump, the pistons are located in a fixed pump body. The center shaft has a cam that drives the pistons in and out of the body as it rotates.

In the "rotating piston" pump, the pistons are located in a rotating cylinder block. As the cylinder rotates, the pistons are thrown out against the outer housing. Since the rotating cylinder is offset in the housing, the pistons are moved back and forth in the block as they follow the housing.

Let's look at the basic operation of each pump.

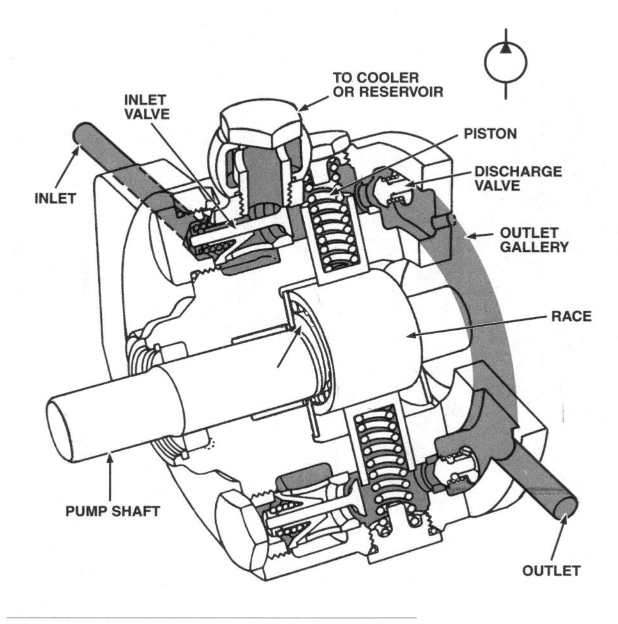

Fig. 20 — Radial Piston Pump in Operation

RADIAL PISTON PUMP (ROTATING CAM TYPE)

The typical radial piston pump shown in Fig. 20 uses the "rotating cam" principle and is normally designed as a four- or eight-piston model. They can also be designed as dual bank pumps, which consists of two pump bodies bolted together. Dual pumps use a common driveshaft and common control valving.

The radial piston pump consists of a housing with piston bores located radially around the center line. The housing also contains an annular passage on each side of the piston bores. One is the inlet passage and is connected to each of the piston bores through a spring loaded inlet check valve. The other is an outlet oil passage and is connected to the piston bores through a spring loaded discharge check valve.

Pistons are located in the bores and contact an eccentric cam on the driveshaft. They are held on the cam with springs on top of the piston.

As the driveshaft is turned, the cam will allow the spring to move the piston down. Pressurized oil in the outlet passage will hold the discharge valve on its seat. A partial vacuum is created in the piston cavity. Oil to fill the cavity will be forced past the inlet valve by charge or atmospheric pressure.

As the cam begins to push the piston up, the pressure will rise. This will force the inlet check valve on its seat and open the discharge valve to allow oil to enter the outlet cavity.

The cycle of each piston works in rapid sequence as the cam rotates. This produces a constant flow of oil which is determined only by the pump displacement (piston diameter, cam lift and number of pistons) and the driveshaft speed.

In the operation we have just described, the pump would have a fixed displacement. But if that were all we needed a cheaper gear or vane pump would probably do the job as well.

To vary the delivery of this pump, we would need to vary the displacement or the speed. Speed variation is not very practical so lets look at displacement. Here we can change the piston diameter or the length of stroke. Changing the diameter would be difficult, which leaves the piston stroke.

To shorten the piston stroke, the crankcase cavity will be pressurized to hold the pistons out away from the cam. Remember that it will take a small pressure to start to compress the spring and a much higher pressure to completely compress it. This means that varying the crankcase pressure will vary the distance the pistons can be pushed toward the cam by the springs. This will change the length of the piston stroke and vary pump delivery from zero to maximum.

USE OF STROKE CONTROL MECHANISM

This pump is used with closed-center valves. Its control system reacts to the pump outlet pressure. It provides standby (maximum system) pressure to the system control valves when there is no oil requirement. It will try to supply enough oil to maintain that pressure at all other times.

The stroke control mechanism (Fig. 21) is normally contained in the pump end housing. It contains two valves. The stroke control valve connects the pump outlet passage to the crankcase. The crankcase outlet valve connects the crankcase to the pump inlet passage. There is also a bleed hole (fixed orifice) connecting the crankcase to the inlet.

When the machine is started, the crankcase outlet valve is held open by a spring. This dumps any oil in the crankcase to the inlet, not allowing crankcase pressure to build up. The stroke control valve is held on its seat by a spring, not allowing pressurized oil into the crankcase.

As the cam is turned, the pistons will travel through a full stroke. The pump output will be maximum. As the needs of the hydraulic system are met, the excess oil will cause the system pressure to rise.

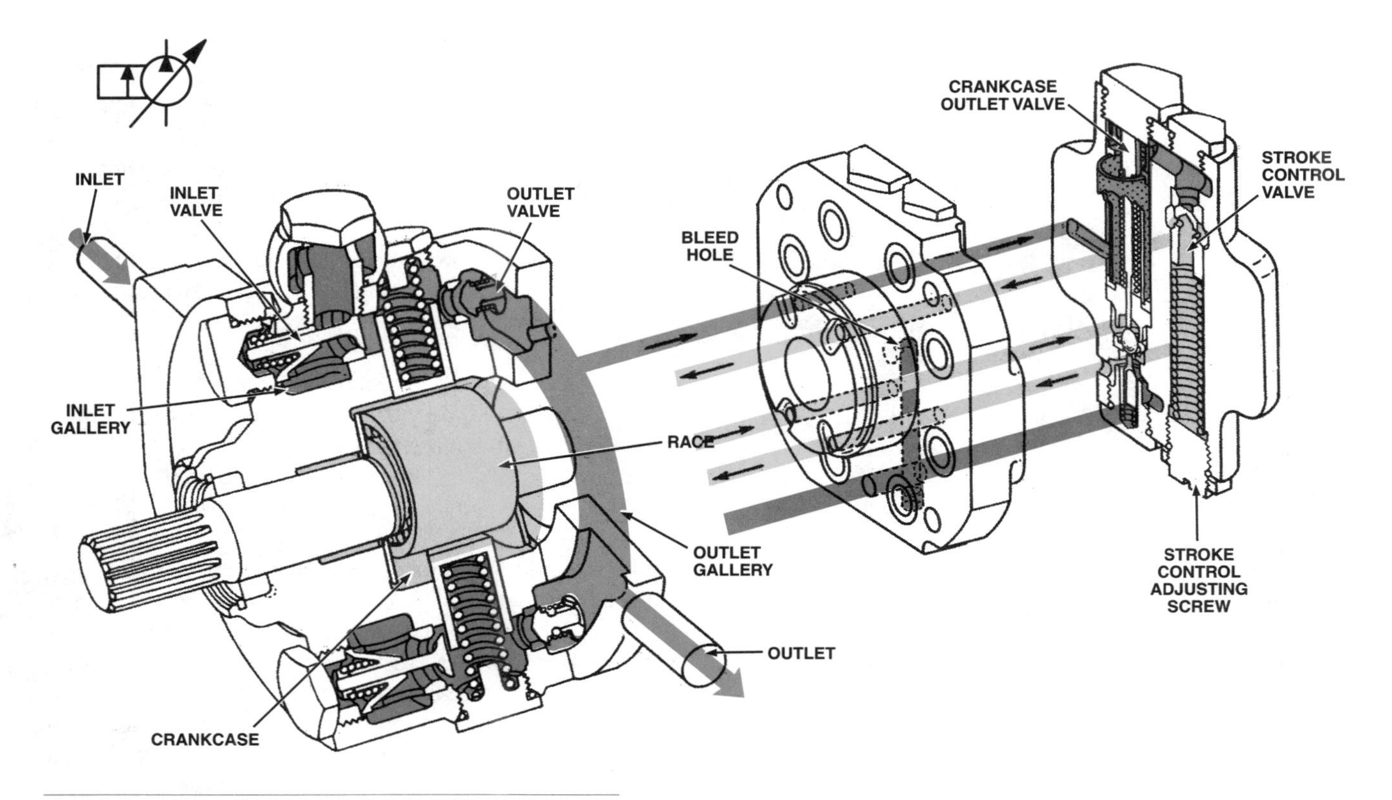

Fig. 21 — Stroke Control Mechanism on Radial Piston Pump

When the outlet pressure reaches about 75% of the maximum (standby) pressure, the crankcase outlet valve will close. The pump will still be in full stroke because the stroke control valve is still closed, not allowing oil into the crankcase.

When outlet pressure approaches standby, the stroke control valve will begin to open and send oil to the crankcase. The higher the outlet pressure, the more oil will flow to the crankcase. Because the orifice is continually bleeding oil out of the crankcase, it requires flow to build and maintain crankcase pressure.

When pressure is enough to overcome the force of the extended springs the pistons will have a slightly shorter stroke.

As outlet pressure builds, more oil flows past the stroke control valve into the crankcase. This builds more pressure, shortening the piston stroke even more.

Standby pressure will be reached when no oil flow is required by the system. In this standby mode, the pump maintains standby pressure to the valves and will pump only enough oil to maintain crankcase oil pressure.

The orifice is continually leaking oil from the crankcase to the inlet to provide cooling and lubrication for the pump. It also allows the pump to get back into stroke.

When any valve requiring less than full pump output is activated, the system outlet pressure drops slightly. The flow of oil into the crankcase decreases. Because oil is continually bleeding through the orifice, the crankcase pressure will drop slightly. This allows the springs to push the pistons down a slight amount. The pump will go into partial stroke to satisfy the needs of the system.

As the system oil needs are met, the system pressure rises, flow increases into the crankcase thus raising the pressure, and the pump will return to standby.

When a valve requiring the full output of the pump is opened, the system pressure will drop considerably. This will open the crankcase outlet valve (sometimes called a quick dump valve), dumping the crankcase to the pump inlet passage. This allows all the pistons to drop to the cam. This puts the pump immediately into full stroke.

The pump returns to standby when no flow is required by the system.

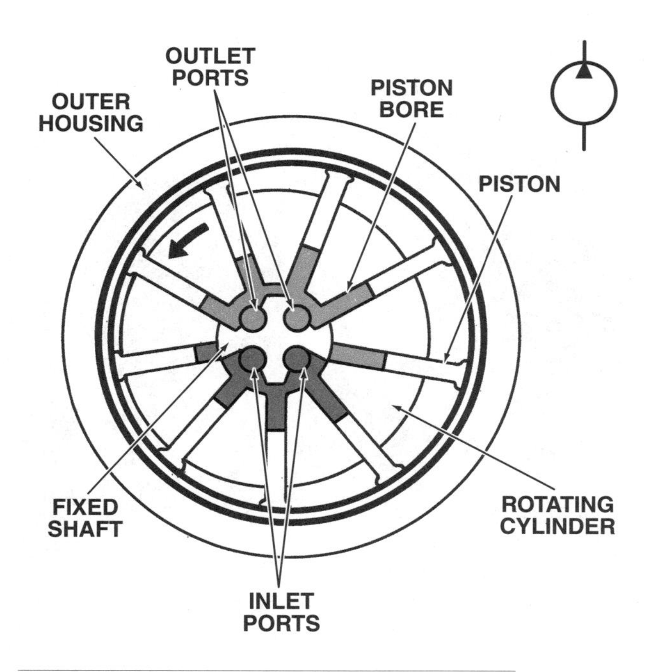

Fig. 22 — Radial Piston Pump (Rotating PistonType)

RADIAL PISTON PUMP (ROTATING PISTON TYPE)

The rotating piston version of the radial piston pump is shown in Fig. 22. It has a rotating cylinder block, which contains the pistons. The cylinder block is offset from the pumps circular housing. The housing contains a cam ring on which the pistons will ride.

As the cylinder block is rotated, centrifugal force holds the pistons against the cam ring. As the pump turns, the pistons are pushed into the block during one half of the rotation and allowed to come out of the block during the other.

During the part of the cycle that the pistons are being pushed into the block, the piston bores are aligned with the pump outlet (pressure) passage.

During the part of the cycle the pistons are being forced out of the block, the bore is aligned with the pump inlet passage. As they are being forced out by centrifugal force, a partial vacuum is created and oils flows into the piston cavity from the inlet port.

The pump can be made variable displacement by adjusting the relation of the cylinder block to the outlet housing. This changes the length of the piston stroke and thus the pump outlet flow.

AXIAL PISTON PUMPS

Axial piston pumps usually fall into one of two types:

- **Inline** - which have the pistons operating in line with the input shaft.
- **Bent-Axis** - which have pistons operating at an angle to the input shaft.

INLINE AXIAL PISTON PUMPS

The pump in Fig. 23, like most in line pumps, has a cylinder block assembly (rotating group) which is shown in detail in Fig. 24. It also contains a swashplate, on which the pistons ride, and the valve plate that directs oil from the pistons to the pump ports.

The rotating group is the assembly that is driven by the pump-input shaft. It consists of a cylinder block which has piston bores, the pistons with slipper plates, a valve plate, a piston retainer and a spring arrangement to load the piston retainer and hold the slippers on the swashplate. In some cases, a bearing plate is used with the valve plate.

Fig. 25 shows an inline axial piston pump. The cylinder block is splined to the input shaft and turns with it. The piston bores of the cylinder block are parallel to the axis of the block. The swashplate is a tilted plate that is attached to the pump housing.

Slipper plates are attached to the piston ends. They are held against the swashplate by a spring-loaded piston retainer.

As the input shaft turns the cylinder block, the pistons are pushed into the block as the slipper plates ride up the ramp of the swashplate.

The pistons are pushed out of the block by the spring-loaded piston retainer as the slippers follow the swashplate ramp down.

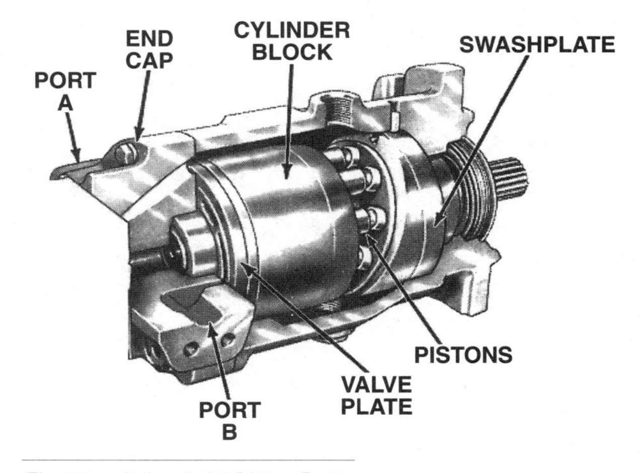

Fig. 23 — Inline Axial Piston Pump

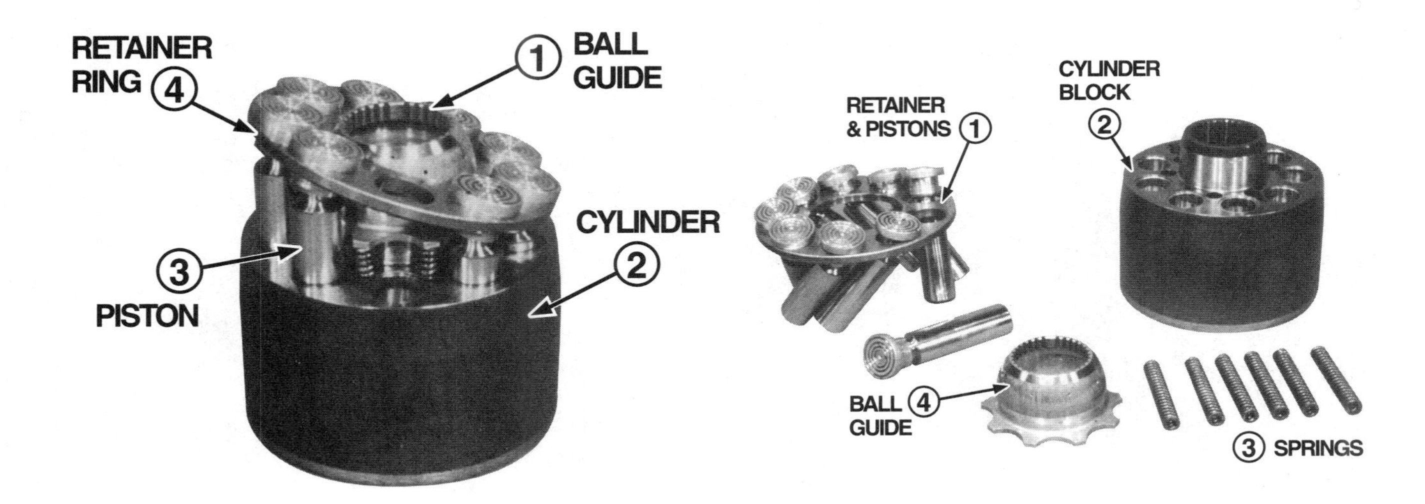

Fig. 24 — Rotating Group

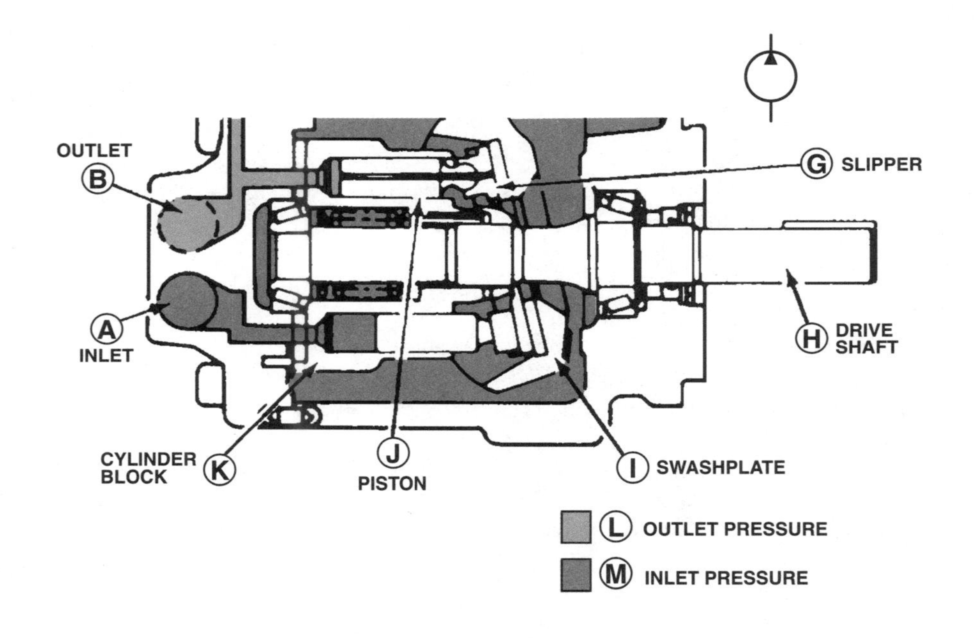

Fig. 25 — Inline Axial Piston Pump

The piston bores, through the valve plate, are aligned with the pump outlet port while the pistons are being pushed into the block. They are aligned with the inlet port while they are being pushed out of the block.

Oil is therefore drawn into the piston cavity while the piston is being pushed out of the block. Oil is forced into the outlet cavity while the pistons are being forced into the block.

If the swashplate was attached to the pump housing and not allowed to move, the pump would be a fixed displacement pump. While there are many of these is use, a majority of piston pumps are variable displacement.

Fig. 26 shows a variable displacement inline axial piston pump on which the swashplate is allowed to pivot. The swashplate is spring loaded to a maximum angle.

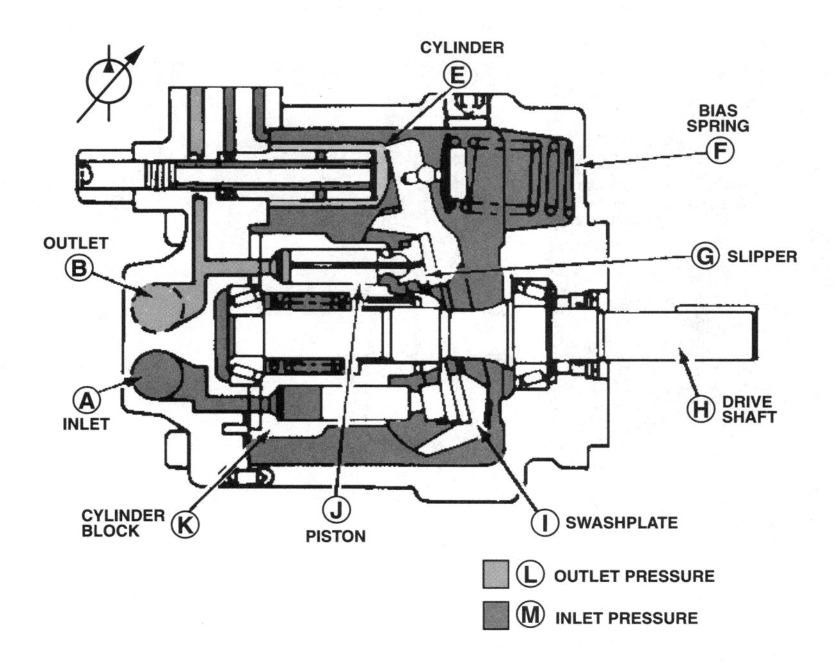

Fig. 26 — Variable Displacement Pump

It also has a hydraulic cylinder, called a servo cylinder, which if pressurized, could move the swashplate, against the spring, to a lesser-angled position.

The force required to start compressing the spring is much less than to completely compress it. By controlling the pressure in the servo cylinder, the swashplate can be moved and held at any desired angle.

As the swashplate angle is decreased, the pistons will have a shorter stroke in the block and result in a decrease in pump output flow. When the swashplate is at right angles to the block, there will be no piston movement in the block and the pump will stop pumping oil.

Most axial piston pumps operate as has been described. The great number of applications in which these pumps are now being used comes from the versatility of the pumps and their ability to adapt to the many types of control systems. Let's look at a few.

PRESSURE SENSING PUMPS

The pressure sensing system uses closed-center valves. The pump maintains a standby pressure to the valves when no flow is required. This standby pressure is the maximum operating pressure in the system. When valves are operated, the pump furnishes enough oil to try to maintain that pressure.

The pump in Fig. 27 has controls that adapted it to the pressure sensing system. When the machine starts, the spring holds the swashplate at a maximum angle. The pump is in full stroke. As the pressure approaches standby pressure, the pressure valve begins to pass oil to the swashplate piston (servo piston).

The servo piston has a small controlled leak so the amount of flow past the pressure valve will determine the servo pressure. As system pressure approaches standby, flow will increase, gradually raising the servo pressure and moving the swashplate against the springs to the no delivery position.

When a control valve on the machine is operated, the system pressure will drop slightly. The pressure valve will send less oil to the servo piston. Because of the controlled leak in the piston, the pressure will drop slightly. This will allow the spring to angle the swashplate and start to pump oil. When the machine hydraulics no longer need oil, the system pressure will reach standby and the pump will go out of stroke.

If a valve is opened which requires almost full pump output, the system pressure will drop considerably. This allows the

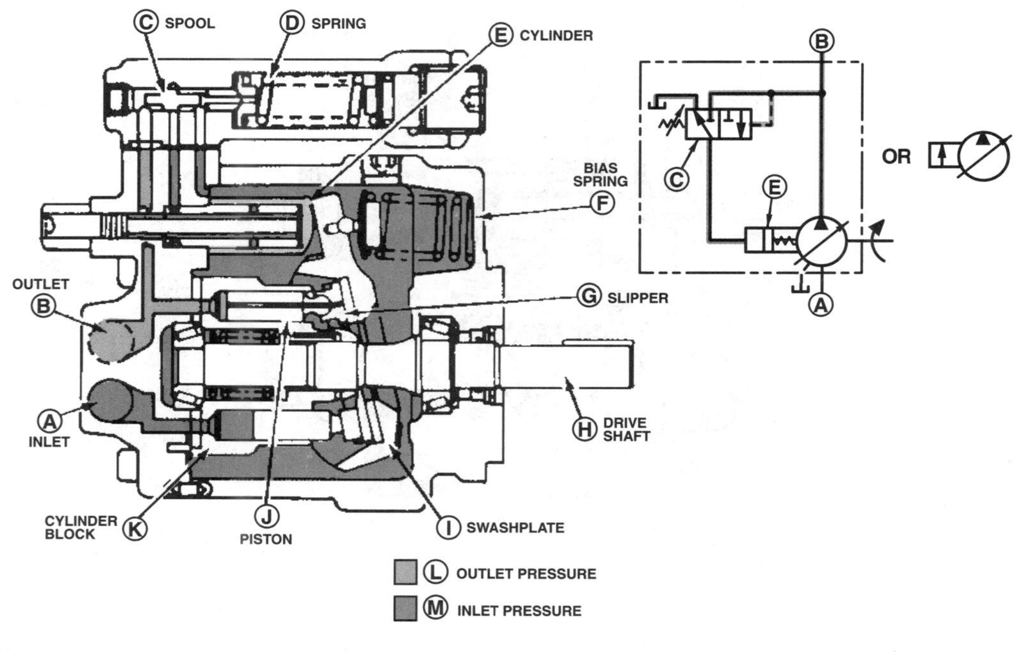

Fig. 27 — Pressure Compensated Pump

spring to move the pressure valve to a position where it dumps the swashplate piston oil to return. This allows the pump to go into full stroke immediately.

The pump just described is used in the same applications as the radial piston pump described earlier in this chapter.

LOAD SENSING PUMPS

The load sensing systems also use closed-center valves. When the valves are in neutral, the pump maintains a low standby pressure. When valves are operated, the pump will then maintain a pressure slightly higher than that needed to lift the load.

Fig. 28 shows a load pump. The pump is almost identical to the one shown in Fig. 27 except for the addition of a load-sensing valve. This valve operates the same as the stroke control valve except that it is set at a much lower pressure.

When there is no requirement for oil, the load-sensing valve will maintain a low standby pressure to the control valve.

When a control valve(s) is activated, a load sensing system will pressurize the cavity behind the load-sensing valve to the highest pressure needed in any of the control valve workports.

Pump outlet pressure will now have to overcoming the working pressure required by the control valve plus the spring force of the load-sensing valve. Pump outlet pressure will now be maintained at a pressure higher than that needed to lift the load by the amount of the load sensing valve pressure setting.

Let's take a system on which the spring adjustment of the load sensing valve will require 300 psi (2100 kPa) to destroke the pump. When all of the machines control valves are in neutral, the pilot circuit is connected to the reservoir so there can be no pressure in the load sensing valves spring area. The pump will destroke and maintain 300 psi (2100 kPa) pressure to the control valves.

In our example, we'll activate a control valve to lift a load that requires 1500 psi (10,345 kPa). That pressure will be built in the sensing system and in the spring area of the load-sensing valve.

In order to destroke the pump, system pressure will now have to reach 1800 psi (12,425 kPa) to overcome working pressure plus the spring force.

The control valve contains check valves to insure that the pump senses only the highest pressure when more than one valve is used at the same time.

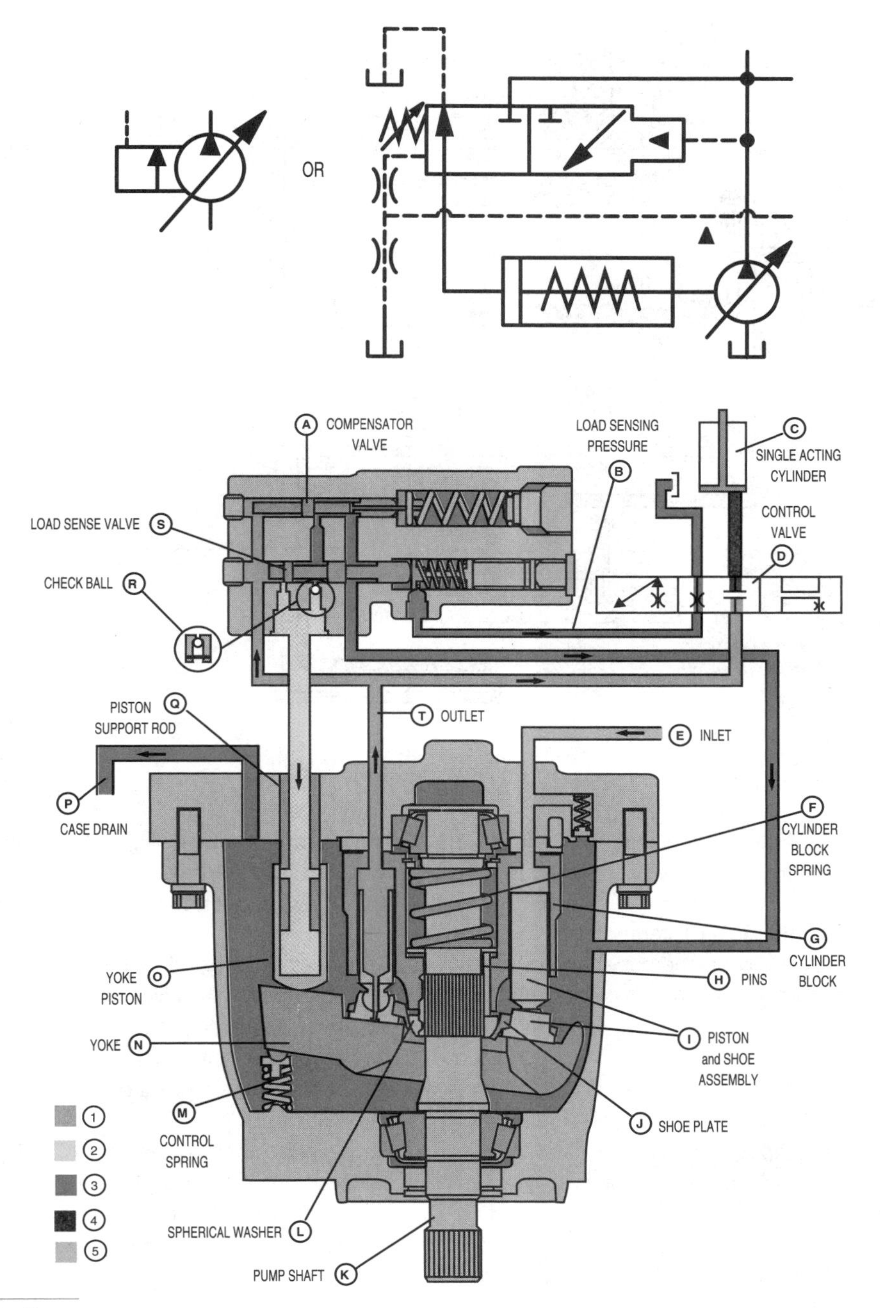

Fig. 28 — Load Sensing Pump

Because the load-sensing valve is always trying to supply more pressure than it senses, it cannot be used for limiting the maximum system pressure. For this reason, the high-pressure standby (compensator) valve destrokes the pump when maximum pressure is reached. It operates the same as on the pressure sensing pumps.

REVERSIBLE AXIAL PISTON PUMPS

The reversible piston pump is normally used in a closed-loop system. This means that the two outlets of the pump are connected to the two inlets of a hydraulic motor. There is no need for valves to direct the high-pressure oil.

The pump shown in Fig. 29 is reversible. All of the parts of the rotating group are identical to those shown in the previous

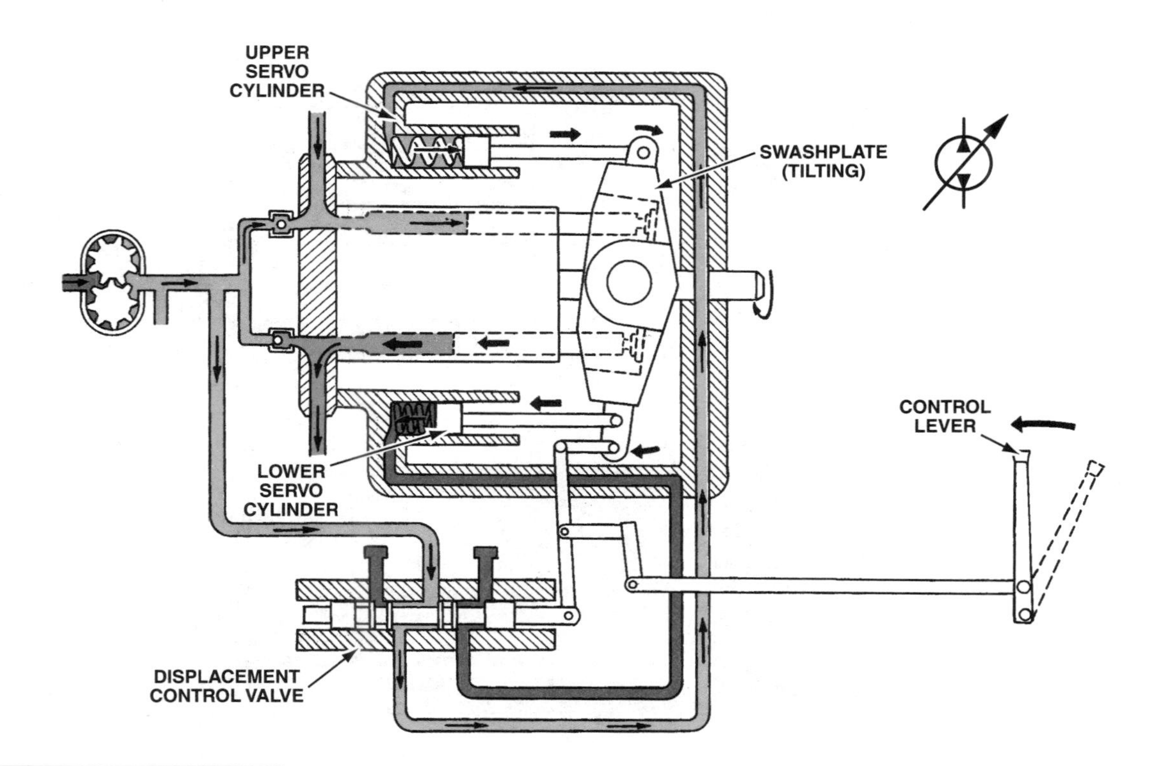

Fig. 29 — Reversible Axial Piston Pump

axial piston pumps. It has a swashplate that can be tilted in either direction. To do this, it requires two servo cylinders.

When the swashplate is rotated past center in a clockwise direction as shown, the lower port is the outlet and oil is sent to the motor. The upper port is the intake, receiving outlet oil from the motor.

When the swashplate is rotated past center in the counter-clockwise direction, the oil flow is reversed. This would result in the motor turning in the opposite direction. The swashplate can be stopped and held in any position, making this a variable displacement pump.

Because this is a closed loop, there is a need to supply make-up oil to the low-pressure circuit in case any leakage should occur. For this reason, most of these pumps have a small supply pump. Low-pressure oil is supplied to a check valve in each of the pressure circuits. When the circuit is pressurized, the check valve will be held on its seat. When the circuit is the return side, charge oil can unseat the check valve and supply make-up oil as needed.

The position of the swashplate (pump output) on most closed-loop systems is determined by the operator and has little to do with the system pressure.

The control shown in Fig. 29 is typical. The control lever is attached to the displacement control valve through linkage, which also senses the swashplate position.

As the lever is moved to the left, the spool will move to the left. This opens the valve and sends oil from the charge pump to the upper servo cylinder. As the swashplate is rotated, it will rotate the vertical link counterclockwise. This will pull the valve back to the neutral position and trap oil in both servos.

For every position of the control lever, there is a corresponding position of the swashplate. This means the operator will have infinite control of the speed and direction of the hydraulic motor driven by this pump.

In automated systems, the displacement control valve is connected to an electric force motor. The force motor will move the linkage based on the voltage sent by an electronic control unit (ECU)

REVOLVING SWASHPLATE PUMP

The revolving swashplate pump shown in Fig. 30 can only be fixed displacement. The cylinder block is stationary while the swashplate rotates. The pistons are spring loaded against the swashplate. As the swashplate rotates, the pistons slide back and forth in their bores, pumping oil.

Check valves are used to control the flow of oil through the pistons. Each piston has an inlet and outlet valve. As the piston is being pushed out of the block, pressurized oil holds the outlet valve on its seat while oil is being drawn in through the inlet valve. When the piston is being forced into the block, the pressure being produced will close the inlet valve. The pressure will build to overcome system pressure and be forced out the discharge valve.

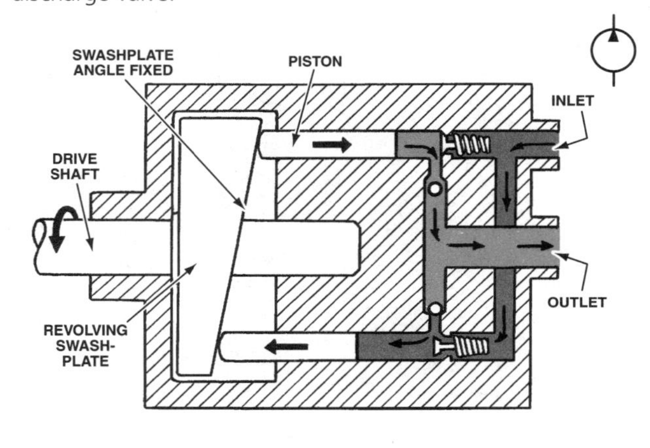

Fig. 30 — Revolving Swashplate Pump

BENT-AXIS AXIAL PISTON PUMP

Another type of axial piston pump is the bent-axis type. Fig. 31 shows a fixed displacement model of this pump.

In this pump, the pump housing is slanted in relation to the driving member. The piston heads are connected to the drive member. Both the cylinder block and drive member are driven by the drive shaft and rotate in the pump housing.

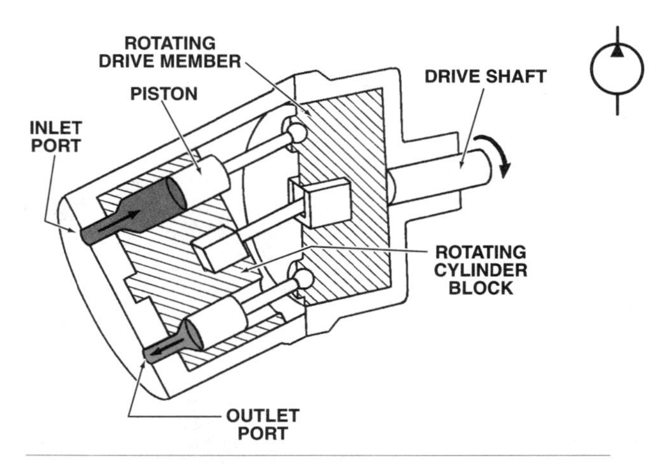

Fig. 31 — Bent-Axis Axial Piston Pump - Fixed Displacement

As the two units revolve, the tilted drive member forces the pistons in and out of their bores. During the one half revolution that a piston is being pushed into the block, the piston bore is aligned with a pressure port. During the time it is being pulled out of the block, it is aligned with the inlet port.

Fig. 32 shows a variable displacement version of the bent-axis pump. In this version, the block (rotor) is moved in the housing by a servo piston. When the servo is down as shown, the pistons will have maximum travel in the block so the pump will have maximum delivery.

As the servo piston is moved up, the angle between the rotor and drive member is reduced. This shortens the piston stroke and reduces the delivery.

The servo piston is moved with oil from the regulating valve. On this pump, the servo is not spring loaded so regulating the pressure to the servo will not control pump output as was done on other axial piston pumps.

On this pump, the regulator valve adjusts the position of the servo and rotor. For that reason it requires a follow-up link to let the regulator know the position of the rotor.

The pump shown in Fig. 32 is a load-sensing pump. With electronic control, it can respond to working pressures, line flows and engine speeds to vary its output.

This concludes our descriptions of the types of pumps used in modern hydraulic systems. There are many other pumps and applications but most will use the principles of the pumps covered in this chapter.

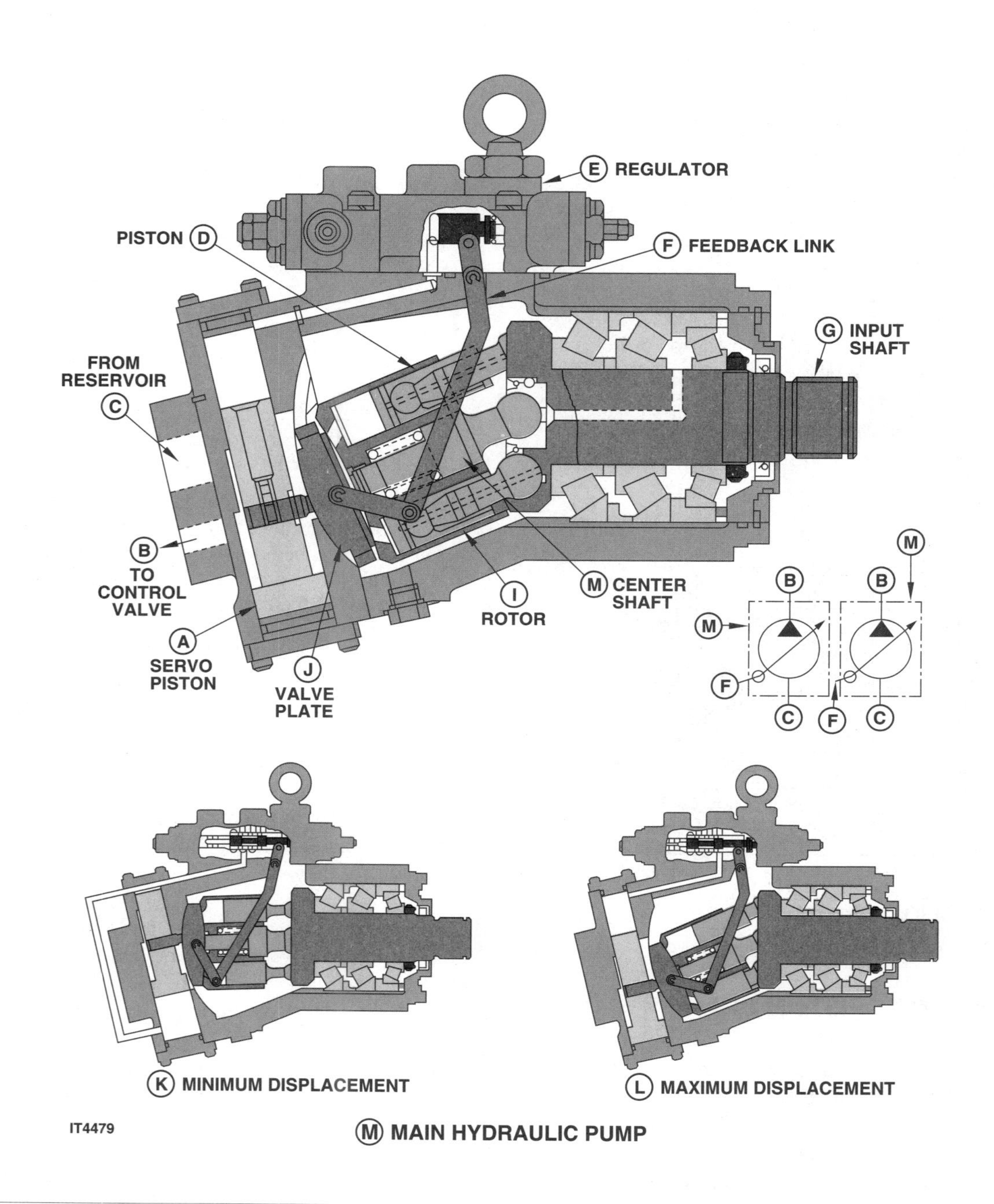

Fig. 32 — Bent-Axis Axial Piston Pump - Variable Displacement

SUMMARY OF PUMP TYPES

Before going into the application and efficiency of hydraulic pumps, let's review some of the points that we have just covered.

In summary:

1. A hydraulic pump converts mechanical force into hydraulic energy or fluid power.
2. Of positive and non-positive displacement pumps, the positive displacement type is the only one suited for modern mobile hydraulics because it can produce a controlled flow against high pressures.
3. A hydraulic pump can be designed to produce either a specific volume of fluid at a specific speed, or to produce a variable volume of fluid at a constant speed ... fixed displacement or variable displacement.
4. The three types of pumps most often used in machine hydraulic systems are gear, vane, and piston.
5. These three pumps are ideal for machine hydraulics because of their ability to produce a great volume of oil for their relatively small size.
6. The preceding text covers only basic hydraulic pumps. There are a great number of variations on all of the pumps selected. While many are designed similar to those covered, the design can vary greatly depending on the pressures, volumes and applications of the pumps.
7. Because of the wide variety of applications and systems in which hydraulic pumps are expected to perform, there is no one pump better than another for all applications.

HYDRAULIC PUMP QUALITIES

1. Pump Capacity - also called delivery rate, discharge rate or size, is expressed in two ways. The first is the theoretical delivery and is expressed in cubic inches (cubic centimeters) per revolution. This is the quantity of oil the pump would deliver when turned one revolution if there were no internal leakage from the outlet to the inlet cavities.

The other is an actual delivery and is expressed in gallons (liters) per minute at a specific pressure and a specific speed. In pump specifications, the viscosity and temperature of the oil will also be specified. This is what the pump will actually do.

Variable displacement pumps are always rated at full delivery.

2. Pump Efficiency - All pumps pressurize and pump their theoretical capacity to the outlet. The amount that does not come out has leaked past the gears, vanes or pistons inside of the pump.

Pump efficiency is the comparison of the amount of oil a pump actually can deliver to what it could deliver if there were no internal leakage. It is expressed as a percentage.

Leakage from the outlet to the inlet side of a pump can be considered like an orifice that connects the two. We know from Chapter 1, that flow through an orifice is increased as pressure is increased. *A pump efficiency rating, therefore, must be at a specific pressure.*

We also know that on a pump with a certain leakage from outlet to inlet that the amount of oil leaking would be about the same whether the pump were running fast (higher output) or slow (low output).

The percentage of that leakage would be higher if compared to the low output than to the high output. *A pump efficiency rating must, therefore, be made at a specific rpm.*

Because leakage is affected by oil viscosity and temperature, it is important that - *A pump efficiency rating be made with oil of a specific viscosity and temperature.*

Pumps selected by a manufacturer for an application will normally have 80-95% efficiency rating at the pressures at which they will be used.

Consult the technical manual when testing a pump or system for flow. It is important to use the specifications for that machine. *Tests for minimum flows should then be made at the speed, pressure and temperature in those specifications.* Machine specifications normally assume that you have already checked for proper oil usage.

3. Pump Comparisons

Efficiency - The gear pump has the largest range of quality of construction. For this reason, it is usually the one chosen as the most economical for low pressure applications.

In its current highest quality construction, the gear pump can be used at the same pressures as the vane pump.

The piston pump is much more efficient when compared at the same pressures as the gear and vane pumps.

Pressure - The gear and vane pumps are about the same with about a 4000 psi (28,000 kPa) upper pressure limit. The piston pump can be used on systems up to 10,000 psi (70,000 kPa).

With the increased popularity of piston pumps, they are being manufactured at a variety of quality levels to meet a broader range of pressure requirements.

Speed - The axial piston pump can be used at speeds up to 6000 rpm, the vane pump to 4000 rpm, and the gear and radial piston pumps up to 3500 rpm.

Size - For equal flow capacity, the piston pump is larger and more expensive than the gear or vane pumps.

MALFUNCTIONS OF PUMPS

THE HUMAN ERROR FACTOR

Hydraulic pumps can wear out through normal use, but very few pump failures are attributed to "old age." Let's look at some of the factors that can speed the aging process.

The majority of hydraulic pump failures are due to: poor maintenance, bad repair, exceeding operating limits, and the greatest cause - the use of fluid which is the wrong type, is dirty or is of poor quality.

From the list of causes, you will see that the majority of failures are caused by the HUMAN ERROR factor.

PUMP FAILURES

It is impossible to list all of the possible reasons for pump failures; however, these are a few general items.

Breakage of parts - Most likely exceeding designed pressures. Could also be the result of a lubrication failure caused by a low oil supply or improper oil.

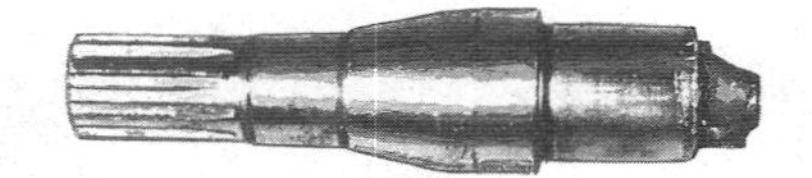

Fig. 33 — Broken Pump Drive Shaft

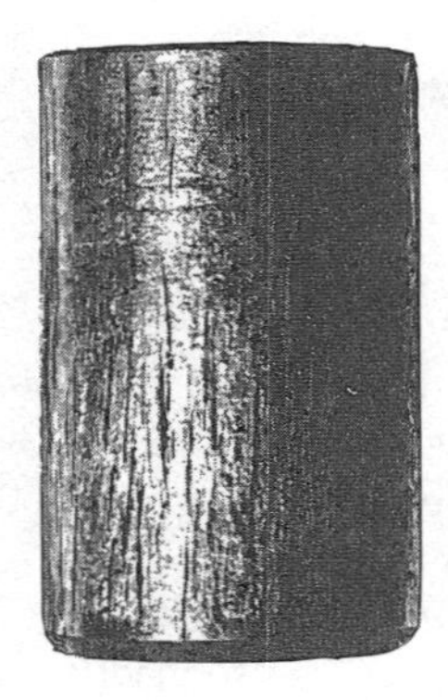

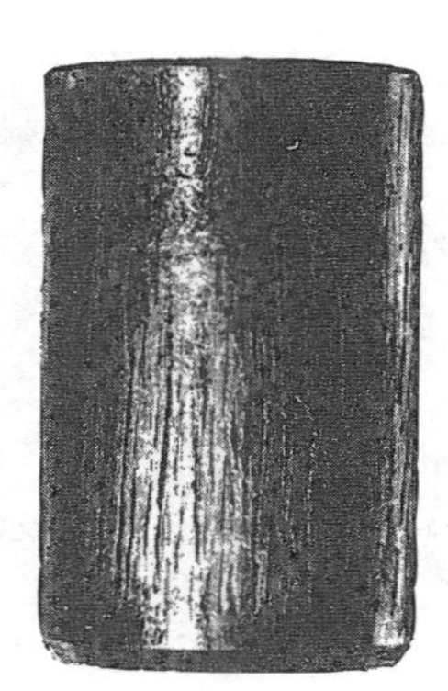

Fig. 34 — Scored Pistons

Fig. 35 — Abnormal Wear

Scoring of moving parts - Larger pieces of contamination in the oil caused by other failures in the system, improper filtration or improper handling of oil.

Pitting at inlet passages - Restriction of pump inlet.

Waved wear pattern on vane pump cam ring - Restricted inlet flow.

Rapid wear of parts - Fine particles in the oil caused by improper or poor quality oils, insufficient oil supply, improper servicing, improper filtration or exceeding operating limits.

The pump is designed or selected to operate at certain loads and speeds. Increasing those limits greatly increases the wear rates of the components.

Let's use the example of an operator who rather than operate at the designed 2000 psi, raised the pressure and consistently operated at 4000 psi. Engineering tests had determined that the bearings should last for 4800 hours if operated at the recommended pressure.

What would be the effect on the pump bearing life?

The formula:

$$\text{Bearing Life} = \frac{\text{Old Bearing Life}}{(\text{New Pressure/Old Pressure})^3}$$

In our example:

New Bearing Life =

$$\frac{4800}{(4000\text{ psi}/2000\text{ psi})^3} = \frac{4800}{2 \times 2 \times 2} = \frac{4800}{8} = \textbf{600 hours}$$

One eighth of the service life is the result of doubling the pressure. It would be one third or 1600 hrs. if the pressure in the example were raised to 3000 psi.

Let's see what effect increasing the speed from 2000 rpm to 3000 rpm would have.

The formula:

$$\text{New Bearing Life} = \frac{\text{Old Bearing Life x Old Pump Speed}}{\text{New Pump Speed}}$$

In our example:

$$\text{New Life} = \frac{\text{4800 Hours X 2000 rpm}}{\text{3000 rpm}} = \frac{9{,}600{,}000}{3000} = \textbf{3200 hrs}$$

This is a substantial decrease in service life.

To prevent the failures described above, it is important that the operator know the hydraulic system, maintain the system, operate it as designed, and use the proper fluids. In other words, *follow the operator's manual provided with the machine.*

TEST YOURSELF

QUESTIONS

1. (Fill in the blanks). "A hydraulic pump converts ____________ force into __________ force."
2. (True or false?) "Hydraulic pumps produce flow, not pressure."
3. (True or false?) "Non-positive displacement pumps are the best hydraulic pumps because they produce a continuous pressurized flow of fluid."
4. What three types of hydraulic pumps are generally used in modern farm and industrial hydraulic systems?
5. What is the major difference between an external gear pump and an internal gear pump?
6. Why does a pump cavitate?
7. What does "axial" and "radial" mean in reference to the piston pumps?
8. Of the pumps discussed in this chapter, which is the most adaptable to the load sensing system?
9. What are the parts that make up the "rotating group" of an axial piston pump?
10. What is the pump flow when the pump is in "standby"?
11. What is the most frequent cause of hydraulic pump failures?
12. What effect will doubling the operating pressure have on pump life?

HYDRAULIC VALVES

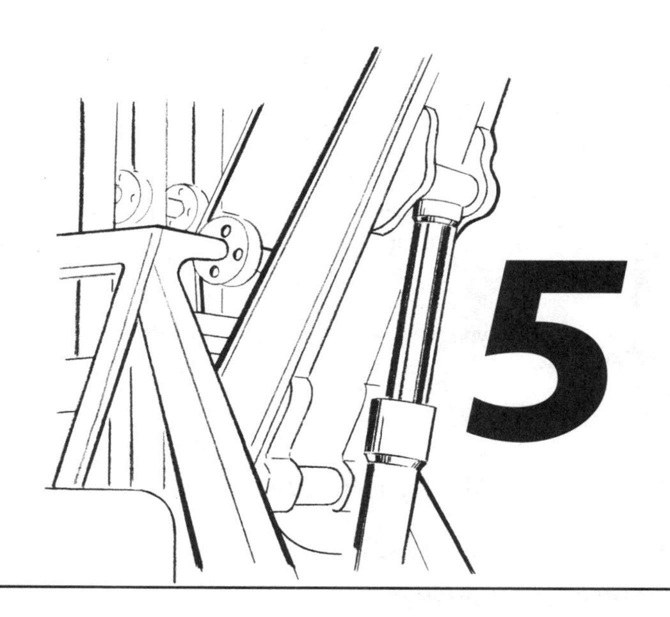

INTRODUCTION

Valves are the controls of the hydraulic system. They regulate the pressure, direction, and volume of oil flow in the hydraulic circuit.

Valves can be divided into three major types:

- **Pressure Control Valves**
- **Directional Control Valves**
- **Volume Control Valves**

Fig. 1 shows the basic operation of the three types of valves.

PRESSURE CONTROL VALVES are used to limit or reduce system pressure, unload a pump, or set the pressure for oil entering a circuit. Pressure control valves include relief valves, pressure reducing valves, pressure sequence valves, brake valves, pilot control valves and unloading valves.

DIRECTIONAL CONTROL VALVES control the direction of oil flow within a hydraulic system. They include check valves, spool valves, rotary valves, pilot controlled poppet valves, flow divider valves and electro-hydraulic valves.

VOLUME CONTROL VALVES regulate the volume of oil flow, usually by throttling or diverting it. They include compensated and non-compensated flow control valves.

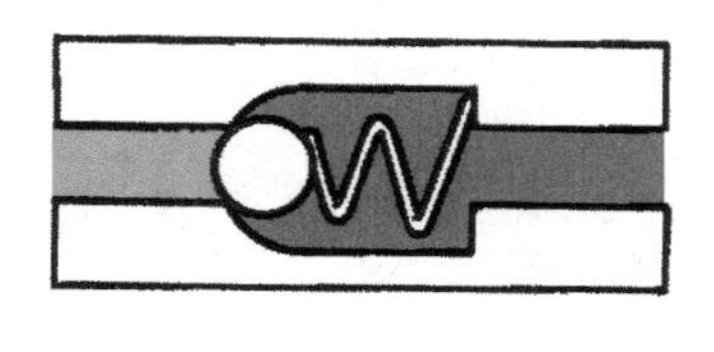
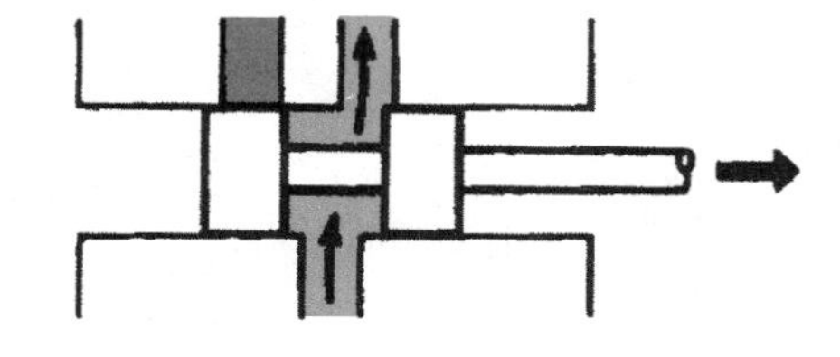
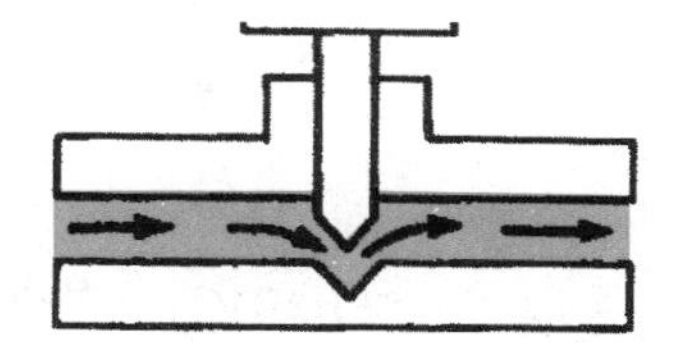
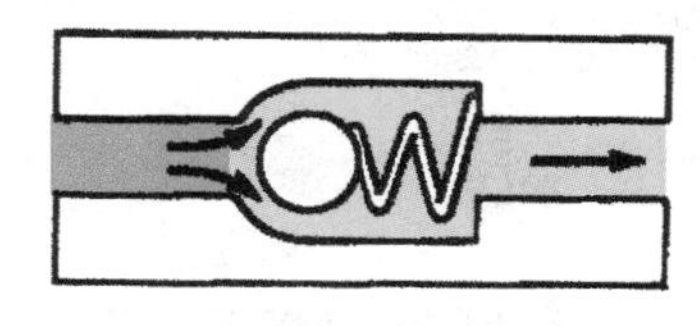
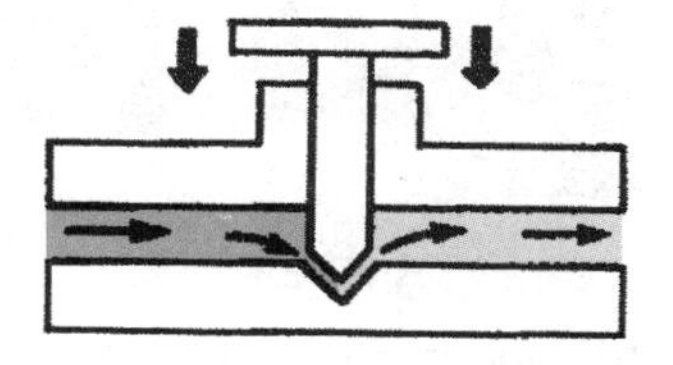

PRESSURE CONTROL — DIRECTIONAL CONTROL — VOLUME CONTROL

Fig. 1 — The Three Types of Valves

Some valve assemblies are variations of the three main types. For example, many direction control valves use a built-in pressure control and flow control valves.

Valves can be controlled in several ways: manually, hydraulically, electrically, or pneumatically. In some modern systems, the entire sequence of operation for a complex machine can be made automatic.

Let's discuss each type of valve in detail, starting with pressure control valves.

PRESSURE CONTROL VALVES

Pressure control valves are used to:

- **Limit system pressure**
- **Reduce pressures**
- **Set pressure for oil entering a circuit**
- **Unload a pump**

RELIEF VALVES

Each hydraulic system is designed to operate in a certain pressure range. Higher pressures can damage the components or develop too great a force on the work to be done.

Relief valves remedy this danger. They are safety valves that release the excess oil when pressures get too high.

Two types of relief valves are used:

DIRECT ACTING relief valves are simple open-closed valves. The pressure oil acts directly on the valve.

PILOT OPERATED relief valves have a "trigger valve" which controls the main relief valve.

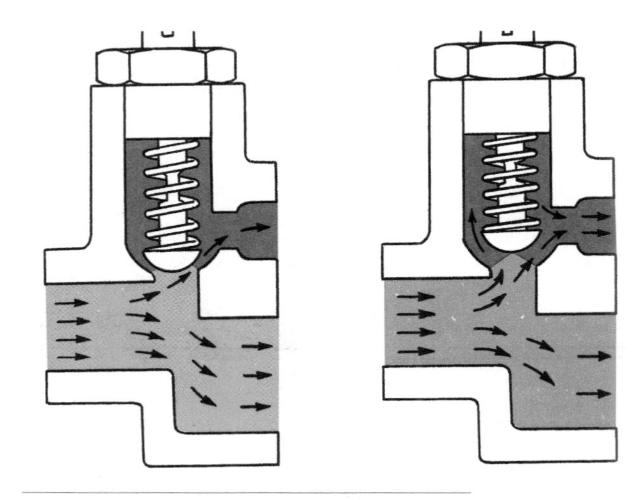

Fig. 2 — Direct Acting Relief Valve in Operation

DIRECT ACTING RELIEF VALVES

Fig. 2 shows the operation of this simple valve. When closed, the spring tension is stronger than the force of the inlet oil pressure, holding the valve on its seat.

When pressure rises at the oil inlet, it pushes on the valve. When the pressure on the valve can overcome the spring force, the valve will open. Oil then flows out to the reservoir, preventing any further rise in pressure.

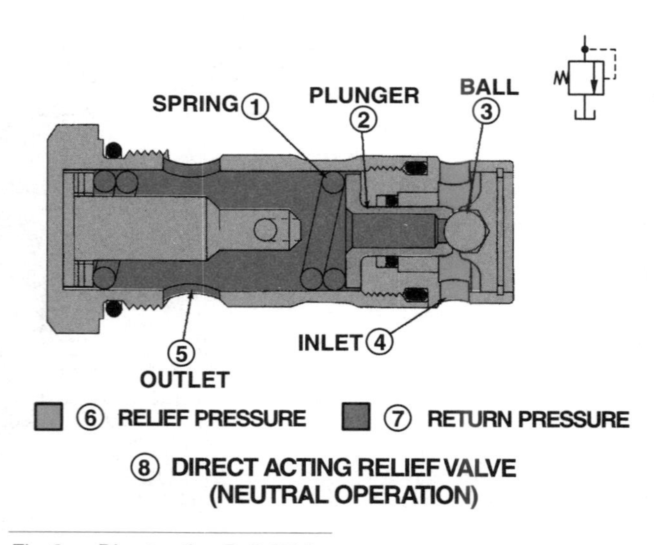

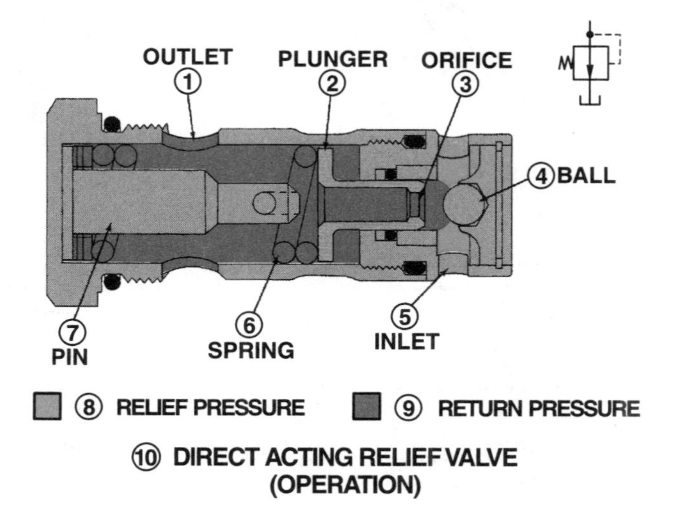

Fig. 3 — Direct acting Relief Valve

When the pressure drops enough to allow the spring force to overcome the pressure, it will push the valve closed and hold it on its seat.

Some relief valves are adjustable. Often a screw is installed behind the spring (see Fig. 2). By turning the screw in or out, the relief valve can be adjusted to open at a certain pressure. Fig. 3 shows a shim adjusted valve.

In most applications, the relief valve poppet must be able to pass the full output flow of the system. This means that when the flow is large, the valve has to be large. Pressure acting on a large surface will exert a large force and require a large spring to overcome that force.

To prevent relief valves from becoming too large, they are usually designed to have the pressure oil act on only a small portion of the poppet and thus require a smaller spring. In Fig. 3, the spring acts against the yellow poppet to hold it against the ball. The ball is part of the valve housing and does not move.

Pressurized oil will now exert its force on the small area of the poppet, which is between the ball seat and the outside diameter of the poppet. This reduces the force acting against the spring and allows it to be smaller.

When pressure increases at the inlet, it exerts a force on the plunger. The plunger moves to the left because the oil pressure force is greater than the spring force. The plunger moves away from the ball allowing oil to go through the poppet and to the outlet.

When pressure is reduced, the spring will close the valve.

USES OF DIRECT ACTING RELIEF VALVES

Direct acting relief valves have a fast response, making them ideal for relieving shock pressures. They are often used in trapped oil circuits (Circuit Relief Valves) to prevent damage to components.

These valves are used mainly where volume is low, and for less frequent operations. For full flow applications, the pilot operated relief valve is normally used. These will be covered next.

PILOT OPERATED RELIEF VALVES

When a relief valve is needed to handle large volumes of oil, and where low pressure override is desirable, the pilot operated relief valve is used.

The pilot operated relief valve shown in Fig. 4 consists of a poppet valve (1) which is hydraulically "triggered" by a pilot valve (2).

The main poppet has a very light spring (3) holding it on a seat. Inlet oil comes into the seat area as shown. The poppet also has a small orifice (4) connecting the pressure oil passage to the backside of the poppet.

The pilot valve is on the backside of the poppet. This pilot valve is simply a miniature version of the direct acting relief valves discussed in previous paragraphs.

When the pressure is low and the pilot valve is closed, the oil is trapped behind the poppet. Because there is no flow

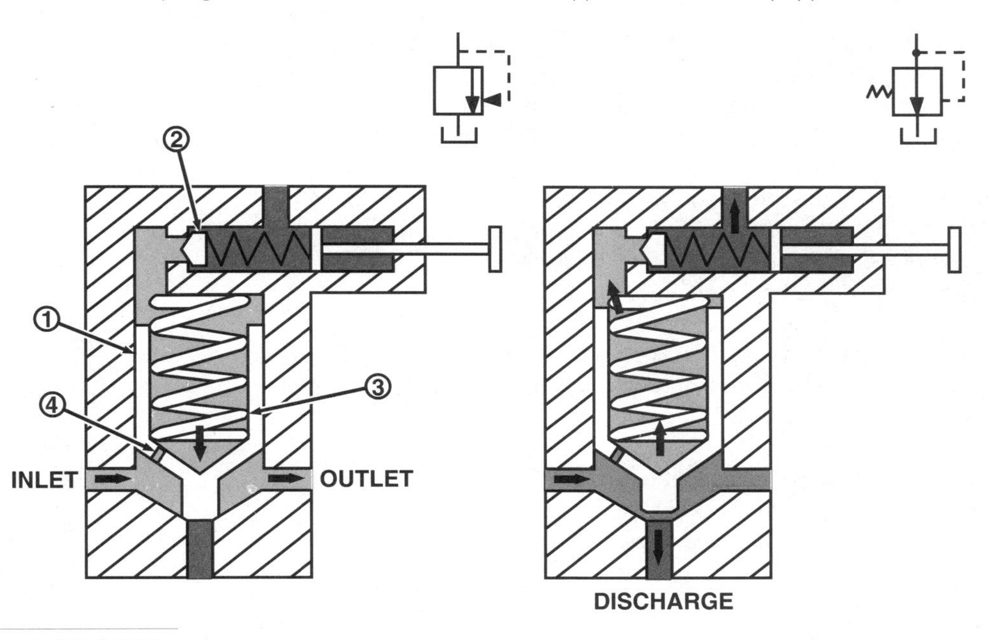

Fig. 4 — Pilot Operated Relief Valve

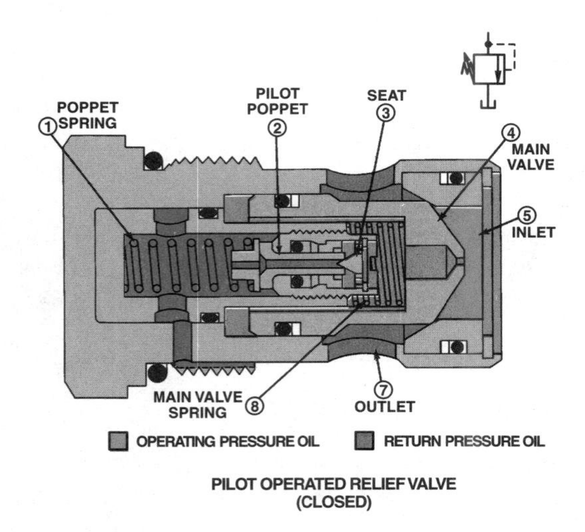

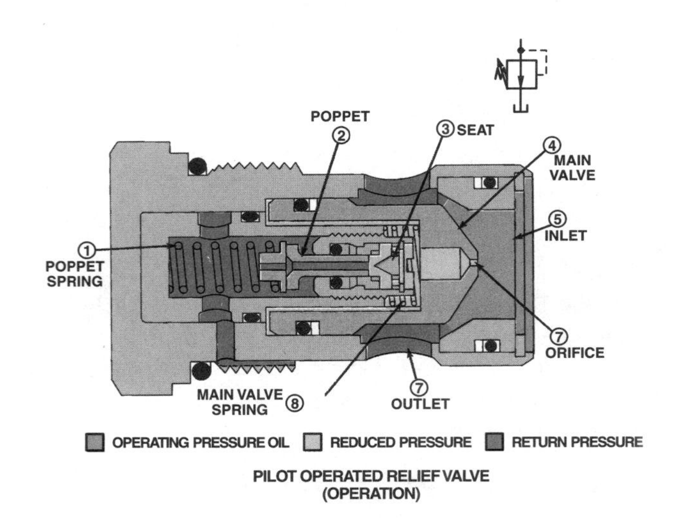

Fig. 5 — Pilot operated Relief Valve

through the orifice, the pressure will be the same on both sides of the poppet. The area exposed to the pressure is greater behind the poppet than on the seat side. This will hydraulically hold the poppet on its seat. The higher the pressure the harder it will be held closed.

When pressure in the system reaches the relief setting, the pilot valve will open. This will cause oil to escape from behind the poppet and create a flow through the orifice. We know that flow through an orifice will cause a pressure drop. The increased pressure on the seat side of the poppet will now hydraulically open the poppet and allow system oil to be discharged.

During discharge, the pressure behind the poppet will remain near the pressure at which the pilot valve opened. The system pressure will rise just enough to maintain a pressure differential across the orifice.

When the system pressure is reduced, the pilot valve will close. There will no longer be flow through the orifice so the pressure will be the same on both sides of the poppet. The poppet will be hydraulically closed and held on its seat.

Fig. 5 shows a pilot operated relief valve. The pilot is a miniature version of the one shown in Fig. 3. The only difference is that a cone seat is used instead of a ball.

The poppet has an orifice in its end. The poppet is exposed to the pressure oil on its entire area on the backside and a lesser seat area on the front. Operation is the same as described in Fig. 4.

Notice that pilot operation has allowed us to have a large bypass valve to handle the flow while allowing the springs to be very small.

CRACKING PRESSURE AND PRESSURE OVERRIDE

Pilot operated relief valves have less pressure override than simple direct acting types. Fig. 6 compares two of these valves. While the direct acting valve in Fig. 6 starts to open at about half its full-flow pressure, the pilot operated valve opens at about 90 percent of its full-flow pressure.

This means the pilot operated valves will have more oil available to the working circuit as pressures are near the relief setting. This will result in less heat buildup.

USES OF RELIEF VALVES

We have seen that in Fig. 6 the pilot operated relief valve has a tighter range of pressure from cracking to full flow, which makes it best suited for system relief applications.

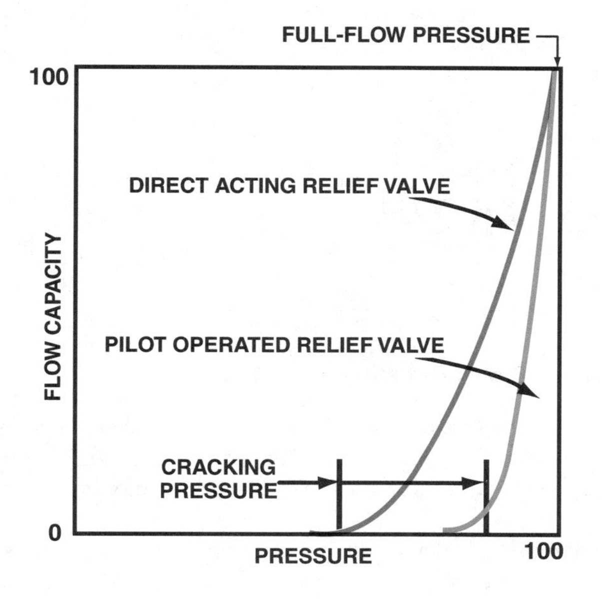

Fig. 6 — Relief Valves Compared at Cracking and Full-Flow Pressures

NOTE: System relief valves are placed in the main circuit and protect everything from the pump to the valves at all times. They also protect the circuits while they are activated.

Circuit relief valves protect the individual circuits when the valve is in neutral and the oil is trapped in the circuit.

The direct operating valves have a greater override but open quicker providing better protection for shock loads. They are normally used in circuit relief applications which seldom require full flow operation.

When hydraulic systems are designed, the relief valves are chosen in each application for their operating characteristics and the effect on the overall system. It is important, therefore, that relief valves be adjusted to or replaced by valves with the same operating characteristics and pressure settings as the original.

It is important to *consult the machine technical manual* for testing and adjusting relief valves. System relief valves are normally tested and rated at a specific flow (rpms) with oil at a specific temperature.

Circuit relief valves are normally tested and rated with a very low flow at a specific temperature. Because of the low flow, they are adjusted very near their cracking pressure and not at full flow.

Circuit relief valves often contain anticavitation valves, which will be covered in the spool valve section of this chapter.

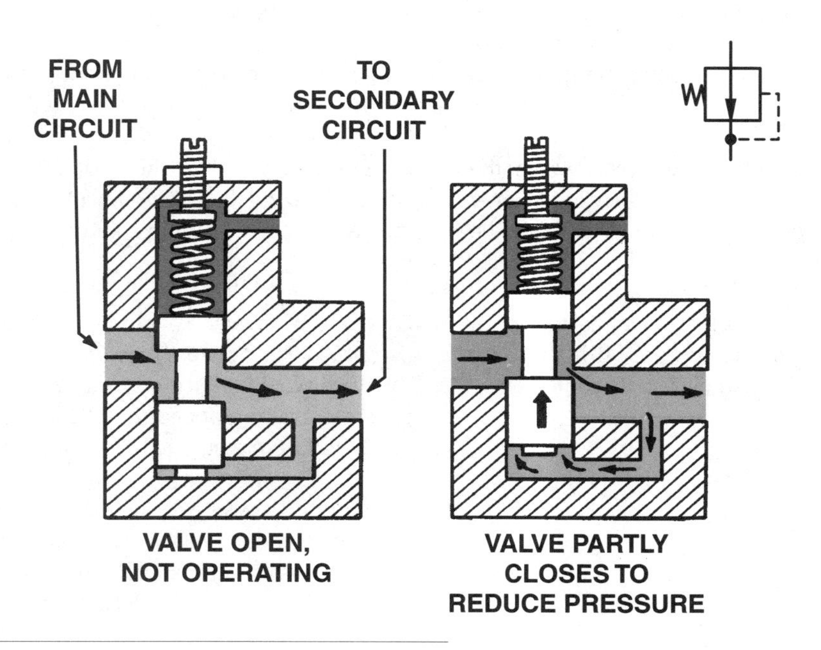

Fig. 7 — Pressure Reducing Valve — Reduced to a Constant Pressure

PRESSURE REDUCING VALVES

A pressure-reducing valve is used to control the pressure in one branch of a circuit. The pressure in that branch will always be at or below the main circuit.

TWO TYPES OF PRESSURE REDUCING VALVES

Pressure reducing valves can operate in two ways:

- **Reduced to a Constant Pressure**
- **Reduce a Fixed Amount**

VALVES WHICH REDUCE TO A CONSTANT PRESSURE maintain a constant pressure in the branch for which it supplies oil, regardless of main system pressure.

Operation of the reducing valve which reduces to a constant pressure in shown in Fig. 7. Until the system pressure is equal to the desired pressure in the secondary circuit (first view) the valve will remain open. The pressure in the secondary circuit will be equal to the main pressure.

When pressure starts to rise in the secondary circuit, that pressure will exert a force on the bottom of the valve spool. This force will act against the end of the valve to compress the adjustable spring. The valve will be partially closed, restricting oil flow to the secondary circuit. When the desired pressure is reached, the valve will automatically adjust the restriction to maintain that pressure.

When secondary pressure lowers, the spring moves the valve to open it a little more. When the pressure rises, the extra force on the bottom of the valve compresses the spring and closes the valve a little more.

If this valve is used in a circuit with a closed-center valve, the reducing valve would completely close when the closed-center valve was closed. It would open just enough to maintain the desired pressure to the circuit.

This valve is used to reduce pressures for differential lock, pilot control circuits, etc. With slight modification, it is the valve used for hydraulic power brakes and pilot valve controllers.

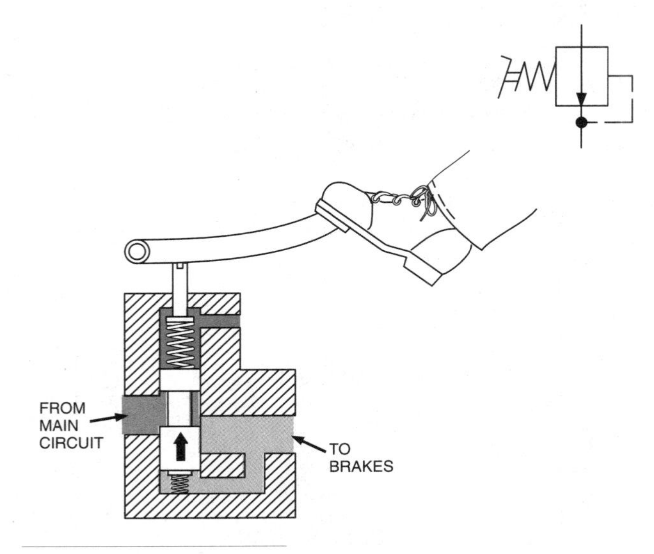

Fig. 8 — Power Brake Valve

In Fig. 8 the adjusting screw used in Fig. 7 has been replaced with the brake foot pedal. With the pedal at rest, a light spring below the valve holds it in the shut off position.

As the pedal is depressed, it will compress the upper spring and open the valve. When pressure is built in the brake circuit, it will force the valve up against the upper spring. When that force is enough to overcome the spring force, the valve will close and maintain that pressure.

The pressure regulated in the brake circuit, and thus the amount of braking, will be in direct proportion to the amount of force put on the pedal by the operator. *See the brake valve section in Chapter 1.*

Add a hand lever instead of a foot pedal and you have a pilot controller for a spool valve. *See spool valves in this chapter.*

FIXED AMOUNT PRESSURE REDUCING VALVES, as shown in Fig. 9, supply a fixed amount of pressure reduction, which means that the outlet pressure will vary with the main circuit pressure.

For example, the valve might be set to give a reduction of 500 psi (3450 kPa). If system pressure were 2000 psi (13,790 kPa), the valve would reduce pressure to 1500 psi (10,345 kPa). If system pressure dropped to 1500 psi (10,345 kPa), the valve would reduce pressure to 1000 psi (6895 kPa)

In Fig. 9, the pressure of the main circuit acts on a portion of the valve trying to push it open against the spring force. Until that pressure is high enough to open the valve, there is no pressure in the secondary circuit or behind the valve.

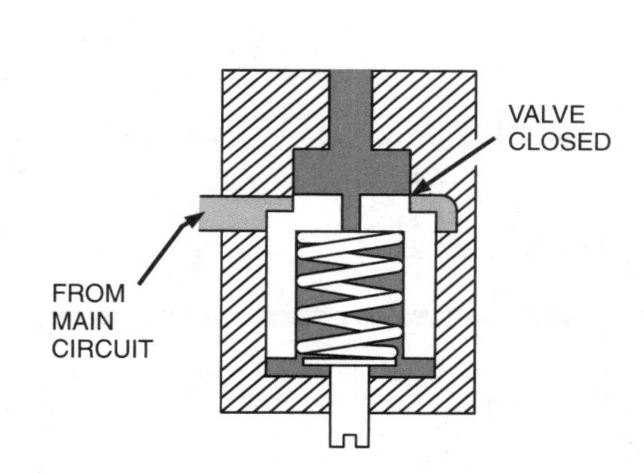

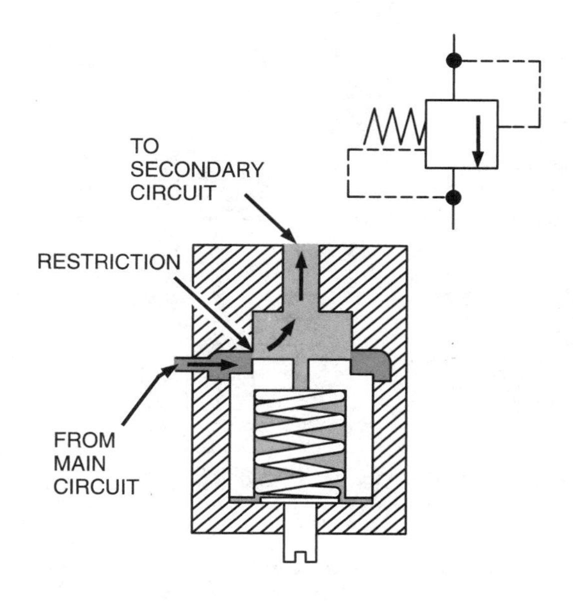

Fig. 9 — Fixed Amount Pressure Reducing Valve

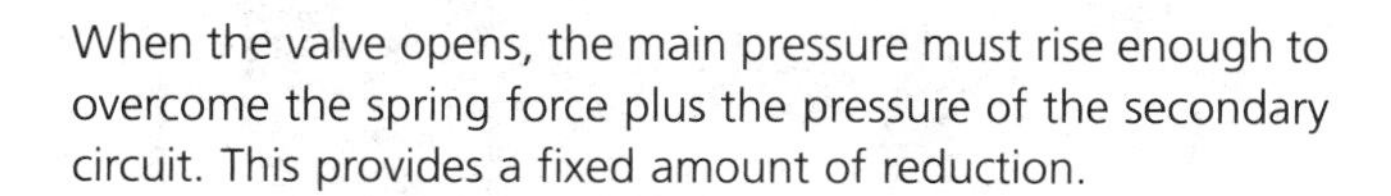

When the valve opens, the main pressure must rise enough to overcome the spring force plus the pressure of the secondary circuit. This provides a fixed amount of reduction.

An example of the fixed amount reduction was found in the stroke control valve of the variable displacement radial piston pump. In Fig. 10, main pressure oil in the pump outlet tries to open the stroke control valve. Spring force prevents this until pump pressure approaches "standby", allowing the pump to remain in full stroke. When the valve opens, oil flows into the pump crankcase. From this point, the outlet and the crankcase pressure will rise together. As the pressure in the crankcase rises, it will hold the pump pistons away from the cam and the pump will stop pumping. *See radial piston pumps in Chapter 4.*

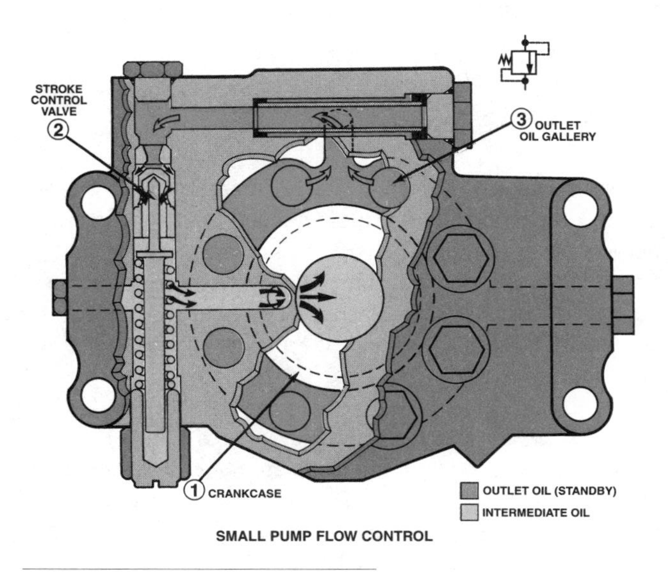

Fig. 10 — Pump Stroke Control Valve

SUMMARY

For **Reduced to a Constant Pressure Valves**, sensing for the valve comes from the outlet side, or the secondary circuit.

For **Reduced a Fixed Amount Valves**, sensing for the valve comes from the inlet side (main circuit) and the outlet side (secondary circuit). The inlet oil pressure tries to open the valve against the spring force. The outlet oil pressure helps the spring to close the valve.

PRESSURE SEQUENCE VALVES

Pressure sequence valves are used to control the sequence of flow to various branches of a circuit. Usually the valves allow flow to a second function only after a first has been fully satisfied.

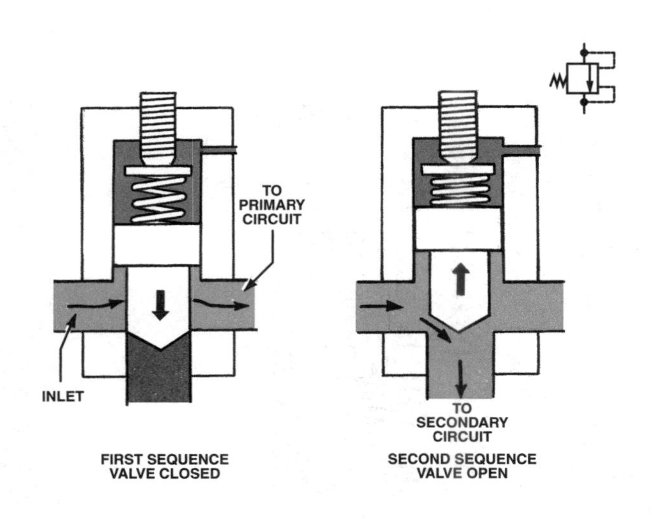

Fig. 11 — Pressure Sequence Valve in Operation

Fig. 11 shows a pressure sequence valve in operation. When closed, the valve directs oil freely to the primary circuit. When opened, the valve diverts oil to a secondary circuit.

The valve opens when pressure oil to the primary circuit reaches a preset point (adjustable at the valve spring). The pressurized oil then lifts the valve off its seat as shown and oil can flow through the lower port to the secondary circuit.

One use of the sequence valve is to regulate the operating sequence of two separate cylinders. The second cylinder begins its stroke when the first completes its stroke. Here the sequence valve keeps pressure on the first cylinder during the operation of the second.

Sequence valves sometimes have check valves that allow a reverse free flow from the secondary to the primary, but sequencing action is provided only when the flow is from primary to secondary.

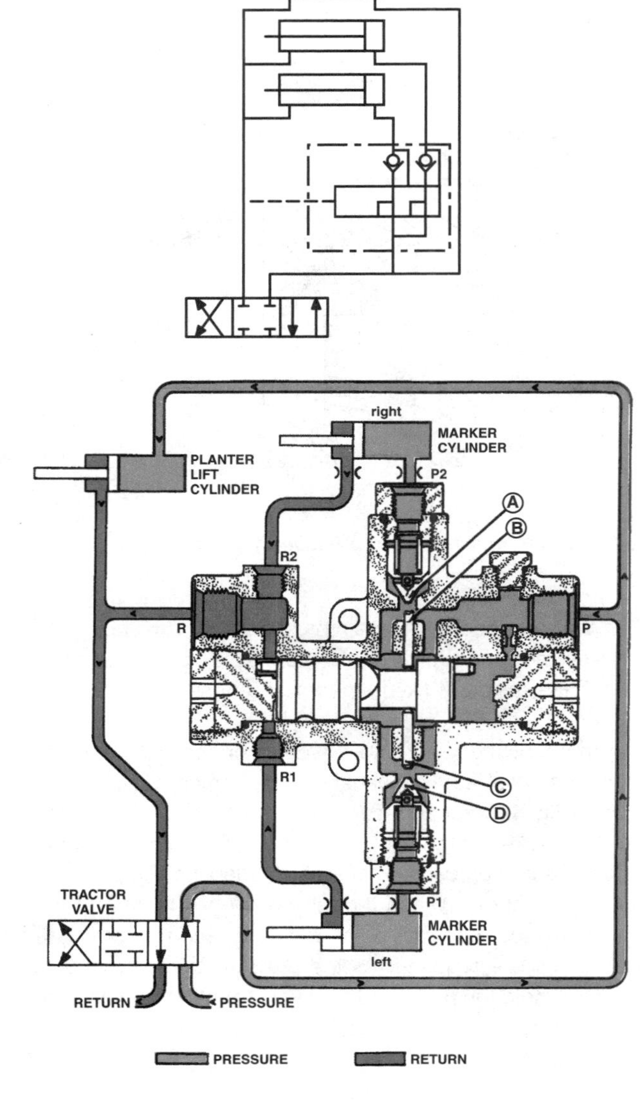

Fig. 12 — A Typical Sequencing Valve

The sequence valve shown in Fig. 12 is used to move one or both secondary cylinders at the same time as a primary cylinder is moved. The logic behind this sequence valve is that if action A occurs, then action B needs to take place. This applies to machines such as corn planters. If the lift lever is actuated for the planter, then the planter markers need to automatically rise.

Fig. 12 shows the hydraulic fluid flow when a planter is raised with a sequence valve. In the first step, the valve spool actuates the planter lift cylinder. Second, poppets A and D are pushed open allowing oil flow to the piston end of both marker cylinders, raising the down marker(s).

When the planter is lowered, the spool valve directs pressure oil to the rod end of all three cylinders. It also pressurizes the sequence valve spool and moves it to the right. The spool contacts pin B and/or C depending on the rotary position of the valve. This moves valves A and/or D to allow oil to return from the marker cylinder(s). The rotated position of the sequence valve spool causes the proper marker to lower. Orifices in the poppets control the drop speed of the markers.

UNLOADING VALVES

The unloading valve is used when a fixed displacement pump and an accumulator are used with closed-center valves. *See Fixed Displacement Pumps with Accumulator System in Chapter 1.*

The unloading valve directs pump output oil to the accumulator until the desired system pressure is reached. It then directs the pump oil back to the reservoir at low pressure until the system again needs charging.

When sensing pressure (accumulator pressure) is low (Fig. 13), spring force holds the valve in the closed position. The pump to reservoir passage is closed and no unloading occurs. Pump output is now forced through a check valve to the accumulator.

As the accumulator pressure rises that pressure acts against the small piston at the end of the sensing line. When the desired pressure is reached, the force on the small piston will move the main valve, against the spring, to the opened position. Oil will now flow from the pump to the reservoir. The check valve in the accumulator circuit will trap oil.

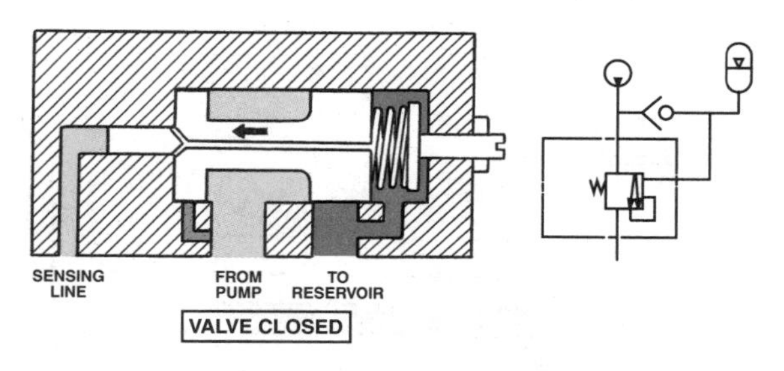

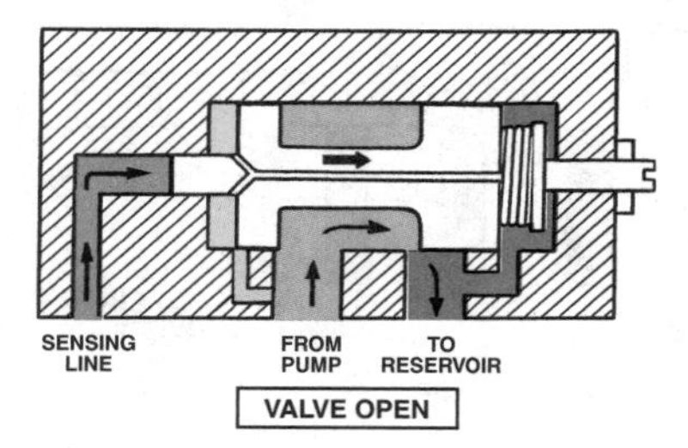

Fig. 13 — Unloading Valve in Operation

There is a slight backpressure as oil flows through the valve. When the valve opened, a small passage opened to allow this backpressure to act on the entire valve face.

To move the valve to the open position, the spring will now have to overcome the accumulator pressure on the pilot piston plus the additional pressure on the valve face.

This means that the valve will close at a lower pressure than the pressure at which it opened. This prevents the valve from cycling when the circuit it serves is in use.

In a typical application, the unloading valve would open when the accumulator was charged to 2500 psi (17,240 kPa) and close to recharge when the pressure was down to 1750 psi (12,065 kPa).

This is normally an independent system used for brakes or differential locks on large machines that have open-center main hydraulic systems. It has, however, been used on some small farm and utility tractors for the main hydraulic system.

DIRECTIONAL CONTROL VALVES

Directional control valves direct the flow of oil in hydraulic systems. They include these types:

- **Spool Valves**
- **Poppet Valves**
- **Rotary Valves**

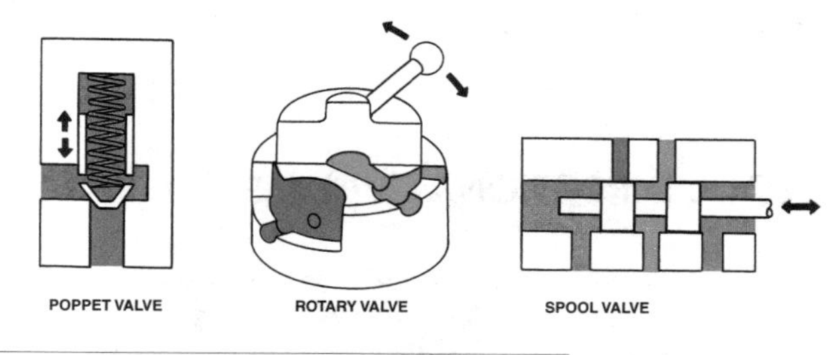

Fig. 14 — Types of Directional Control Valves

The three basic types of directional control valves are compared in Fig. 14. Each uses a different type of valving element to direct oil. The *spool valve* uses a sliding spool that moves back and forth to open and close oil routes. The *poppet valve* uses a poppet that seats and unseats to block or pass oil. The rotary valve uses a *rotary spool* that opens and closes oil routes as it turns.

Let's discuss each valve in more detail.

CHECK VALVES

Before discussing the types of valves, let's look at the check valve, which can be a poppet, ball or spool type. Check valves are simple one-way valves that can be a separate component, but usually are a part of another complex valve. They open to allow flow in one direction, but close to prevent flow in the opposite direction.

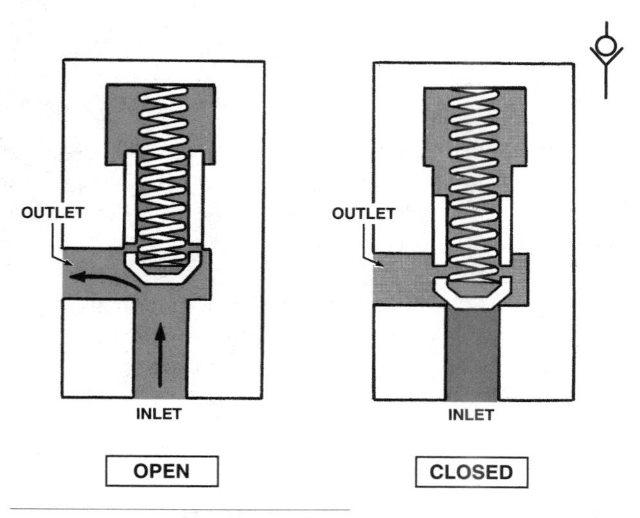

Fig. 15 — Check Valve in Operation

Fig. 15 shows a simple poppet type check valve in operation. The valve is forced opened by pressurized oil, which flows past the valve as shown.

The valve closes when flow stops. This stops reverse flow and traps oil already in the circuit. This valve can be held in the closed position by the spring but often it is only gravity or reverse flow that closes it.

SPOOL DIRECTIONAL VALVES

The sliding spool valve is the most used on modern hydraulic systems. The uses of the spool valve are so great because it is compact, inexpensive to produce and can incorporate almost unlimited options. Some of these options are: system and circuit relief valves, anticavitation valves, lift checks, float position, regenerative position and it can be activated manually, hydraulically or electrically.

Two or more valves can be contained in a single assembly. Each spool controls a separate circuit.

Fig. 16 shows a stack valve in which individual valve sections are bolted together. They have an endcap on each end. The endcaps contain the inlet from pump and outlet to reservoir ports and, where required, the system relief valves. The clearance between the spool and housing is so close that improper tightening of the bolts holding the sections together can distort the housings and cause the spools to bind.

Also shown in Fig. 16 is a unibody valve. A single casting contains all the oil ports and valves.

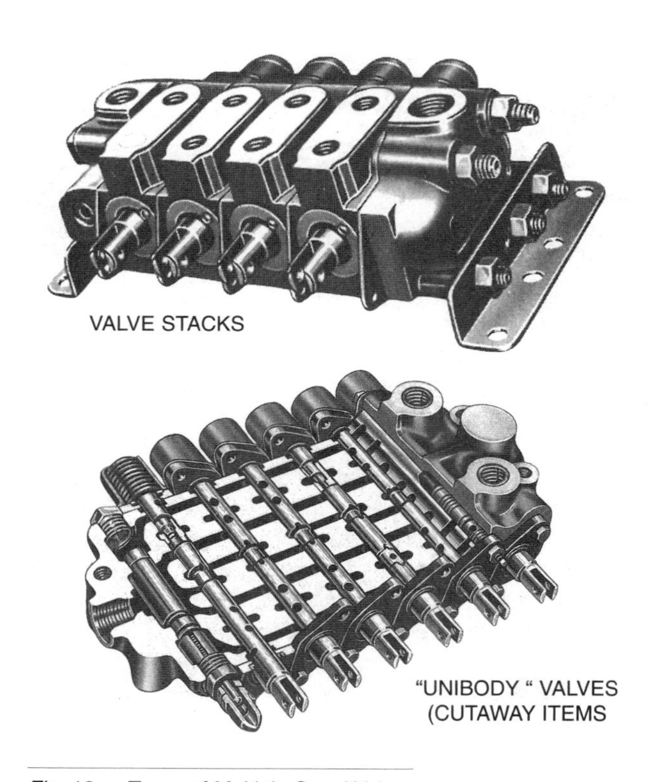

Fig. 16 — Types of Multiple Spool Valves

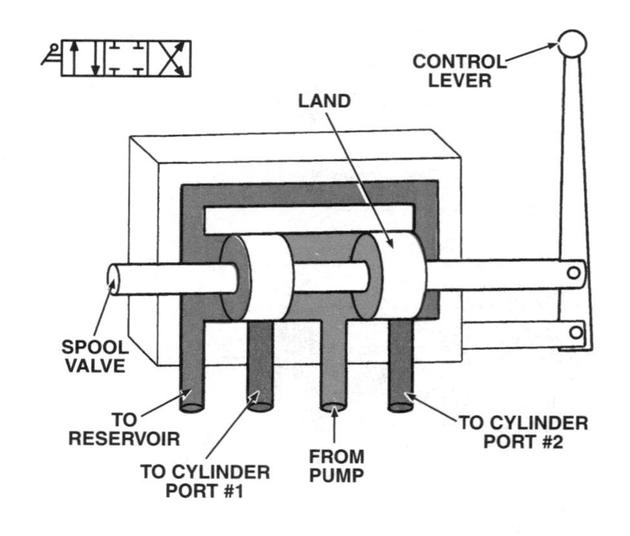

Fig. 17 — Spool Directional Valve

Fig. 17 shows a simple two-land spool valve. Moving the spool from neutral (shown) to the right or left opens up some passages and closes others. In this way, it directs oil to and from the cylinder. The spool lands seal the inlet from the outlet oil passages.

The spool is usually hardened and ground to produce a smooth, accurate, and durable surface. It may also be chrome-plated to resist wear, rust, and corrosion.

The spool valve shown in Fig. 17 is called a "three-position, four-way" valve. The valve has three positions: neutral, left, and right. It is connected to the circuit in four ways: to pump, to reservoir, to cylinder port 1, and to cylinder port 2.

Fig. 18 shows the same spool valve in operation. As the valve is moved to the left, it directs oil from the pump to the left side of the cylinder, actuating it as shown. At the same time, the valve opens a passage that allows oil from the opposite end of the cylinder to return to the reservoir.

When the valve is moved to the right, the flow is reversed and the cylinder operates in the opposite direction.

In neutral (see Fig. 17), the spool lands seal off both cylinder ports, trapping oil to hold the cylinder in place.

Open- and Closed-Center Spool Valves

In Chapter 1, we covered the two kinds of hydraulic valves, open and closed-center. Each uses a different type of spool valve (Fig. 19).

OPEN-CENTER spool valves allow pump oil to flow through the valve during neutral and return to the reservoir

CLOSED-CENTER spool valves stop (dead-end) the flow of oil from the pump during neutral. Note that the valve in Fig. 17 is a closed-center valve.

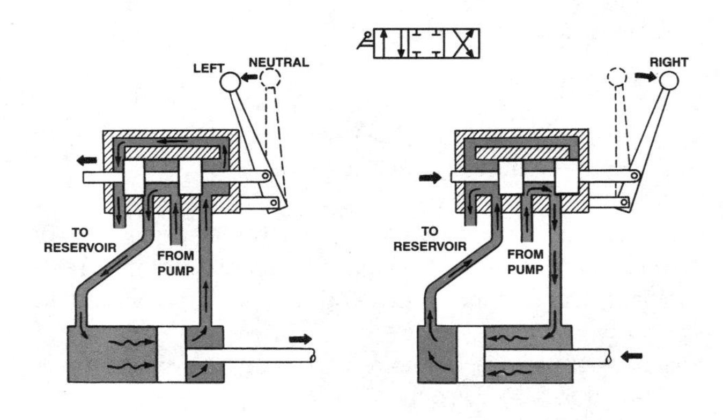

Fig. 18 — Spool Valve Directing Oil to Cylinder

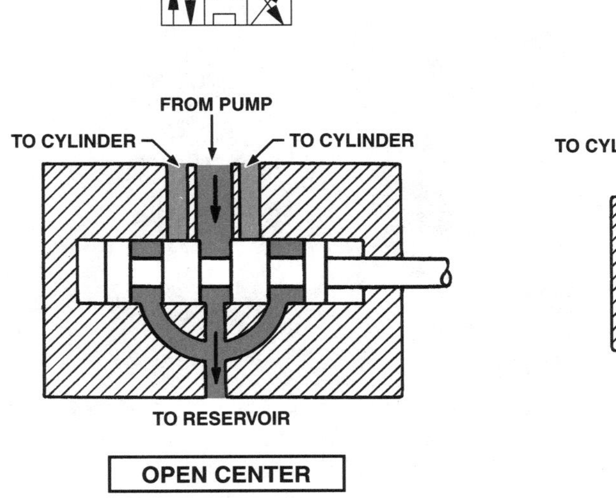

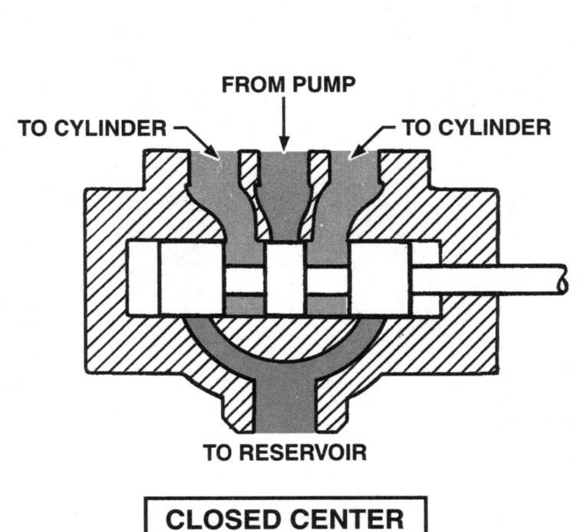

Fig. 19 – Open- and Closed-Center Spool Valves in Neutral

Most multiple stacked spool valves are the series-parallel design.

To better understand the series operation of a multiple spool open-center control valve, the cutaway of a valve is shown in Figs. 20 and 21.

Fig. 20 is a top view of the open-center valve showing the oil flow during neutral operation. Because of the position of the spools in neutral, oil enters the inlet (4) where it flows around the center of the spool (1) and on to the next spool. It exits the valve at the return passage (2). Oil in the work ports (3) is trapped.

With this oil flow through a multiple spool valve, a series connection exists between the spools.

When any spool is activated to engage a function, flow to the other spools is blocked (Fig. 21). Therefore, no flow would be available to the other hydraulic circuits.

A power passage connected to the valve inlet does provide flow to all the control valve functions regardless of their spool positions. This passage is dead-ended at the last valve section.

This parallel passage overcomes the problems, which would be created by a series connection.

With this power passage, more than one function may be operated at a time. Each spool operated will connect the parallel passage to one of its work ports (one end of a cylinder). Therefore, the function, which requires the least amount of pressure, will be the first function to operate.

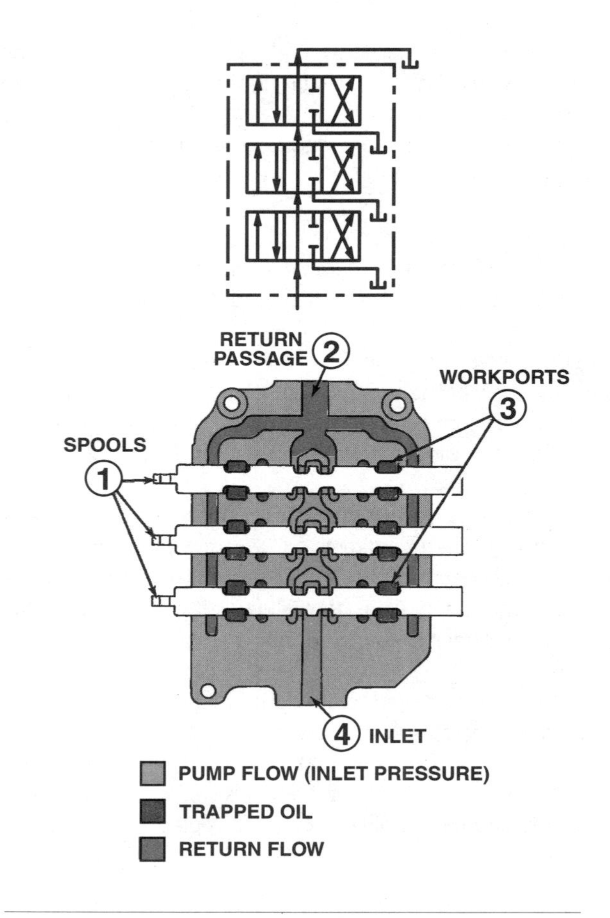

Fig. 20 — Three Spool Open-Center Control Valve in Neutral

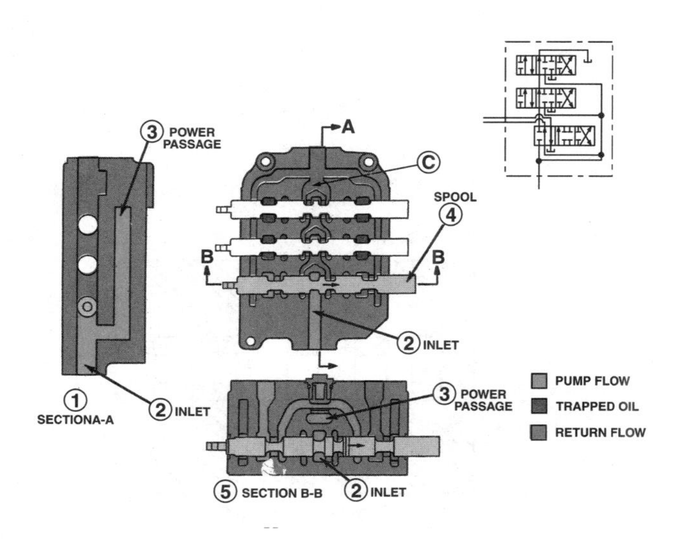

Fig. 21 — Open-Center Control Valve — One Valve Operated

An operator, however, would meter the flow to the cylinders to provide flow to several circuits at the same time.

This valve can be made into a closed-center valve by blocking the neutral passage at point C. This is normally done by installing an endcap which does not have that passage drilled.

Let's look at the operation of individual valve sections.

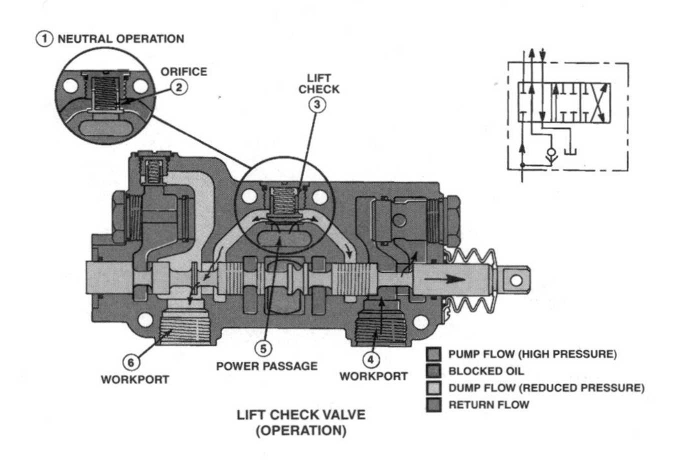

Fig. 22 — Spool Valve in Operation

The spool valve in Fig. 22 has been activated. The spool valve has: 1. blocked the neutral flow passage, 2. connected the power passage to the left workport (6), and 3. connected the right workport (4) to the return passage.

When moving the valve, the workport is opening at the same time the neutral passage is being closed. When moving a cylinder that already has a load on it, there is an instant when oil could flow from the cylinder to the power passage and out the neutral flow passage. This would allow the load to drop a short distance before enough pressure had built in the power passage to lift the load.

For this reason, oil from the power passage must flow through the lift check valve (3) to get to the workports. The lift check valve will not open until oil pressure in the power passage exceeds pressure in the work port. This prevents back flow from the work ports to the power passage.

When a lift check valve on a machine is leaking, the hydraulic function can drop before it starts to rise as the valve is activated.

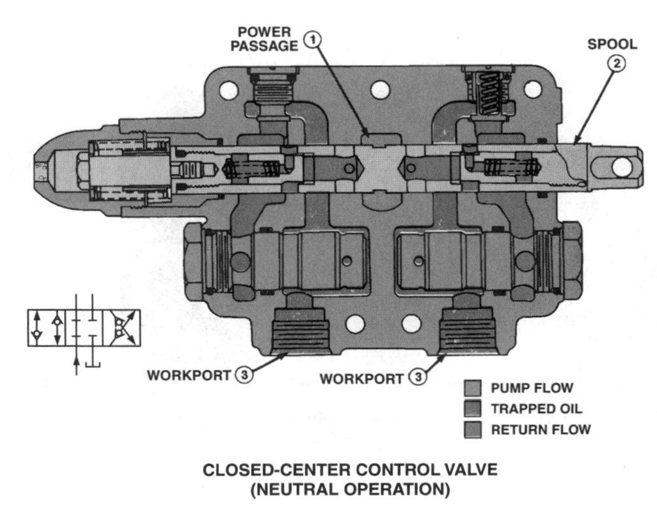

Fig. 23 — Closed-Center Valve in Neutral

Fig. 23 shows a closed-center control valve during neutral operation. This has internal spool passages as opposed to the land type shown in previous examples. The valve is designed specifically for closed-center applications because it does not have a neutral flow passage.

The power passage connecting the spools makes pressure oil available to all of the spools regardless of their position

In neutral, the spool blocks (dead-ends) the inlet flow of oil and traps the oil in the workports (cylinders).

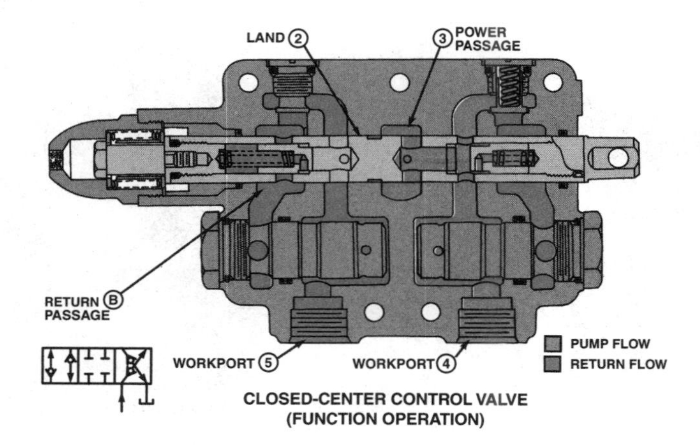

Fig. 24 — Closed-Center Valve in Operation

The spool in Fig. 24 has been moved to the left. Oil from the power passage can now enter the spool, push the check valve off its seat and flow out the right workport. The check valve is the lift check that prevents a backflow of oil until system pressure is enough to lift the load.

At the other end, the spool passage has aligned with the left workport. Oil in the workport can enter the spool, open the check valve, and enter the return passage and flow back to the reservoir.

By moving the spool to the right, the direction of flow through the workports would be reversed.

The closed-center valve shown in Figs. 23-24 has circuit relief valves and anticavitation valves in each circuit.

Let's now look at some of these additional features which can be added to both open- and closed-center valve sections.

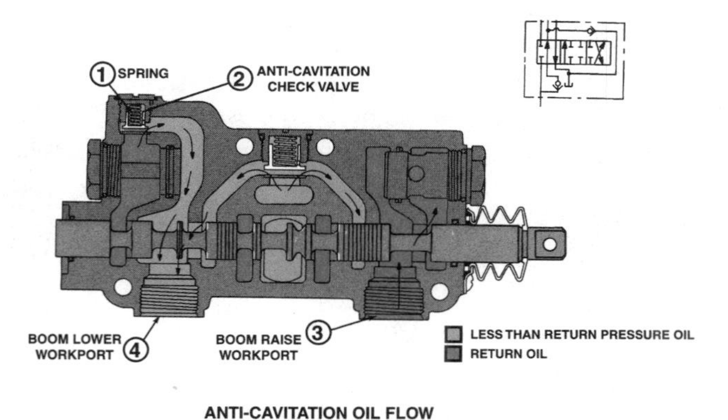

Fig. 25 — Anti-Cavitation Valve Operation

CIRCUIT RELIEF AND ANTI-CAVITATION VALVES

In some applications, the pump cannot supply enough inlet flow to keep up with the requirements of the hydraulic function.

One example would be lowering a loader with a full bucket. Gravity can force the oil out of the lift cylinder faster than the pump can fill the other side. If the valve is closed before the pump can catch up, there will be cavitation trapped in the cylinder. If the next requirement were to put downpressure on the bucket by lowering the loader, there would be a pause before the cylinder would move because the pump would have to fill the cavitation first.

To solve this problem, an anti-cavitation check valve is used. Fig. 25 shows the anti-cavitation valve installed in the boom lower circuit of a loader valve.

As gravity moves the cylinder faster than the pump can supply oil, a vacuum is created in the cylinder. The check valve allows oil from the return passage to enter the work port. This prevents the cavitation. When the valve is closed, there is a solid column of oil from the valve to the cylinder piston.

Fig. 25 shows a circuit relief valve in the boom raise circuit. A circuit relief valve is there to protect that circuit from high pressures when the spool is in the neutral position.

If an external force would cause the pressure in the circuit to rise, the circuit relief would open. This would allow oil from the workport to escape to the return passage.

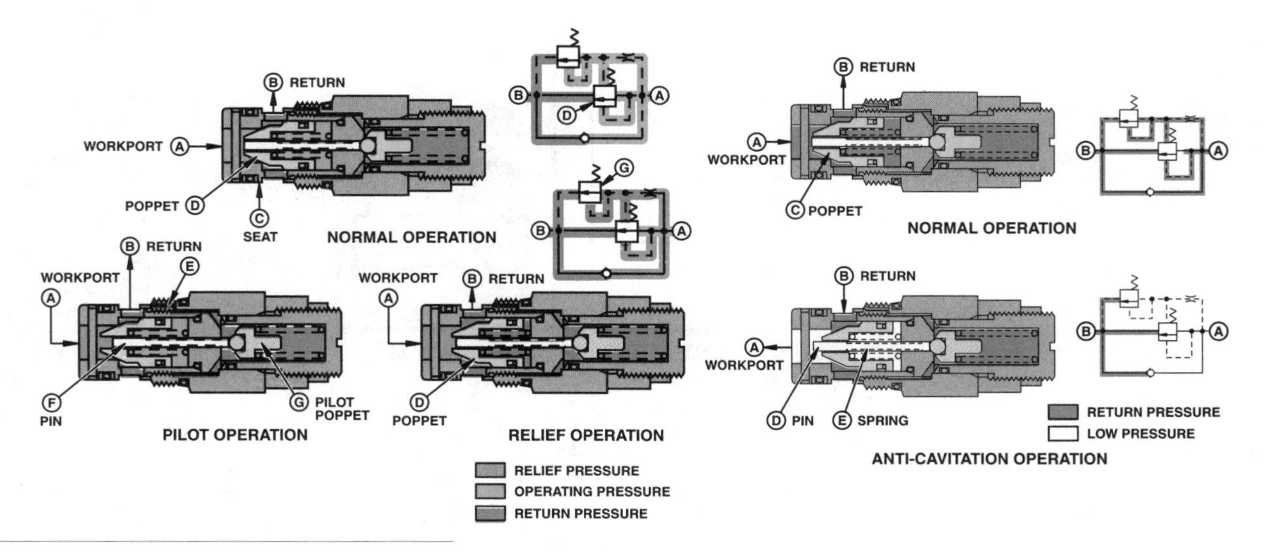

Fig. 26 — Circuit Relief Valve with Anti-Cavitation

As the cylinder is moved to force oil out the circuit relief valve, cavitation will occur on the other side of the cylinder. The anti-cavitation valve will allow oil from the return passage to fill that void and prevent cavitation in the cylinder.

For this reason, anti-cavitation check valves are almost always used in the circuit opposite from one using a circuit relief valve.

Most modern hydraulic systems incorporate the anti-cavitation valve into the circuit relief valve as shown in Fig. 26.

The relief valve is pilot operated. The pilot valve opens when the pressure in the circuit reaches its setting. This creates a flow through the orifice in the poppet, which causes a pressure drop across the orifice. The extra pressure on the poppet then pushes it open and allows circuit oil to flow into the return passage.

When cavitation occurs in the circuit, that negative pressure is felt on the backside of the poppet. This allows the return and atmospheric pressures to push the poppet open and allow oil from the return passage to enter the circuit to satisfy the cavitation.

In certain applications, the operator wishes to have a bucket or blade follow the ground contour. To accomplish this, some valves have a float position. This is a fourth position of the valve spool.

In the float position, shown in Fig. 27, the spool: 1. opens the neutral passage to allow pump oil to flow to return, 2. blocks power passage oil from entering the workports and, 3. connects both workports to the return passages.

This allows the cylinder pistons to move up and down in the cylinders, pushing oil into and drawing oil out of the return passages as it moves.

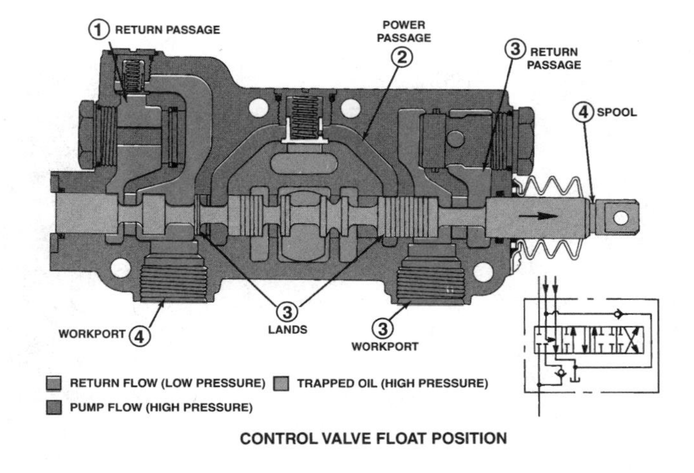

Fig. 27 — Control Valve in Float Position

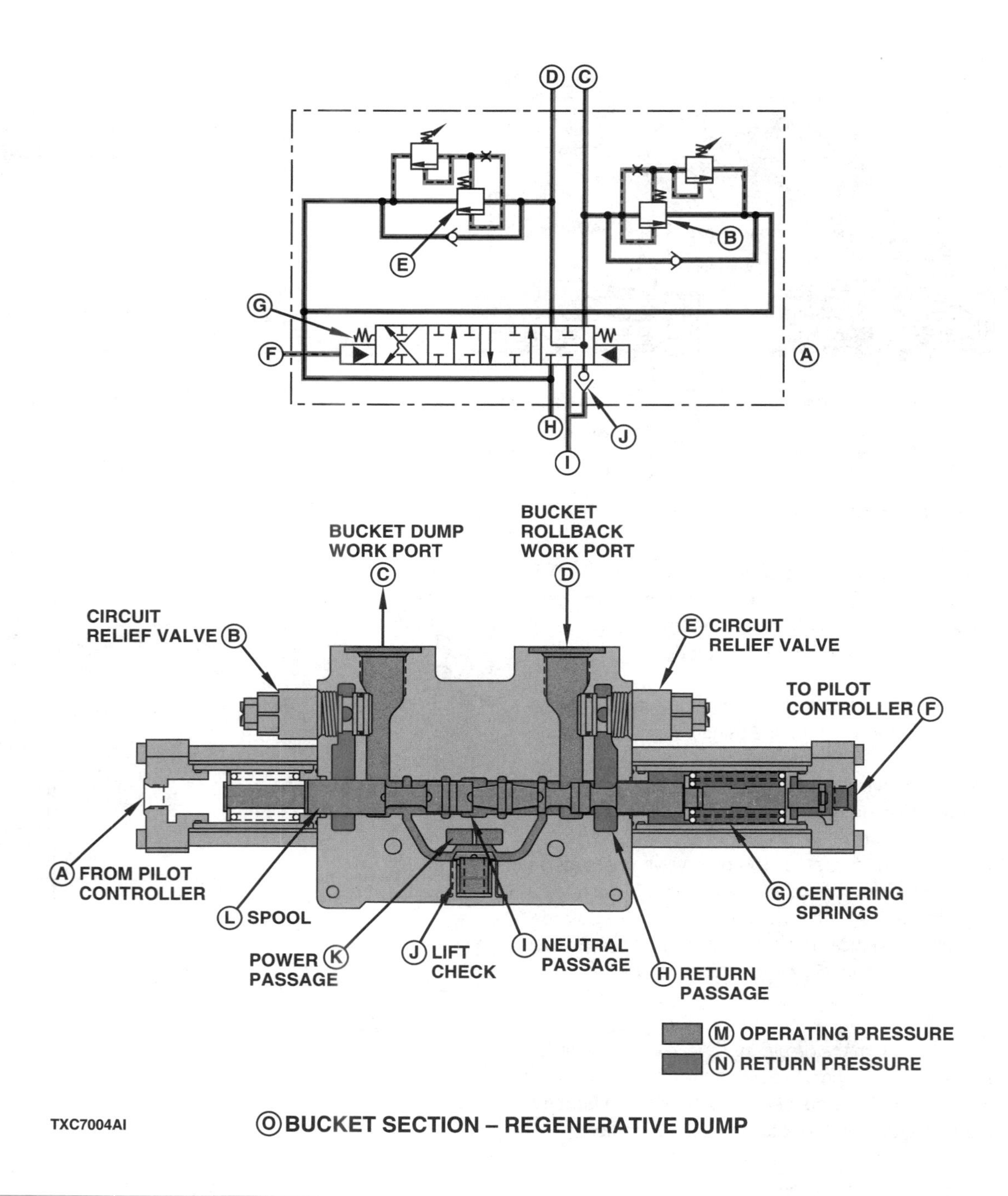

Fig. 28 — Valve in the Regenerative Position

REGENERATIVE POSITION

There are applications, such as dumping a bucket where a faster movement of the hydraulic cylinder is desired. To accomplish this, a regenerative (quick dump) position is added to the valve.

This is a fourth position of the spool. It is in addition to the normal power extend position of the spool.

Fig. 28 shows a valve in the regenerative position. In that position, the spool has blocked the neutral flow and connected the power passage to both workports.

Pressure oil will act against the entire piston area of the piston end of the cylinder. On the rod end of the cylinder, pressure will act on the piston area less the area of the rod. Because of the extra area and extra force on the piston end, oil will be forced out of the rod end. This oil will go back to the valve and out to the piston end of the cylinder.

All of the pump oil will go to making up the additional oil needed at the piston end. The cylinder has now effectively become a small single-acting cylinder the size of the cylinder rod instead of a large double-acting cylinder. This means with all of the pump oil going to this small cylinder, it will extend very fast but will have very little force or power.

The regenerative feature can be used only to extend a cylinder.

CENTERING SPRINGS

Centering springs are used on all directional spool valves. Fig. 30 shows the spring holding the spool in the neutral (centered) position. When the spool was moved to the left, the spring retainer on the end of the spool compressed the spring from the right. The end cap retains the left end of the spring pack. When the engaging force is removed from the spool, the spring will move it back to, and hold it in, the neutral position.

When the spool is moved to the right, the capscrew and retainer will compress the spring from the left while the valve housing retains the right end of the spring pack.

DETENTS

Open- and closed-center directional control valves can be equipped with detent assemblies. Their principle function is to hold the valve spool in an engaged position, allowing the operator to free his hand for other operations.

There are many variations in the design of detents. The following are several general types.

The detent in Fig. 29 consists of a detent ramp and groove attached to the valve spool and detents mounted in the valve endcap. As the spool is moved to the detent position, the detents are forced up the ramps and drop into the groove. The spool is then held in this position.

To release the spool, the operator moves the lever in the opposite direction. This forces the detents out of the groove and allows the centering spring to move the spool back to the neutral position.

This type of detent is commonly used to secure the spool in a "float" position. This frees the operator's hands and gives him/her complete control in activating and releasing the float option.

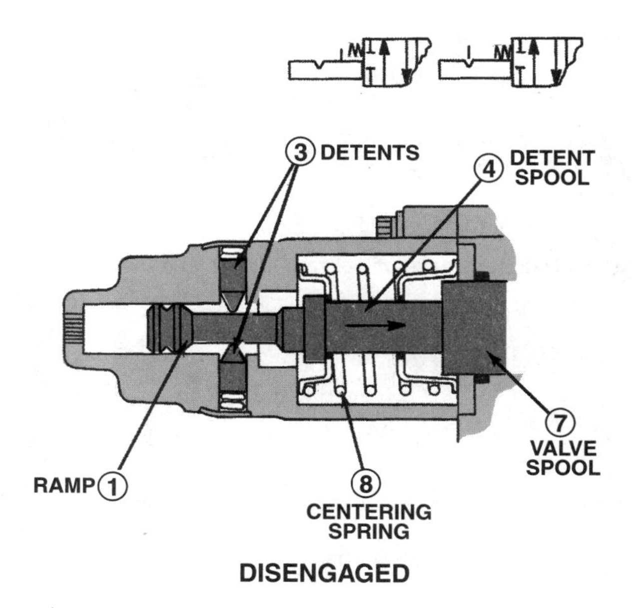

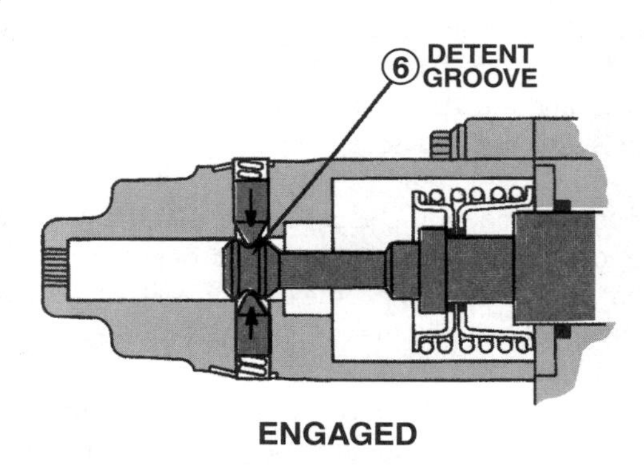

LINEAR DETENT OPERATION

Fig. 29 — Spring Loaded Detent Operation

Many loaders have the features of "Return to Dig" and "Boom Height Control." The return to dig feature allows the operator to move the bucket control lever to the roll-back position, have it detented in that position and automatically release when the bucket is in the desired position for digging or loading.

After filling the bucket, the boom height feature allows the operator to move the control to the boom raise position, have it detented in that position and release when the loader raises to the desired height.

Both options use the same detent and release mechanisms. The following are some of those used.

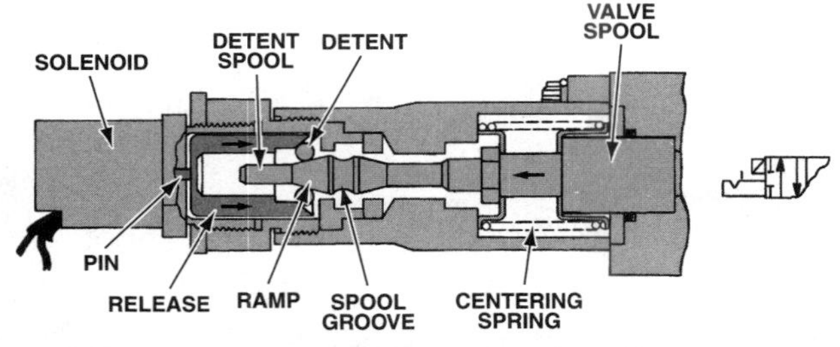

LINEAR SOLENOID OPERATED DETENT (DETENT ENGAGING)

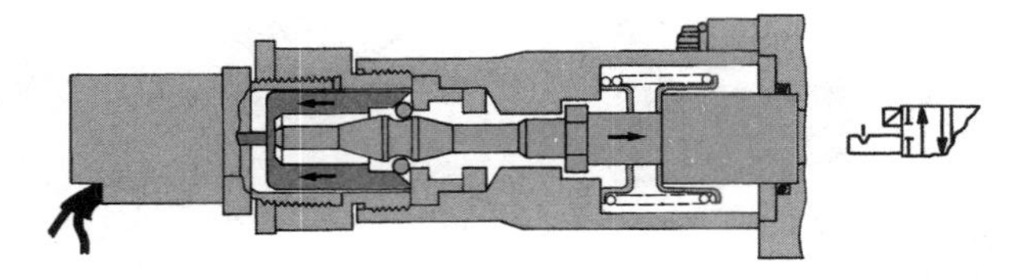

SOLENOID OPERATED LINEAR DETENT (DETENT RELEASING)

Fig. 30 — Solenoid Operated Detent

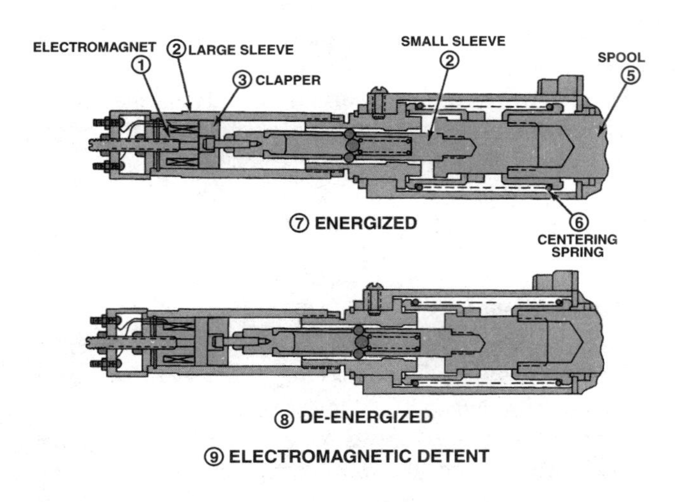

Fig. 31 — Electromagnetic Detent

The detent assembly in Fig. 30 consists of a detent spool with ramp and groove attached to the valve spool. It also has detent balls and a release sleeve that is spring loaded to the right and is attached to a solenoid.

As the operator moves the spool to the left, the valve spool will be moved to a full cylinder extend or retract position. When that position is reached, the detents will be held against the detent spool ramp by the release sleeve.

If the operator now wants to use the automatic feature, he/she can force the detents up the ramp and into the groove. The spool will now be held in position. There is no change of oil flow in the valve as the valve is moved into the detented position.

When the bucket or boom reaches the desired position, a switch attached to the linkage will be closed. This will energize the solenoid. The release sleeve will be pulled by the solenoid to the left. This releases the detents and allows the centering springs to return the valve spool to the neutral position.

The electromagnetic detent is shown in Fig. 31. There is a detent spool attached to the end of the valve spool. On the end of the detent spool is a clapper. At the far left an electromagnet is mounted to the large sleeve.

When the operator moves the spool to the left, the clapper will come in contact with an energized electromagnet. The magnet has enough force to prevent the centering spring from returning the spool to neutral.

When the bucket or boom reach the desired position, a switch on the linkage is opened deenergizing the electromagnet. This allows the centering spring to return the spool to neutral.

Depending on the position of the loader linkage, the electromagnet can be energized when the valve is in neutral. Because of the distance between the clapper and the magnet, there is not enough force to move the spool against the centering springs.

The valve in Fig. 31 also has another type of spring loaded detent to lock the spool in the "Float" position. The detents ride inside a sleeve attached to the valve endcap. They are constantly being forced outward by the spring-loaded larger ball. When the spool is moved to the far right, the balls will be forced over the ramp and lock on the other side. This will hold the spool in position until the operator forces the spool to the left releasing the detents.

LOCKOUT VALVES

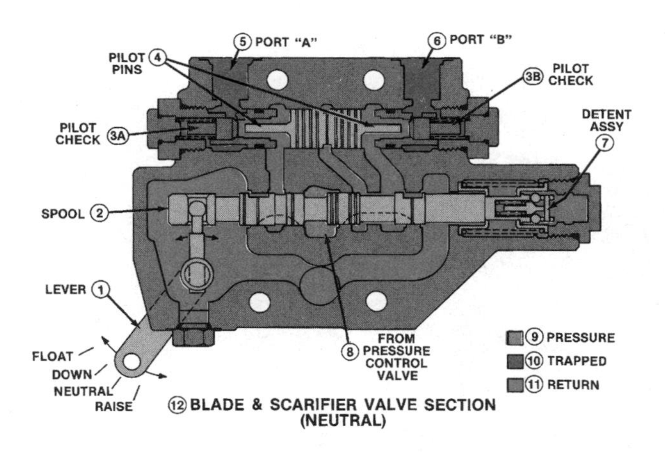

Fig. 32 — Spool Valve with Lockouts

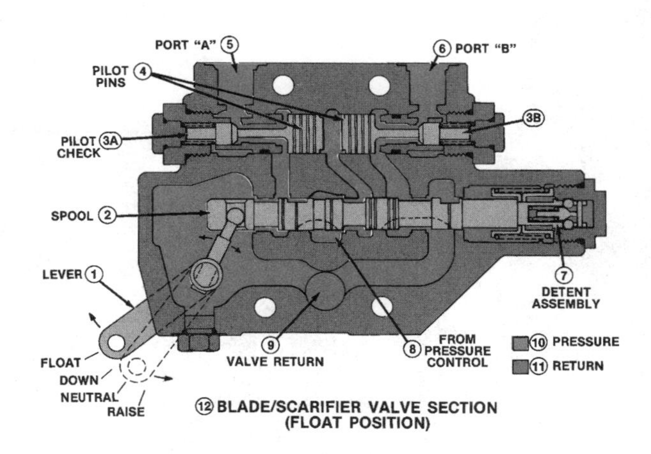

Fig. 33 — Valve with Lockouts in Float Position

There are circuits where it is important that the cylinder be held in position for long periods of time. Because of the tolerance between a valve spool and the housing, it is difficult to insure that leakage from the cylinder will not occur.

Lockout valves, also called loadcheck valves, are placed in the cylinder workports to block the cylinder oil from the spools. Poppets or balls are normally used for lockouts because they can rest on machined seats and will have no leakage.

The lockouts in Fig. 32 are mounted in the valve housing. On hydraulically operated personnel carriers and other applications where safety is a concern, they are mounted between the cylinder ports of the cylinder. This prevents a fall in case of a hose failure.

When the valve in Fig. 32 is in neutral, the lockouts prevent oil from escaping from either end of the cylinder.

When the valve is moved to the right, pressure oil is directed to the left lockout. The oil pushes the lockout open allowing flow to the workport.

Oil returning from the right workport is trapped by the right lockout. Oil pressure will build in the left workport. The piston located between the workports will move to the right opening the right lockout.

When a valve is equipped with a float position both lockouts must be held open while in float. In Fig. 33, the valve has a split piston between the lockouts. When the valve is moved to the float position, both workports are connected to the return passages and pressure oil is directed between the lockout pistons. This holds both lockouts open.

LOAD SENSING VALVES

As we discussed in Chapter 1, the pump in the load sensing system provides the valve with the right amount of oil at the right pressure to satisfy the function(s) being used.

In order to do this, the pump has to receive a signal from the valve as to its needs. The valve also has to properly distribute the oil available.

The valve will sense, and send, to the pump the highest pressure required to lift any of the loads. The pump will then supply a pressure that is a fixed amount higher. *See Chapter 4 for pump operation.*

This pressure will vary depending on the design of the machine and its components. For our example in this discussion, we will have a pump that will maintain a pressure which is 300 psi (2070 kPa) higher than the highest pressure needed in the system.

When the valves are in neutral and there is no requirement, the pump will maintain a standby pressure of 300 psi (2070 kPa).

The valve also controls the oil flow to the cylinders. It supplies full-metered oil flow to all actuated valves until full pump capacity is reached. The valve will then distribute the available flow proportionally to all the valves. This means that all functions will operate but at a slower rate in spite of varying pressures required by the functions.

Let's see how this is accomplished.

Fig. 34 shows a cross section of several valves. The upper right view is a valve in the neutral position. This closed-center valve

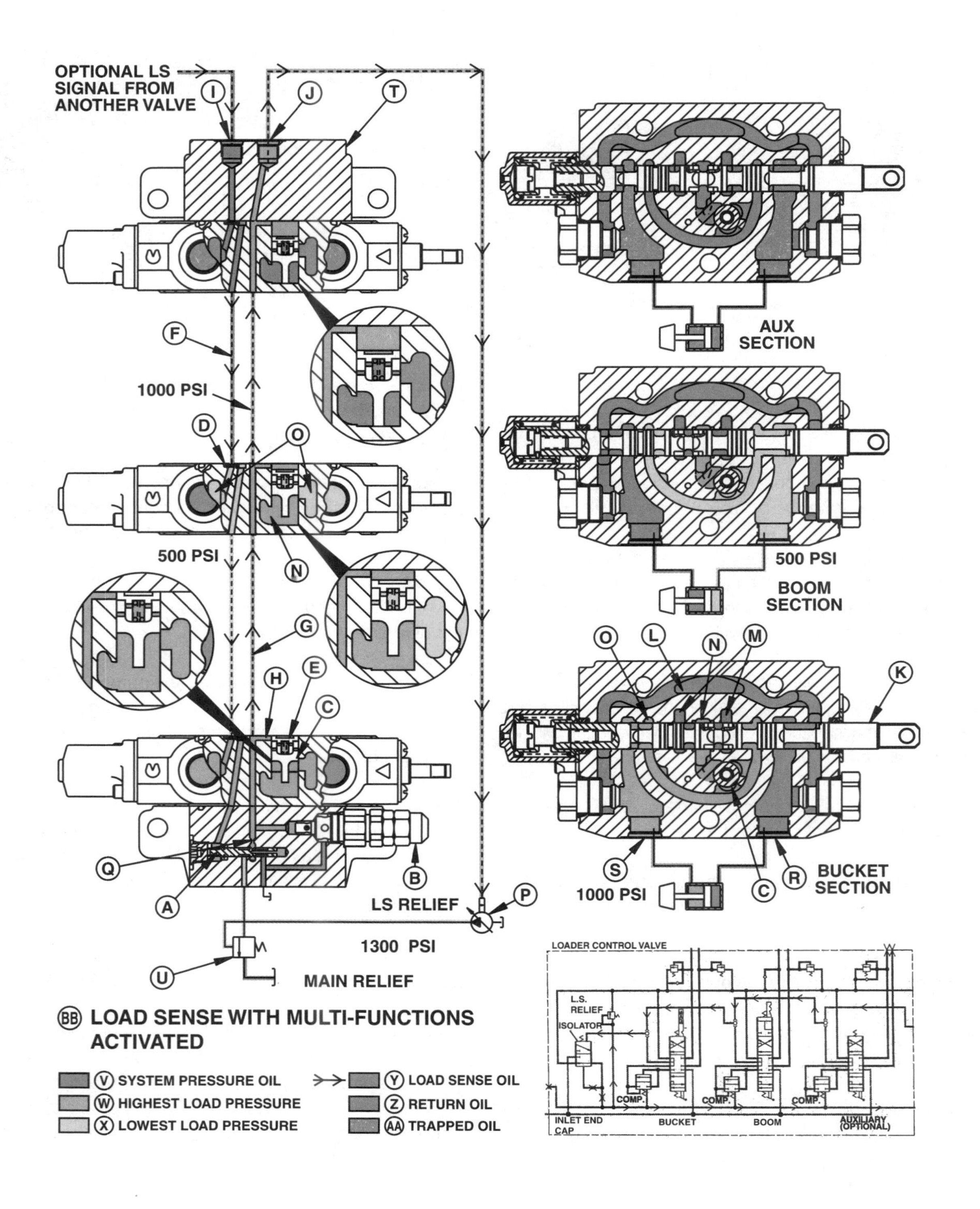

Fig. 34 — Load Sensing Valve Operation

does not allow pump oil to flow to the return. The spool traps oil in the work ports.

This valve has a plug separating the workport from the return passages. In other applications, a circuit relief and/or an anti-cavitation valve could be installed in those openings.

When the valve is actuated as shown on the middle valve, the spool meters pump oil to the compensator valve. Flow will push the compensator valve open allowing oil to go to the bridge passage. The compensator valve acts as a lift check to prevent a load from dropping before it raises.

The spool has connected the bridge passage to one workport and the return passage to the other workport.

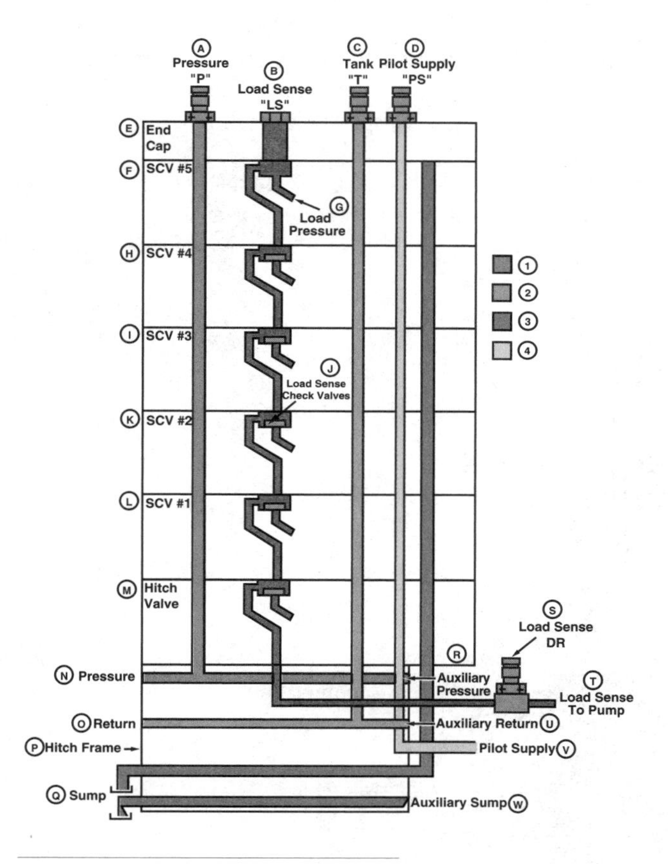

Fig. 35 — Load Sensing Shuttle System

Pilot pressure used to regulate the pump pressure is sensed in the bridge passage. The valves are all connected to the sensing passage through a series of load sensing shuttle checks as shown in Fig 35. Each shuttle check moves away from the highest pressure to block the lower pressure. The highest pressure required in the system is the pressure that will reach the end of the shuttle passage and is the signal the pump receives.

To prevent the sensing line from being a source of leakage, this valve is equipped with a load sense isolator (Fig. 34(A)). This valve is a pressure-reducing valve, which senses the highest bridge pressure and reduces pump pressure to that exact pressure. This oil is then sent to operate the compensator valves and pump pressure control valve.

Oil flowing from the valve inlet passages to the open compensator valve goes through metering grooves in the spool. This acts like an orifice to restrict the flow.

The pump control mechanism maintains pump output pressure at 300 psi (2070 kPa) over the pressure in the bridge area. Flow through the valves metering passages will always be the same because the pressure drop is always the same regardless of the pressure required to lift the load. *Remember in Chapter 1 we learned that if we have a fixed sized orifice and maintain the same pressure differential across that orifice, the flow will always be the same.*

Let's now operate a second valve. This valve is going to have to lift a load which requires more pressure. The pressure that was required on the center valve was 500 psi (3450 kPa). The lower valve will require 1000 psi (6895 kPa).

In the series-parallel valves, which were discussed earlier, pump oil would choose the path of least resistance and all flow would go to the middle valve. Only after the middle cylinder had reached the end of its stroke would the pressure rise and enter the lower valve.

This valve has a passage from the load sensing passage to the top of the compensator valves. This means that the highest pressure required to operate a function (pump pilot pressure) will be pushing down restricting oil flow into the bridge passage.

In the lower valve, the 1000 psi (6895 kPa) is pushing down but the cylinder pressure is the same so the valve will not restrict flow. The pump will supply oil at 1300 psi (8965 kPa) so the pressure drop across the spool lands will be 300 psi (2070 kPa).

The compensator of the middle valve will also be pushed down with the 1000 psi (6895 kPa) force. The pressure on the bottom will only be 500 psi (3450 kPa). This will cause the compensator to move down and restrict the oil flow to the bridge passage. This will cause the pressure to rise on the bottom of the compensator until top and bottom pressures are equal which in this instance is 1000 psi (6895 kPa).

The pump pressure is 1300 psi (8965 kPa) to the valve spool and the compensator is maintaining the pressure between it and the spool at 1000 psi (6895 kPa). The differential pressure across the spool grooves is maintained at 300 psi (2070 kPa). Flow therefore is the same to that function as it was when only one valve was used.

When enough functions are used to exceed the maximum pump flow, the compensators maintain the 300 psi (2070 kPa) pressure drop across all the valves. Available flow will be distributed proportionally to all of the actuated valves.

VALVE CONTROLS

All of the control valves discussed thus far, have been shown to be manually operated. This consists of a handle or handle and mechanical linkage attached to the valve spool.

It is also possible to control the valves hydraulically or electrically. When valves are large and require hydraulic assist or are located in a remote location, these pilot operated valves are used.

When valves are located near the functions they control, it eliminates the need for long hydraulic hoses and pipes to be routed to each function. It is necessary only to use one pressure and one return line to connect the valve to the pump and reservoir.

Another advantage in mounting the valve away from the operator's station is that, with no mechanical connection to the hydraulic system, cab noise is reduced.

Many of the valves we have already discussed can be hydraulically operated and some can be electrically operated.

HYDRAULIC CONTROLS

The valve shown in Fig. 36 is a hydraulically operated spool valve. The valve is the same as a manually operated valve except for a sealed endcap on each end of the spool.

When pilot pressure oil is sent to the left endcap, it will push on the end of the spool. It will overcome the centering spring and move the spool to the right. When the activating oil is released, the centering spring moves the spool back to the neutral position.

If the oil source used to activate the valve was supplied by a constant pressure, "on-off" type source, the valve would have a full speed extend, a full speed retract or a neutral position.

It is normally necessary to be able to partially engage the valve as you can with a manually operated valve.

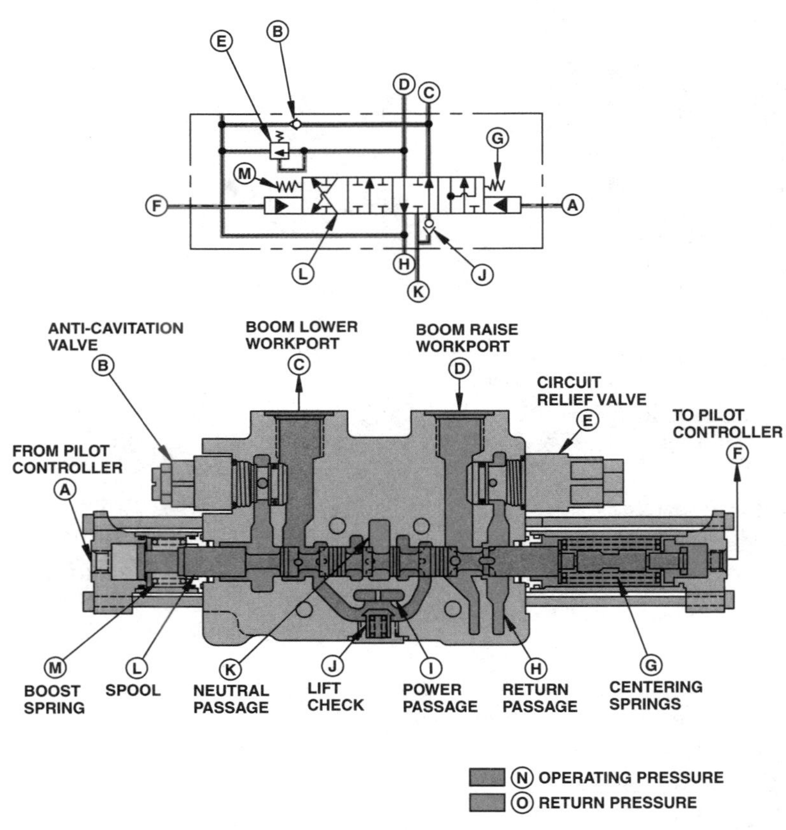

Fig. 36 — Hydraulically Operated Valve

By controlling the pressure to the endcap, the centering spring can be compressed any desired amount.

To accomplish this, a valve controller is used. The controller consists of a small pressure-regulating valve for each endcap in the valve stack. *See the explanation of pressure reducing valves (Figs. 7 & 8) in this chapter.*

The controller in Fig. 37 has four valves controlled by one lever, so two valves can be operated without changing levers.

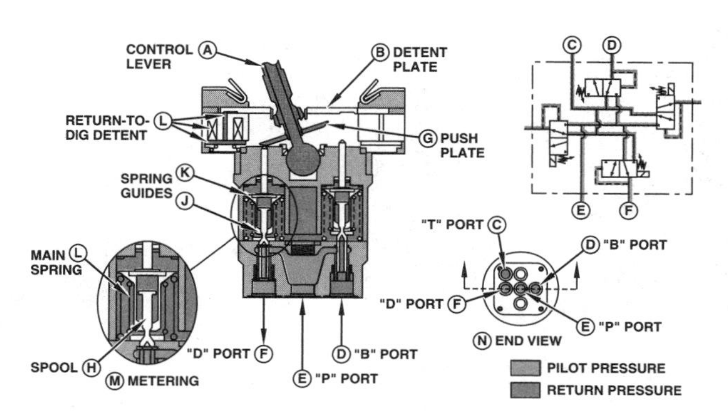

Fig. 37 — Pilot Controller Operation

The lever is moved, it compresses a spring which in turn pushes down on the valve. Oil from the pilot source is sent to the main control valve. When the pressure below the controller valve is high enough to overcome the force of the spring above it, the valve will close and maintain that pressure.

For each amount of lever movement, there is a specific force applied to the spring, which results in a specific pressure applied to the end of the spool in the control valve.

For every pressure applied to the spool, there is a specific valve movement by compressing the centering spring.

The pilot controller will therefore provide the same degree of control and feel that is experienced with manual control.

ELECTRO-HYDRAULIC VALVE CONTROL

Hydraulic control valves may be actuated by an electric solenoid. Solenoids are designed to do mechanical jobs by means of electromagnets (see Chapter 1 or the FOS Electronic and Electrical Systems book).

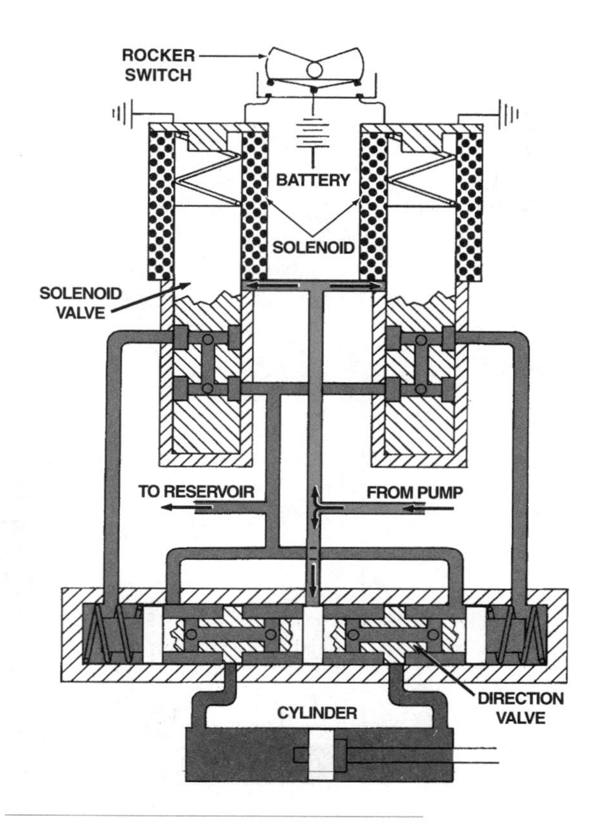

Fig. 38 — Electro-Hydraulic Valve in Neutral

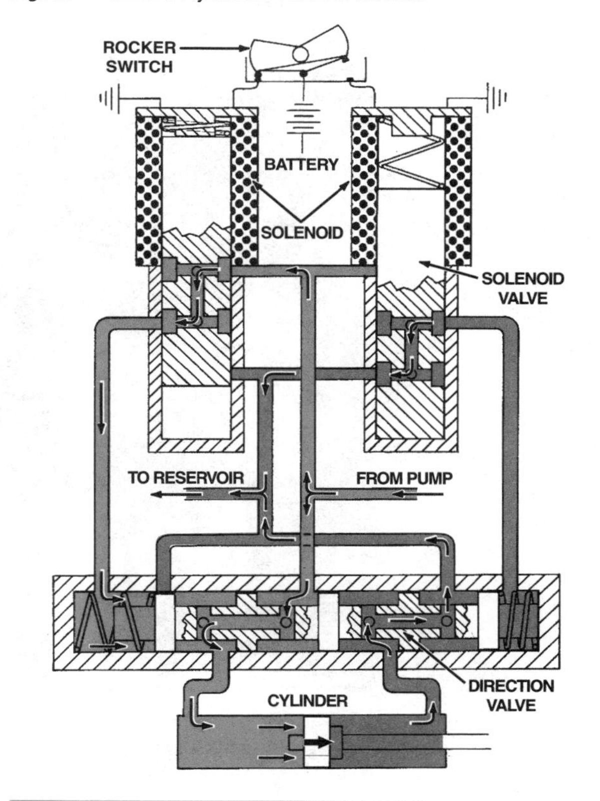

Fig. 39 — Electro-Hydraulic Valve Activated

When the closed-center valve is in neutral (Fig. 38), neither solenoids are activated. The solenoid springs are holding the plunger down which aligns passages to connect both ends of the spool valve to the reservoir and blocking pump inlet oil.

With no pressure at either end of the spool valve, the centering springs at each end will locate the spool in the neutral position. This blocks the inlet oil passage from the pump and traps oil on both sides of the cylinder.

When the left rocker switch is pressed down, the left solenoid becomes energized (Fig. 39). Battery current flows through the windings that surround the hollow core. The magnetic field created by the current will move the solenoid core and valve upward against the spring.

This allows pressure oil from the main hydraulic pump to enter the hollow center of the solenoid valve. Oil is directed through the solenoid valve to the left end of the spool valve.

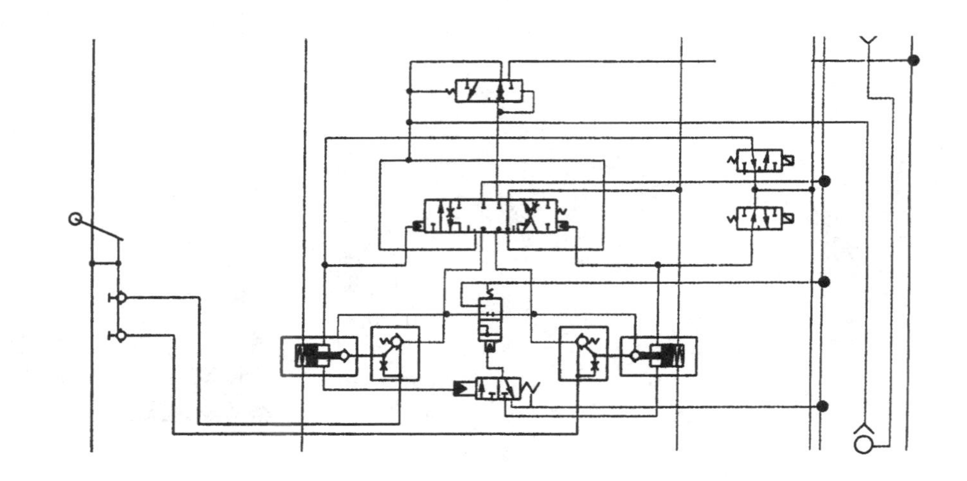

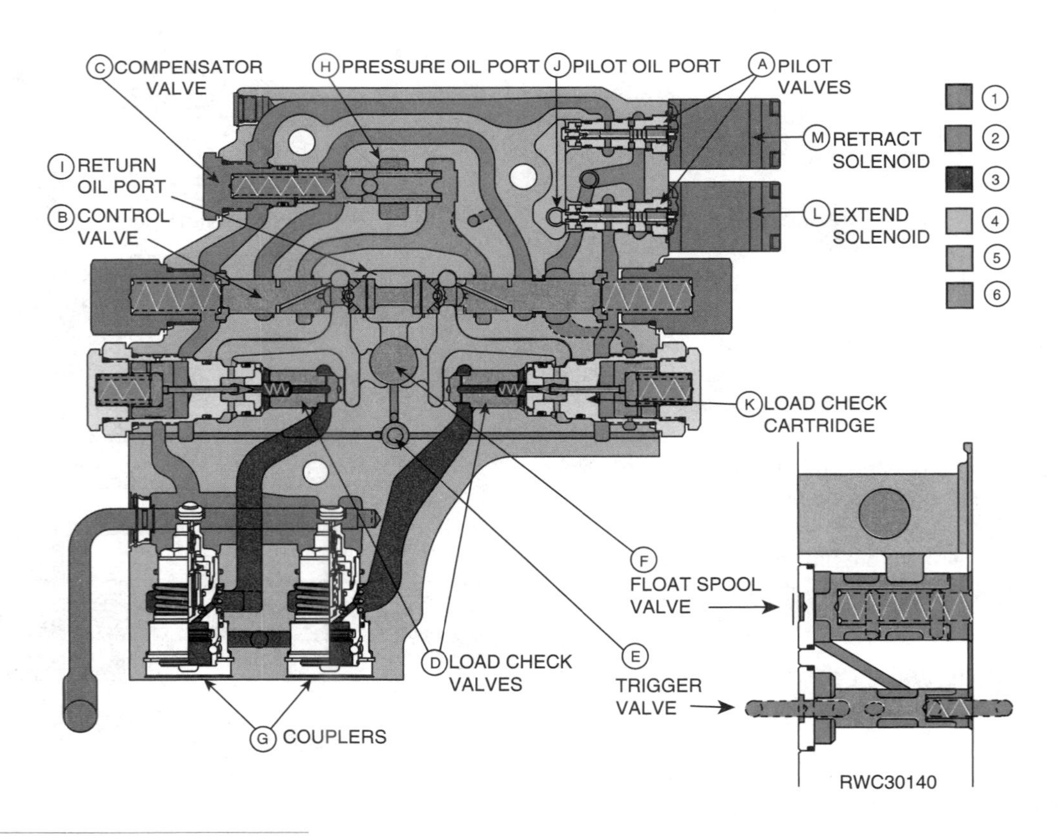

Fig. 40 — Control Valve Using Microprocessor

The spool is moved to the right, sending pressure oil to the piston end of the cylinder and connecting the rod end to reservoir. Oil from the right end of the spool valve flows to the reservoir through the right solenoid valve.

When the rocker switch is released, the left solenoid is deenergized and the spring pushes the core and valve down. This blocks the pressure oil and connects the left end of the spool to the reservoir. With both ends of the spool valve connected to the reservoir, the centering springs will move the spool to the neutral position.

When the right rocker switch is pressed down, the right solenoid becomes energized. The action of the valves and oil flow will be reversed and the rod end of the cylinder will be pressurized.

The solenoid-operated valve does not have the capability to be partially engaged. It is either in neutral or full engagement.

VARIABLE FLOW VALVE USING MICROPROCESSOR

The valve shown in Fig. 40 is closed center and is used in a load sensing system. It is one of several stacked valves. *See Fig. 34 for the load sensing shuttle system.*

Two hydraulic pilot valves direct pilot oil to the ends of the spool valve. In this case, the pilot valves are proportional solenoids connected to pressure reducing valves.

The solenoid receives a variable electrical signal from an electronic control unit (microprocessor). The force exerted by the solenoid core is proportional to the current the windings receive from the microprocessor. This varying force is exerted on the pressure-reducing valve. The solenoid force then determines the output pressure of the pilot valve.

This varying pressure acting against the centering springs will control the amount of opening and, therefore, the flow past the spool valve to the cylinder. *See the explanation of pressure reducing valves (Figs. 7 & 8) in this chapter.*

To make certain that flow past the spool valve is predictable, each valve section has a compensator valve. This valve, located in the valve inlet, is also a pressure-reducing valve. By sensing the load pressure, it drops the inlet pressure to maintain a constant pressure drop across the metering area of the spool regardless of the implement load or the pump pressure. *Same orifice size, same pressure drop equals same flow.*

Each workport has a pilot operated load check (lockout) cartridge with thermal relief (circuit relief) valve.

When the retract pilot is activated, pressure oil is sent to the left end of the valve spool moving it to the right. This connects the pressure oil passage to the right coupler. Oil pushes the right lockout open. The pilot oil was also applied to the back of the left lockout. This opens the lockout and allows oil to return from the left coupler.

The float function is obtained by activating both pilots. This opens both lockouts and leaves the valve spool in neutral. Activating both pilots also sends pressure through the trigger valve to move the float spool to the open position. This connects the two workports.

PILOT-CONTROLLED POPPET VALVES

The poppet valve is a floating valve, which is normally held on its seat by the pressure of the system. When activated by a pilot valve, the system pressure opens the poppet.

Poppet valves can be manually or electrically operated. They can therefore be used when remote mounting of the valve is required.

Poppet valves have little or no leakage eliminating the need for lockout valves.

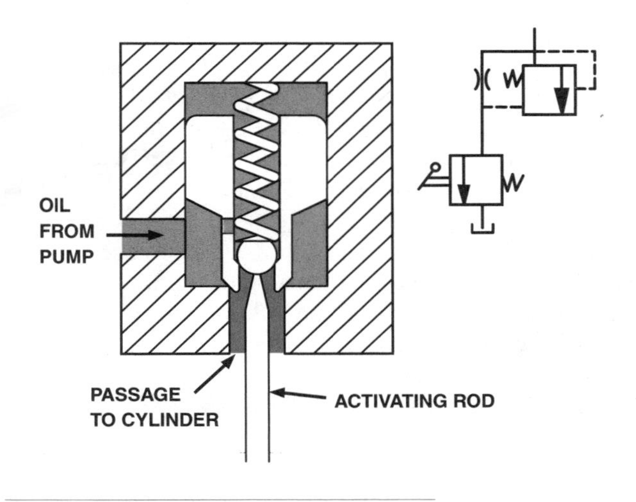

Fig. 41 — Poppet Valve (Steering Valve Type)

In the closed position, the poppet in Fig. 41 is hydraulically held on its seat. Because of the orifice, the pressure behind the poppet is equal to inlet pressure in the poppet. The area behind the poppet is greater than the inlet area, which hydraulically holds the valve on its seat.

The valve is operated by pushing the activating rod and moving the ball off its seat. This allows oil to flow from behind the poppet. Oil will flow across the orifice causing a pressure drop. This allows the inlet pressure to lift the poppet and pass oil to the outlet.

This design is good when precise and variable metering is required. The poppet will open only as far as the ball is moved. When the poppet catches up to the ball, the pressure behind the poppet will increase and hold the valve in that position.

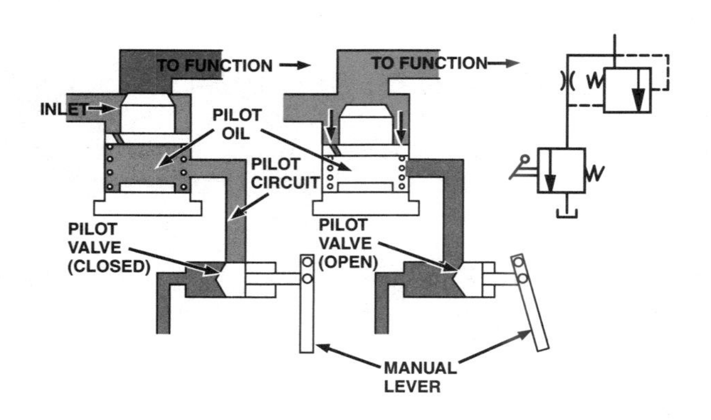

Fig. 42 — Poppet Valve

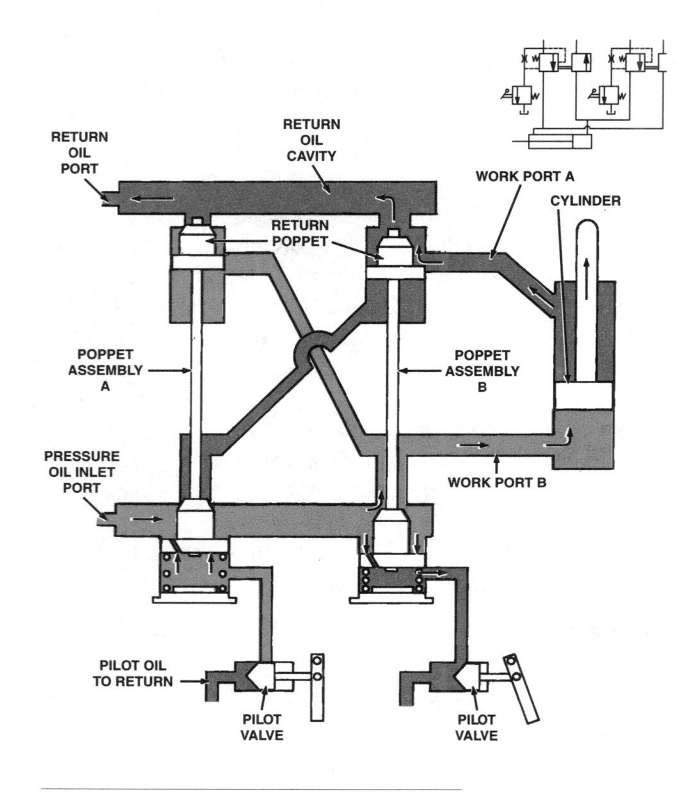

Fig. 43 — Poppet Valves Extending Cylinder

The poppet valve in Fig. 42 is also hydraulically held on its seat when not actuated by a pilot valve. When the pilot valve is opened, oil escapes from behind the poppet. This creates a flow and pressure drop across the poppet orifice. The inlet pressure can now push the poppet down to open it. The valve will remain open as long as the pilot valve is open.

When the pilot is closed, there will be no flow through the orifice. The pressure will be the same on both sides of the poppet. Because of the extra area on the backside of the poppet, it will be hydraulically closed and held on its seat.

The pilot valve can be opened either manually or by an electric solenoid.

The preceding discussion shows that a single poppet valve can control the flow of pressure oil in only one direction. Therefore, to control a function in two directions, two poppet valves are needed.

If the oil is to be trapped in the cylinder when not operating, another set of poppets or lockouts will be required for oil returning from the cylinder.

The valve shown in Fig. 43 is set up to operate a double acting cylinder. There is a pressure and return poppet for each side of the cylinder. The pressure poppets are hydraulically operated. When they open, a rod will manually open the return poppet.

When the right pilot valve is opened, the right poppet will hydraulically open. The right-return poppet is also pulled open.

Pressurized oil is sent to the piston end of the cylinder through workport B. Oil returning in workport A enters the return passage through the return poppet.

The left poppets are operated when the left pilot is opened. Pressurized oil is sent to the rod end of the cylinder retracting it.

In the example used in Fig. 43, the pressure poppets pulled the return poppets open. Fig. 44 shows an example where the pressure and return poppets have their own pilots and operate independently. In this case, there are two hingeplates, each operating a pressure and a return poppet.

When the hingeplate is in the mid-position, the valves are all closed. When the top of the hingeplate is rotated down, the bottom pressure and return poppets are opened. When moved up, the top poppets are opened. The individual poppet used in this valve is shown in Fig. 41.

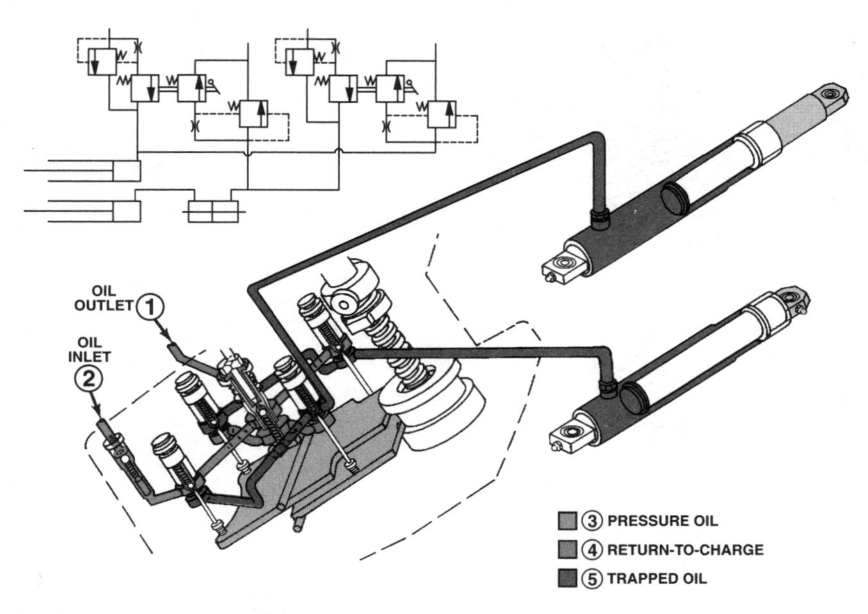

Fig. 44 — Individually Controlled Poppets

ROTARY DIRECTIONAL VALVES

Rotary valves are commonly used as pilot valves to direct flow to other valves.

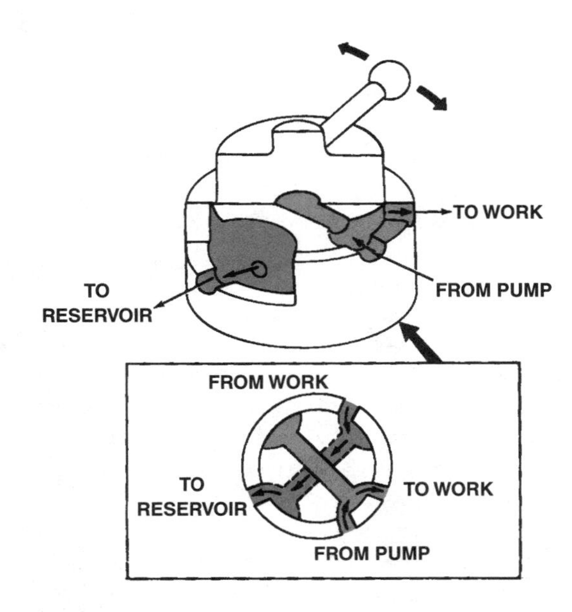

Fig. 45 — Rotary Directional Valve

Fig. 45 shows a four-way rotary valve. The valve has holes that match with holes in the main body as the valve turns. A hand lever turns the valve; other models may be operated hydraulically or electrically.

Fig. 45 shows the valve positioned to allow pressure oil from pump to enter one port, flow through the valve, and out another port to the work. Meanwhile, oil is returning from another work port through the valve to the reservoir. The drilled ports in the valve are actually on two levels to separate them.

Rotary valves can be designed to operate as two-, three-, or four-way valves. Relocating the ports, altering the passages, or adding and removing oil routes does this.

Fig. 46 shows the components of a rotary steering valve. The steering wheel is connected to the valve spool with a splined shaft. The inner gear of the gerotor motor assembly is connected to the valve sleeve by the shaft and pin.

When the steering wheel is turned, the valve spool is turned inside the valve sleeve. Trapped oil in the gerotor assembly prevents the sleeve from turning. This aligns passages to send pressure oil to the gerotor and routes oil returning from the gerotor to a work port connected to the steering cylinder. It also connects the return passage from the steering cylinders to the return to reservoir port.

A—Valve Spool
B—Valve Sleeve
C—Gerotor Assembly
D—Shaft
E—Pin

Fig. 46 — Rotary Valve for Hydrostatic Steering

As the gerotor turns, it also turns the sleeve. As long as the steering wheel is turned, the steering will continue. When the steering wheel stops, the gerotor will rotate the sleeve until it catches up to the spool to neutralize the valve. This causes steering as long as the steering wheel is turned and stops when the desired steering position is reached.

The pin goes through a slot in the valve spool. The slot allows only about 8° rotation between the valve and sleeve. When no hydraulic power is available, turning the steering wheel turns the valve, sleeve and the gerotor. The gerotor acts as a pump, pressurizing oil and sending it to the right or left steering cylinders, thus providing emergency manual steering.

VOLUME CONTROL VALVES

FLOW CONTROL VALVES

Volume (flow) control valves regulate the oil flow rate to a hydraulic circuit. They can provide:

1. Specific Flow
2. Proportional Divided Flow
3. Priority Flow

SPECIFIC FLOW WITH FIXED DISPLACEMENT PUMPS

When a system requires only a portion of the output of a fixed displacement pump, it is necessary to provide another path for the excess oil. The bypass flow regulator valve shown in Fig. 47 provides a constant flow of oil to the priority port. Any additional oil from the pump goes to the secondary port. Oil flow to the priority port is determined by the orifice size in the valve and the spring force.

The secondary port can be connected to another hydraulic circuit or be returned to the reservoir.

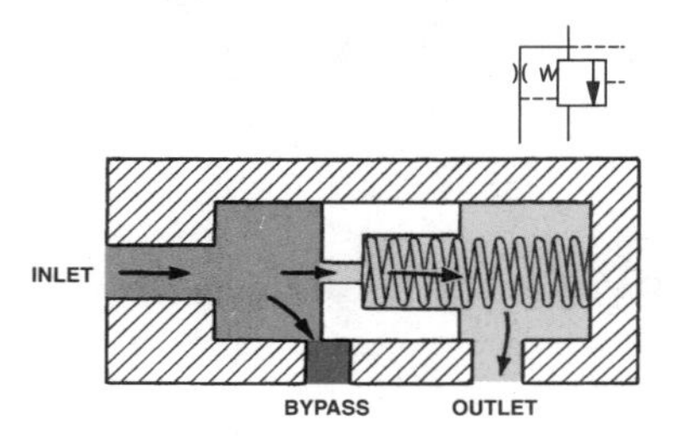

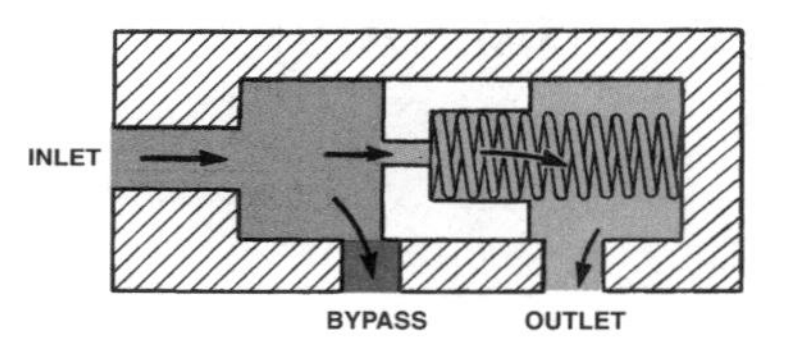

Fig. 47 — Bypass Flow Regulator

As the machine is started, spring force is holding the valve to close the bypass port. Pump flow passes through the valve orifice to the outlet. As flow increases, the pressure drop across the orifice will increase.

When the pressure drop is enough to overcome the spring force, the valve will be pushed back by the inlet pressure and allow the excess oil to flow out the bypass. The spring will maintain the same pressure drop across the orifice at all times regardless of outlet pressure. *With the same pressure drop and the same orifice size, the flow will also be the same.*

When the pressure requirement of the secondary circuit is higher than the pressure required in the priority circuit, the pressure will move the spool back. This restricts the priority opening to build pressure behind the valve. This pressure will build to maintain the same pressure drop across the orifice and thus, the same flow.

Because priority and secondary flow are not affected by the priority or secondary pressures, this is called a **compensated flow control valve.**

This valve can be used as a priority valve because it supplies all oil to the priority port until the predetermined flow is reached.

Changing the spring tension or orifice size can change the volume of oil flowing to the priority port. On the valve shown in Fig. 47, adjusting the spring tension with an adjusting screw or shims would be a possibility. More spring tension will increase flow. Less tension will decrease flow. This is normally adjusted to a specification by a technician.

When it is necessary for the operator to make frequent flow adjustments, an adjustable orifice is more practical. The valve shown in Fig. 48 has a variable orifice that is controlled by the operator. On this valve, the orifice routes oil around the valve rather than through it as shown in Fig. 47.

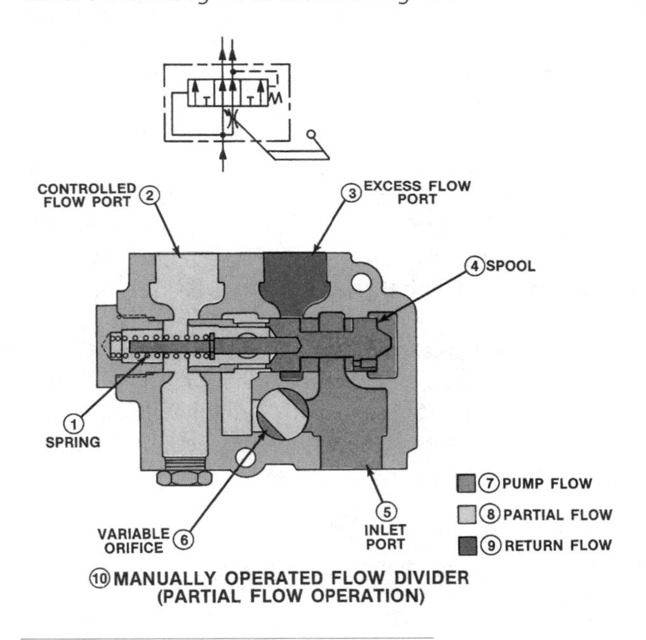

Fig. 48 — Manually Controlled Flow Divider

Inlet pressure is sensed on the right end of the spool through a small orifice. The spring (1) will determine the constant pressure drop across the variable orifice (6). The spool (4) will move to either bypass oil to the excess flow port (3) or restrict flow to the controlled flow port (2).

This will always maintain the same pressure drop. By changing the orifice opening, therefore, the flow out the controlled port can be changed. Increasing the orifice size will increase the flow. Decreasing the orifice size will decrease flow.

If total pump flow is desired at the controlled flow port, the variable orifice can be fully opened. With no pressure drop, the pressure will be the same on both ends of the spool. The spring will move the spool to the right, closing the excess flow port.

WITH VARIABLE DISPLACEMENT PUMPS

The simplest flow control on a system with variable displacement pumps is an orifice or a needle valve (Fig. 49). The problem is that with a constant inlet and a variable outlet pressure, the pressure drop across the orifice is not constant. The flow will, therefore, decrease as the load (outlet pressure) increases. This is called a **noncompensated valve** because it does not compensate for varying pressures.

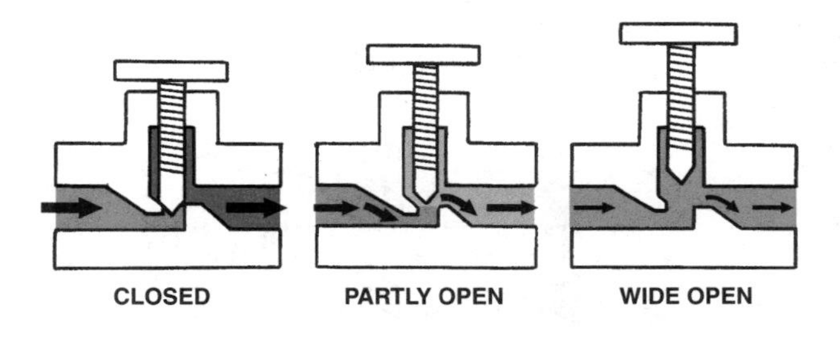

Fig. 49 — Needle Valve

The valve shown in Fig. 50 operates on the same principle as the bypass type. Because the pump is variable displacement, it is not necessary to bypass the excess oil.

As desired flow is reached, the pressure drop across the fixed orifice will move the spool against the spring. This will restrict the flow to the outlet.

When more than the predetermined flow tries to go through the fixed orifice, the pressure difference between the front and back of the valve increases. This compresses the spring and moves the spool to further restrict the outlet flow. This increases the pressure inside the spool and reduces flow to the predetermined amount.

Regardless of changes in inlet or outlet pressures, the spring will maintain the same pressure drop and therefore the same outlet flow.

This type of valve can also have an operator adjusted outlet flow.

The valve in Fig. 51 has an adjustable orifice to control the flow rate to the outlet port. It operates the same way as the previous valve. Inlet oil acts on the right end of the valve. As it passes through the adjustable orifice there is a pressure drop. That reduced pressure acts on the spring end of the valve, through an orifice at the middle of the valve.

The spring will insure the same pressure drop across the variable orifice by restricting the flow to the outlet. The spring insures the same pressure drop, therefore, varying the orifice size will vary the flow across it.

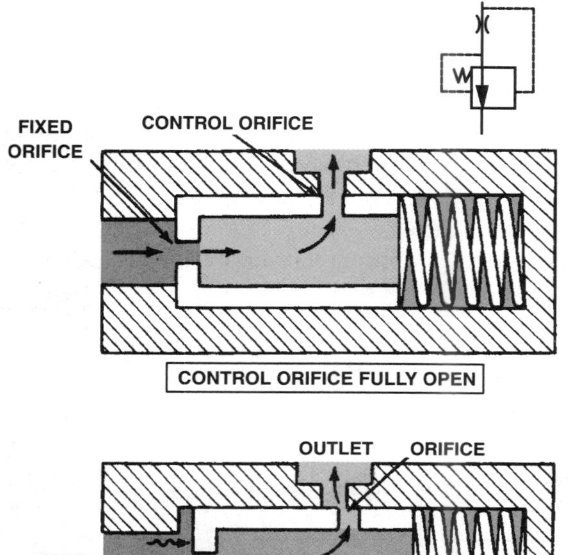

Fig. 50 — Flow Valve-Variable Displacement Pump

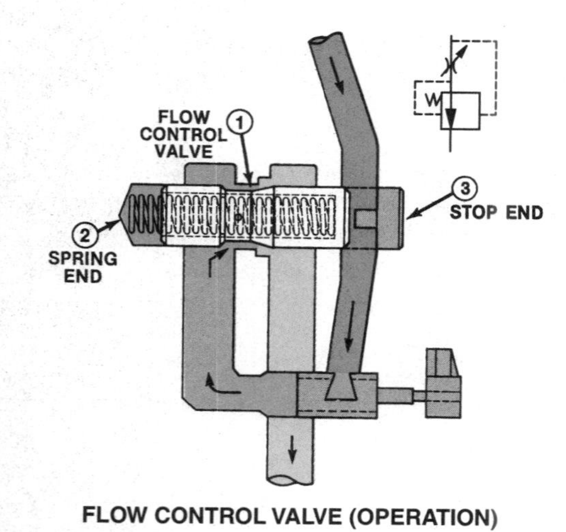

Fig. 51 — Variable Flow Control Valve

PROPORTIONAL DIVIDED FLOW

This type of valve divides the oil flow, sending it to two circuits on a percentage basis. This valve can be used with a fixed or variable displacement pump.

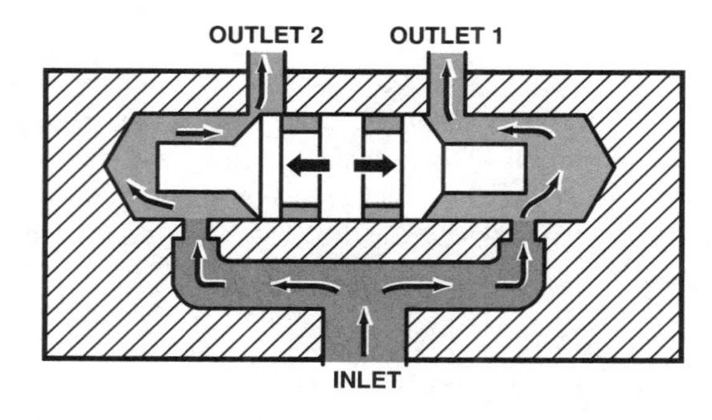

Fig. 52 — Proportional Flow Divider

The flow divider valve shown in Fig. 52 is one that always sends the same flow to each of the outlet ports. Having the two orifices between the inlet passage and the spool ends equal in size does this.

When the function served by outlet No. 1 is activated, the pressure backup from the control valve will move the spool to the left. It will restrict the oil flow to the left outlet port enough to establish the same pressure on each end of the spool. Because the spool is free floating, this balance will always be maintained.

The inlet pressure to each orifice is the same and the pressure at each end of the spool is always the same. Because the flow through an orifice is the same with the same pressure drop, there will always be equal flow to each outlet port regardless of the pressures required at either port.

To divide the flow other than on a 50-50 basis, it is necessary only to vary the proportional size of the orifices between inlet and spool ends.

PRIORITY FLOW

Priority valves make certain that the needs of one circuit are satisfied before any oil is allowed to go to other circuits.

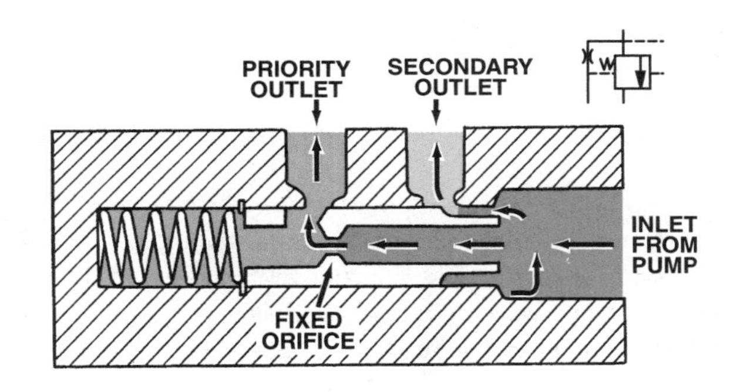

Fig. 53 — Priority Flow Valve

One type of priority valve is the volume control valve shown in Fig. 53. This valve supplies a flow of oil to the priority outlet before any oil is allowed to flow to the secondary outlet. As with the valve shown in Fig. 47, the flow is determined by the size and the pressure drop across the fixed orifice. The spring force keeps the pressure drop constant so the flow is constant. Excess oil is allowed to flow to the secondary outlet.

The priority valve shown in Fig. 54 is used in systems with variable displacement pumps. It does not let oil go to secondary circuits if the inlet pressure drops below a predetermined amount. Thus, a predetermined minimum pressure is assured to the priority circuits such as steering and brakes.

As an example, the priority valve is used in a system with a 2500 psi (17,240 kPa) standby pressure. It is determined that no oil will be allowed to flow to the secondary circuits unless the steering and brakes have at least 1800 psi (12,410 kPa) available to them.

The valve spool has a smaller diameter on the spring end than on the inlet end. A small orifice in the end of the spool allows inlet pressure to act on both ends of the spool. With the differential in areas, pressurized oil will try to move the spool to the left. The spring overcomes that force and holds the spool to the right until the pressure reaches 1800 psi (12,410 kPa). Inlet pressure acting on the extra area of the right end of the spool will move it against the spring, opening the secondary passage and allowing oil to flow to the other circuits.

Note the cavity at the stepped land on the valve is connected to reservoir. If this area was to become pressurized, it would eliminate the differential area on the spool and the spring would hold the valve closed at all pressures.

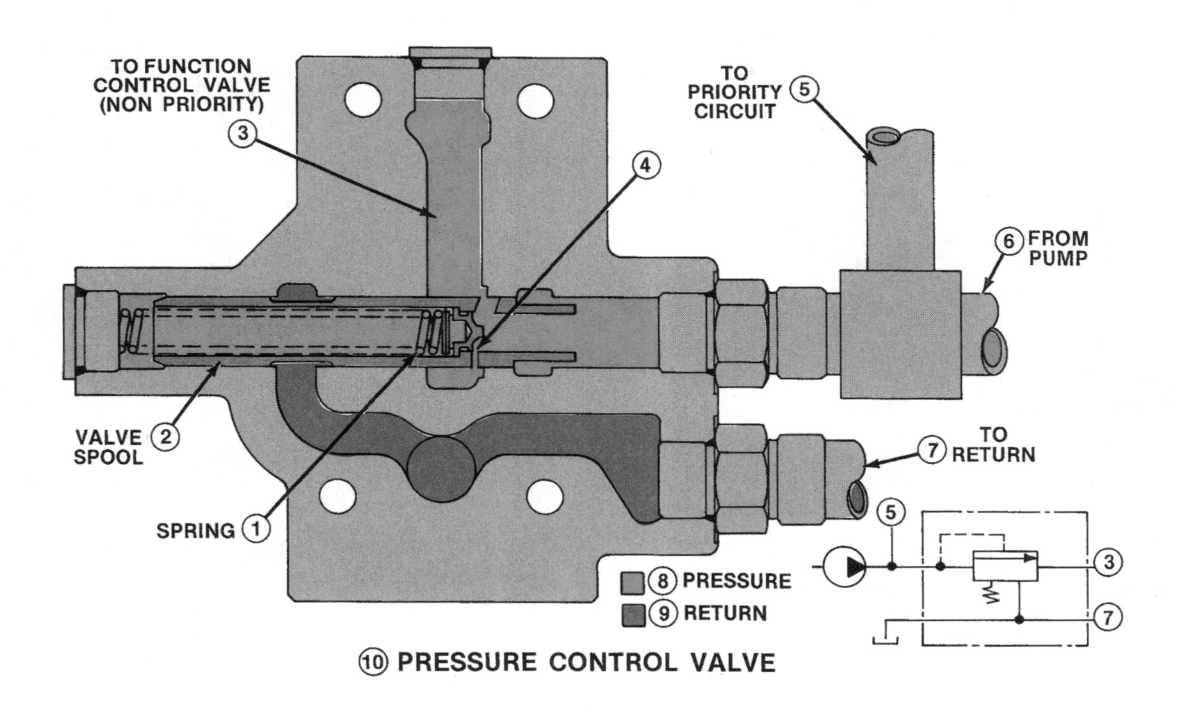

Fig. 54 — Pressure Type Priority Valve

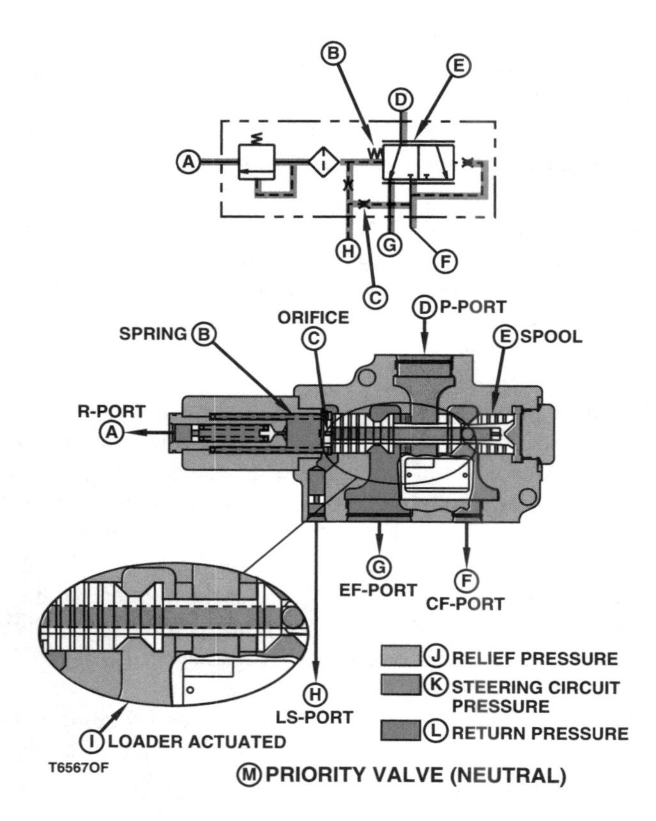

Fig. 55 — Load Sensing Priority Valve-Steering in Neutral

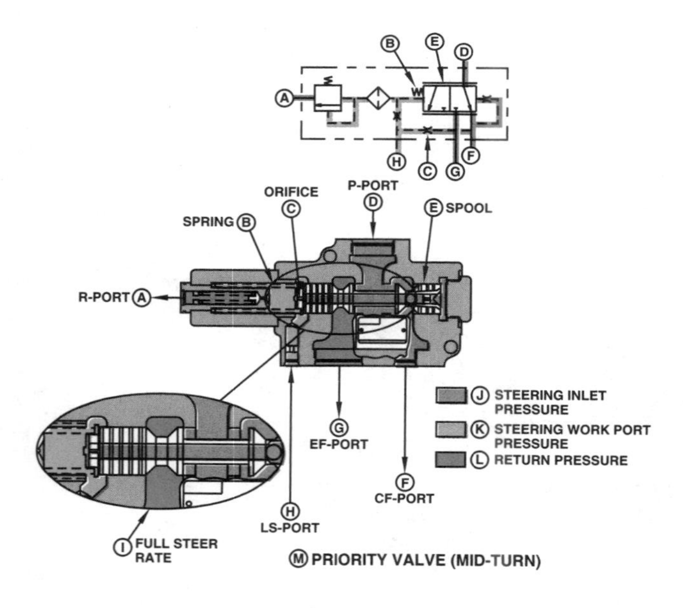

Fig. 56 — Load Sensing Priority Valve-Steering Activated

The priority valve shown in Fig. 55 receives oil from a fixed displacement pump. It first supplies oil to a closed-center steering valve and sends the rest to the secondary circuits.

The priority valve will maintain a low standby pressure to the steering valve when it is in neutral. For our example, the standby will be 100 psi (690 kPa). When the steering is activated, the priority valve will supply enough oil to maintain a pressure which is 100 psi (690 kPa) higher than the pressure of the oil flowing from the steering valve to the cylinders.

A spring holds the valve spool to the right blocking the secondary outlet passage and opening the steering passage. The steering pilot sensing line connects the right end of the spool to the steering valve workports. This line senses the pressure that is actually required to steer the machine.

The spring force is such that 100 psi (690 kPa) acting on the right end of the spool will overcome the spring and open the secondary outlet passage. When the steering valve is in neutral, there is no pilot pressure so the priority valve will restrict oil flow to the secondary outlet enough to maintain 100 psi (690 kPa) to the steering.

When the steering is activated (Fig. 56), the pressure at the steering workports is sensed on the spring end of the spool. It will now require additional pressure on the right end of the spool to open the secondary port.

Again for our example, it will require 800 psi (5520 kPa) at the steering valve workport to steer the machine. This pressure will be sent to the spring end of the spool by the sensing line. The pressure that is required on the right end of the spool to open the secondary port would be 800 psi (5520 kPa) to overcome the pilot pressure plus 100 psi (690 kPa) to overcome the spring force. That pressure will be 900 psi (6205 kPa). This is the pressure maintained to the steering valve, always 100 psi (690 kPa) more than that needed to turn the machine.

When the steering cylinders reach the end of their stroke, the pressure to the steering valve would keep rising. To limit the maximum pressure in the system, a pilot relief valve is added to the pilot circuit. If the relief valve limits the pilot pressure to 2400 psi (16,550 kPa) the pressure maintained to the steering valve would be limited to 100 psi (690 kPa) more, or 2500psi (17,240 kPa).

When the secondary functions require a higher pressure than that required by the steering, the spool will move to the right to restrict the pressure and flow to the steering valve.

MISCELLANEOUS VALVES

GATE VALVES

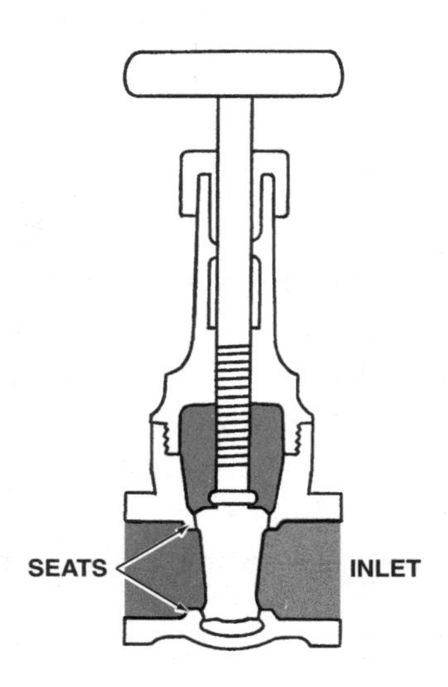

Fig. 57 — Gate Valve in Closed Position

A gate valve is used to open or close a line to flow. The valve element is a wedge-shaped gate, which is raised or lowered by screw action (Fig. 57). This valve is designed to completely open or close a line, but not to throttle flow when partly open.

Although a gate valve offers very little resistance to flow when completely open, it is difficult to open and close under high pressure.

COCK VALVES

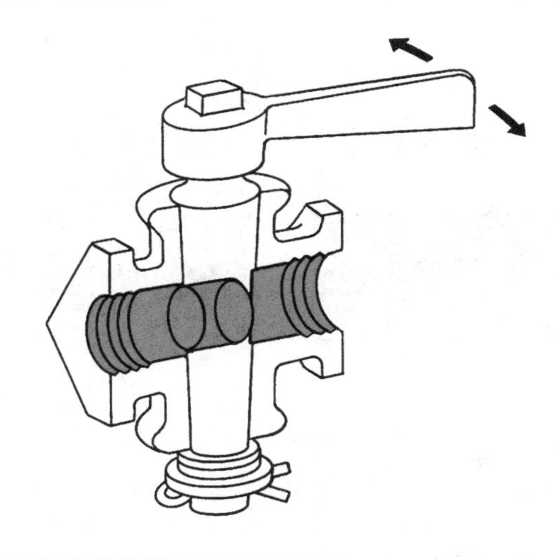

Fig. 58 — Cock Valve in Open Position

Cock valves are very simple and are usually small in size. They are used to bleed air out of a system, turn gauges on and off, or drain the system. Fig. 58 shows a cock valve in the open position. A quarter turn of the handle will shut it off. This type of valve is manufactured to be used in a wide range of pressures. When selecting a valve for use, make certain it will withstand the pressures of the system.

FLAPPER VALVES

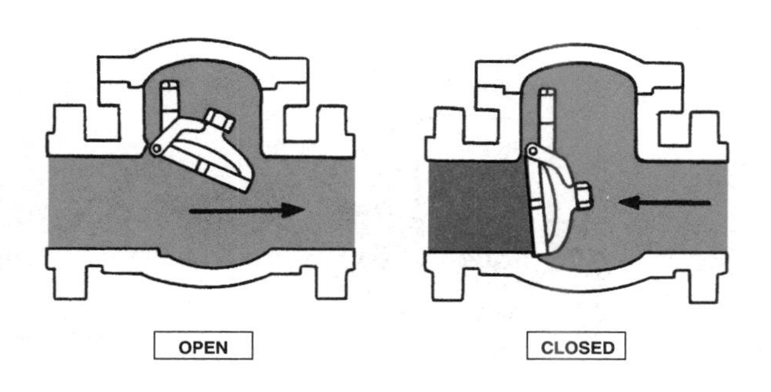

Fig. 59 — Flapper Valve

Flapper valves are essentially check valves. They permit flow in only one direction. They are made in a wide range of sizes. They offer little resistance to flow when fully open. Although they are usually installed so that gravity and pressure close them, they sometimes have a spring. Backpressure causes the flapper valve to seal tightly. These valves are manufactured in a wide range of pressure ranges so be sure to check the rating before using.

VALVE SERVICING AND CARE

Hydraulic valves are manufactured with great precision to accurately control the pressure, direction, and volume of fluid within a system. Generally, valve leakage is controlled with the close fit of the parts.

Contaminants in the oil are the major villains in valve failures. Small amounts of dirt, lint, rust, sludge or metal particles can cause annoying malfunctions and extensive damage to valve parts. Such material will cause the valve to stick, plug small openings, or abrade the mating surfaces.

Any of these conditions will result in poor machine operation, or even complete stoppage. This damage may be eliminated if operators use care in keeping out dirt.

Follow the recommendations in the machine operator's manual. Use only the specified oils in the hydraulic system. The oil and filters should be changed at the proper intervals.

For successful valve service, follow the recommendations in the technical manual for the machine on which the valve is used.

TEST YOURSELF

QUESTIONS

1. Name the three basic valve types.
2. Give the reasons why pressure control valves are used in hydraulic systems.
3. Name the two general types of relief valves.
4. (True or False) "On the direct-acting relief valve, the cracking pressure is closer to the full flow pressure than on the pilot operated relief valve."
5. (True or False) "The pressure to operate a relief valve comes from the valve outlet oil."
6. Which type of directional control valve is the most often used?
7. What are the two basic types of valves?
8. What is the difference in oil flow through the two types of valves in neutral?
9. "When a valve is in the "float" position, both workports are connected to the __________________ port."
   inlet or return
10. (True or False) "A needle valve is a noncompensated flow control valve."
11. (True or False) "With a pressure compensated flow control valve, the flow will be greater when a function requires a high pressure rather than when a low pressure is required."
12. "A "fixed amount" pressure-reducing valve responds to the ______________ pressure."
    inlet or outlet

HYDRAULIC CYLINDERS

INTRODUCTION

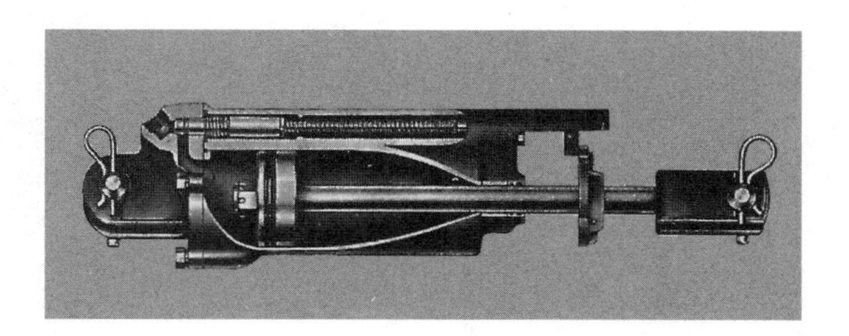

Fig. 1 - Piston Type Cylinder

The cylinder does the work of the hydraulic system. It converts fluid power from the pump back to mechanical power. Cylinders are the "arms" of the hydraulic circuit.

Chapter 1 explained the uses of hydraulics and how cylinders can be used to actuate both mounted equipment and drawn implements (remote uses). In either case, the basic design of the cylinder is the same.

PISTON-TYPE CYLINDERS

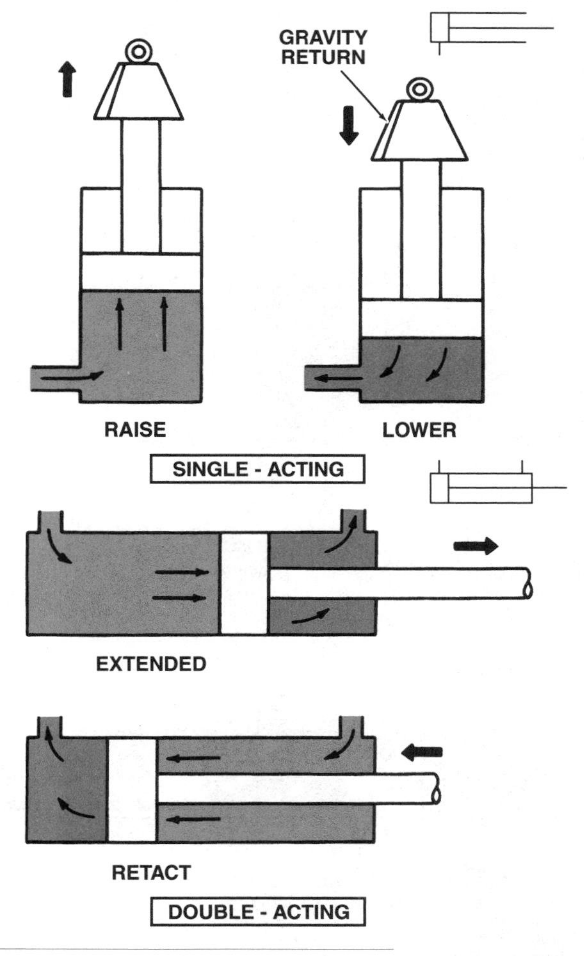

Fig. 2 – Single- and Double-Acting Cylinders

Piston-type cylinders fall into two categories:

- **SINGLE-ACTING CYLINDERS-give force in only one direction (Fig. 2). Pressurized oil is sent to only one port on the cylinder, raising the load. An outside force such as gravity or a spring must return the cylinder when the oil is allowed to escape.**
- **DOUBLE-ACTING CYLINDERS-give force in both directions (Fig. 2). Pressurized oil is admitted first at one end of the cylinder, then at the other, providing power in both directions.**

In both types of cylinders, a movable piston and rod slides in a cylinder housing or barrel in response to pressurized oil admitted to the cylinder.

The cylinder mount holds the cylinder in place while it works (Fig. 4).

The piston uses various packings or seals to prevent leakage.

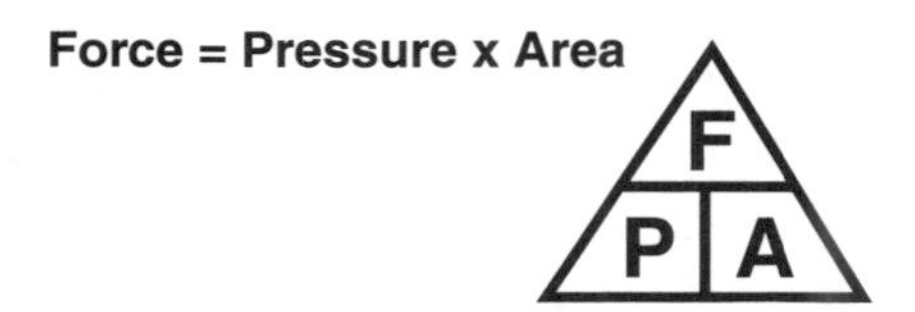

Fig. 3 - Force Triangle (Fig. 4-3 FOS)

The force (F) exerted by a piston can be determined by multiplying the piston area (A) by the pressure (P) applied.

Force = Pressure x Area (See Chapter 1)

SINGLE-ACTING CYLINDER

In a single-acting cylinder, shown in Fig. 4, pressurized oil is sent to only one side of the piston. The piston and rod are forced out of the housing, as shown moving the load. When the oil is released, mechanical force from another cylinder, the weight of the load or a spring device retracts the cylinder.

A seal on the piston prevents leakage of oil into the dry side of the cylinder. A wiper seal in the rod end of the cylinder cleans the rod as it moves in and out of the housing.

The rod end of the cylinder is dry. An air vent is required to release air when the piston rod extends, and to let in air when the rod retracts. To keep out dirt and moisture, a filtered breather is often used in the air vent.

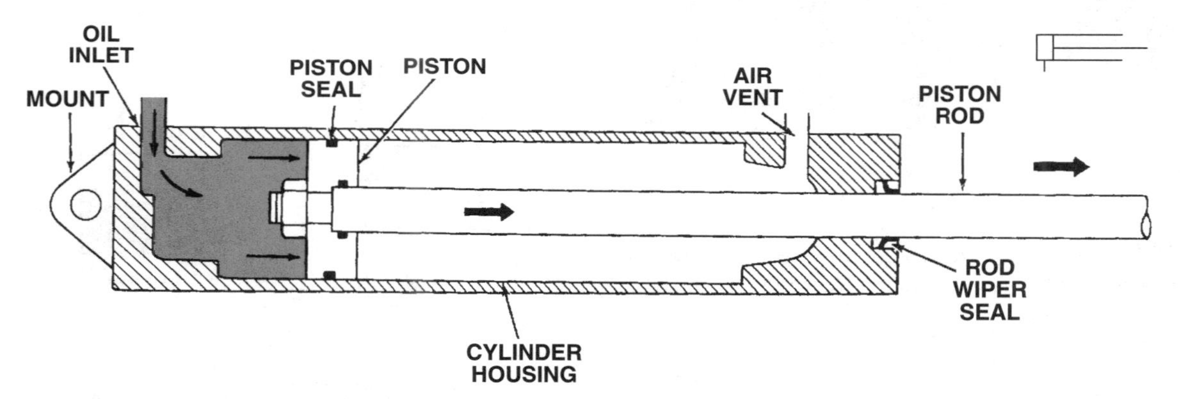

Fig. 4 - Piston Type Single-Acting Cylinder

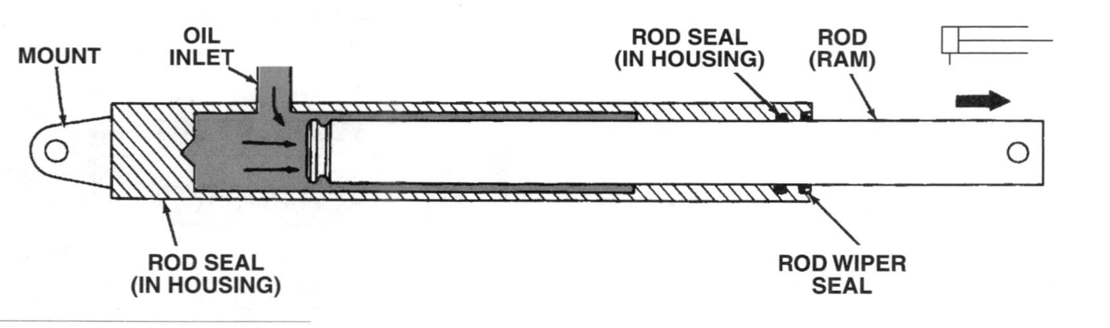

Fig. 5 - Ram Type Single-Acting Cylinder

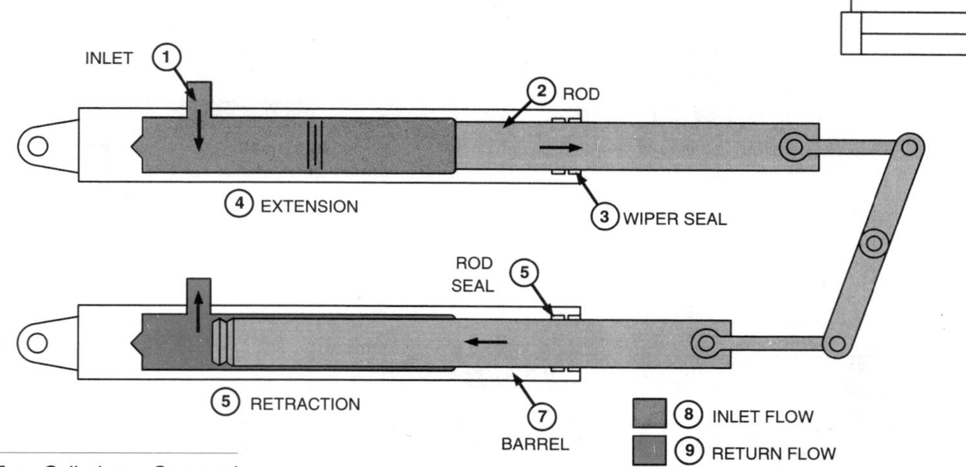

Fig. 6 - Ram Type Cylinders - Opposed

The ram type cylinder shown in Fig. 5 provides a sealed system, eliminating the problems of moisture and dirt in the dry end of a single-acting cylinder. This cylinder does not have a piston on the rod. Instead, the end of the rod serves as the piston.

When calculating the force exerted by the cylinder, the area of the rod end, not the area of the cylinder bore, are multiplied by the pressure.

The rod is slightly smaller than the inside of the cylinder. A shoulder or ring on the end of the rod keeps the rod from being pushed out of the cylinder.

Single-acting cylinders are used on some mobile equipment where the weight of the working unit will lower itself.

They are more often used in pairs that are connected to linkage as shown in Fig. 6. One cylinder provides power in one direction and the other cylinder powers the other direction. A common application of this is for steering, where operating at the same speed in both directions is important.

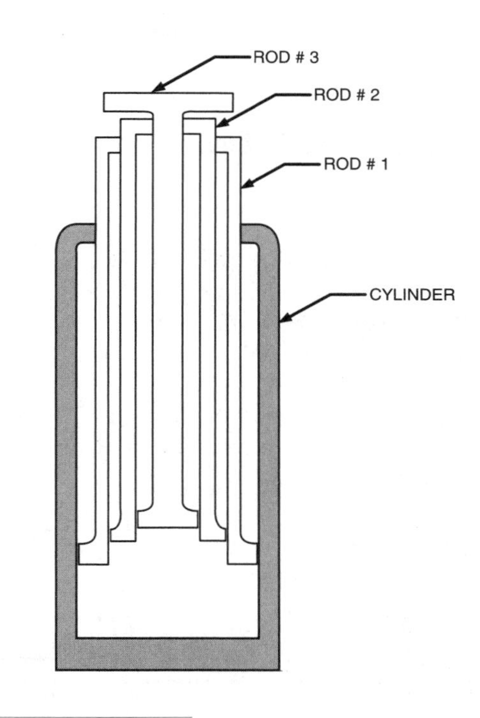

Fig. 7 - Telescoping Cylinder

The telescoping cylinder shown in Fig. 7 is single acting. This type of cylinder would commonly be used in a forklift or crane.

Oil enters at the base of the cylinder and acts against the end of the rods. The three rods will move out together. The force generated will be the pressure against the end of all three rods.

When the outer stop on the largest rod contacts the cylinder, the two inside rods will move out. The force will be the pressure acting against the ends of the two rods. When the second rod reaches the stop, the inside rod will continue to move the load.

Because the area on which the pressure acts changes as the cylinder extends, the lift specifications are greater for the first part of the travel than at maximum extension.

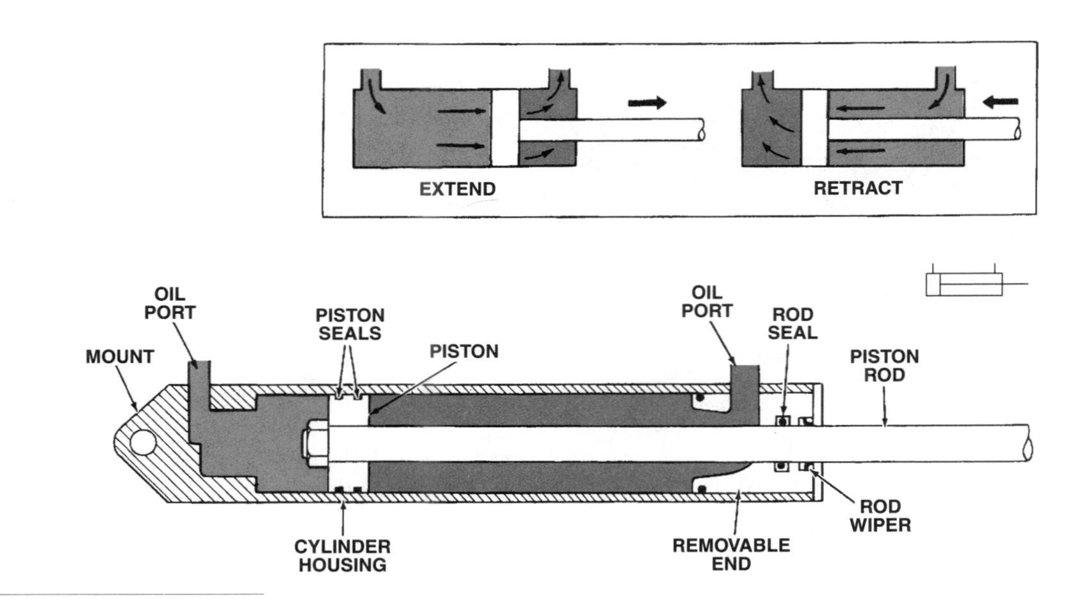

Fig. 8 - Double-Acting Cylinder

DOUBLE-ACTING CYLINDER

Double-acting cylinders provide force in both directions. Pressurized oil enters at one end of the cylinder to extend it, at the other to retract it (Fig. 8). Oil from the non-powered end of the cylinder is returned to the reservoir.

With the double-acting cylinder, both the piston head and the piston rod must be sealed to prevent oil leakage.

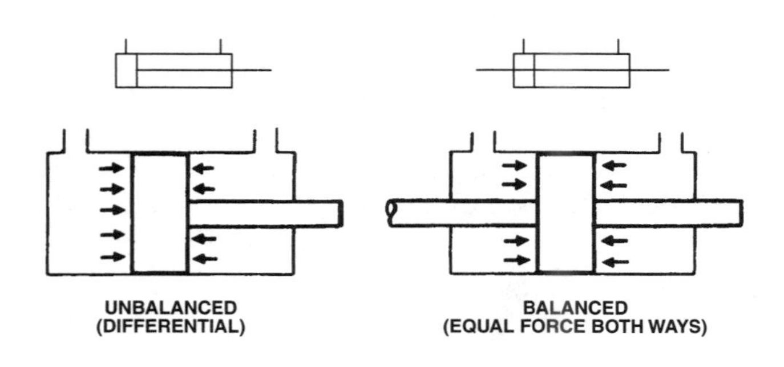

Fig. 9 - Balanced vs. Unbalanced Cylinders

Two types of double-acting cylinders, **balanced** and **unbalanced**, are shown in Fig. 9.

In the unbalanced type, total force when pressurizing the rod end of the cylinder is less than when pressurizing the piston end. This is because when the cylinder is extended, pressurized oil acts on the entire piston area. When retracted, pressure acts on an area that is less because of the rod. This cylinder will, therefore have a slower, more powerful stroke when it extends than when it retracts.

In the balanced cylinder, the piston rod extends through the piston head on both ends. This gives equal working area on both sides of the piston and balances the working force and speed regardless of the direction of travel.

EXTRA FEATURES OF PISTON-TYPE CYLINDERS

Many piston-type cylinders have extra features that add functions or adapt them to different uses.

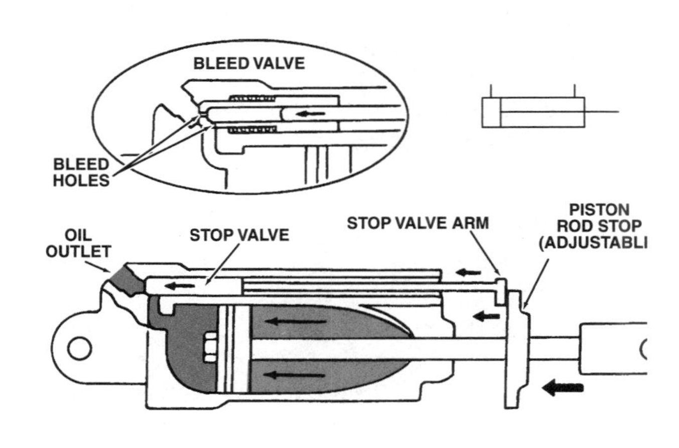

Fig. 10 - Hydraulic Stop in Cylinder

STROKE CONTROL DEVICES

The hydraulic stop cylinder (Fig. 10) can be adjusted to stop at a precise position every time the cylinder is retracted. It enables the operator to lower an implement to the same depth each time it is lowered.

It must be used with a detented valve on which the detent releases when the pressure reaches the relief setting.

The piston rod stop is adjusted on the rod to contact the valve stop arm when the implement is at proper depth. The stop arm moves the stop valve to shut off the return at the piston end of the cylinder. Oil flow is blocked causing the pressure to rise and open the system relief valve. This pressure releases the control valve detent allowing the control lever to go to neutral.

An "override" feature is built into the cylinder shown in Fig. 10. After the stop valve closes, two small bleed holes in the valve (see inset) allow a limited flow of oil out of the cylinder. By holding the control lever in retract position, the piston can be retracted further.

A spring mechanism allows the stop arm to continue travel after the valve is seated. It also allows incoming oil to open the valves when the cylinder is extended.

The rate of cylinder travel is usually adjustable at the control valve by means of a volume control device (see Chapter 5 for details).

CUSHION STOPS

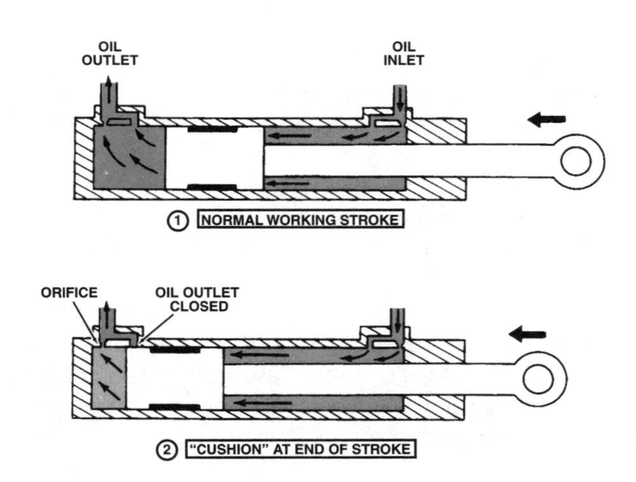

Fig. 11 - Cushion Stops

A cushion is built into some cylinders to slow them down gradually at the end of their strokes. This cushion prevents damage that would occur, if the travel of a heavy load were stopped by having the piston hit the end of the cylinder.

In Fig. 11, the cylinder works normally during its main stroke (top), but slows down as the piston covers the oil outlet (bottom). The small orifice now restricts outlet oil. The momentum of the load causes a very high pressure to build in the end of the cylinder. As the piston slows, the pressure will gradually reduce, providing a smooth gradual slowing of the piston. *Remember - a high pressure drop across an orifice equals high flow—low pressure drop equals low flow.*

STEPPED PISTONS

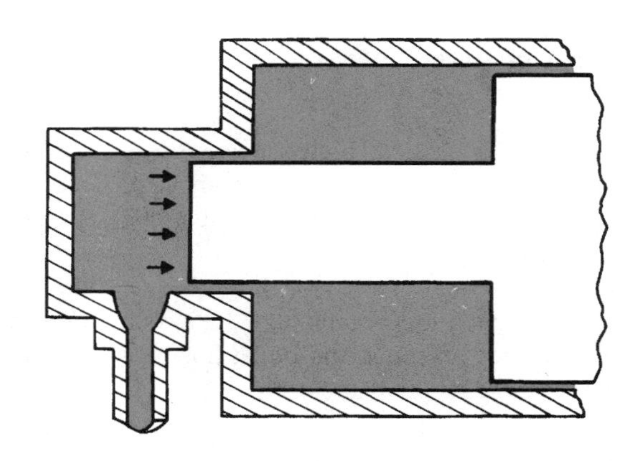

Fig. 12 - Stepped Piston

A stepped piston allows a cylinder to provide a rapid starting stroke with low force and a slower, more powerful working stroke. This is done by admitting oil first against the smaller part of the piston, which moves rapidly until the work is contacted (Fig. 12). Then the entire piston surface takes over for the power stroke.

VANE-TYPE ACTUATOR

A vane-type actuator provides limited rotary motion.

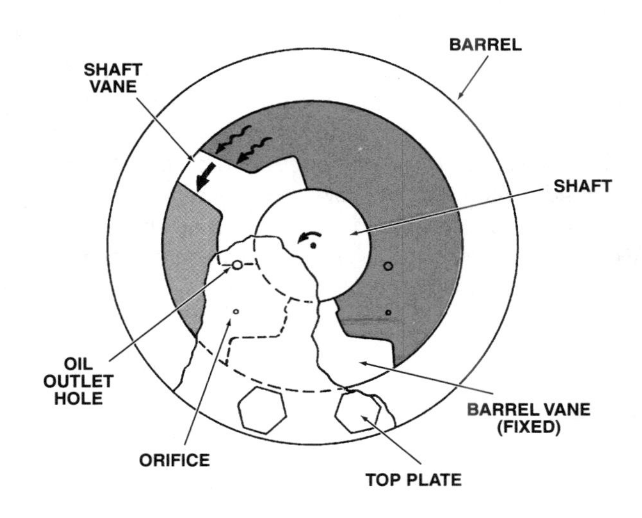

Fig. 13 - Vane-Type Actuator

The vane-type double-acting actuator shown in Fig. 13 consists of a barrel, shaft and two vanes. A movable vane is attached to the shaft and a fixed vane to the barrel. They separate the pressure cavities.

Oil entering the right cavity rotates the shaft counterclockwise. Oil is discharged through the outlet hole in the other side of the cylinder.

A "cushion" or "hydraulic brake" slows the rotating vane as it comes to the end of its stroke. The shaft vane shuts off the oil outlet hole in the top plate forcing the oil to flow through the small orifice.

Pressurizing the other cavity reverses the rotation.

SEALS

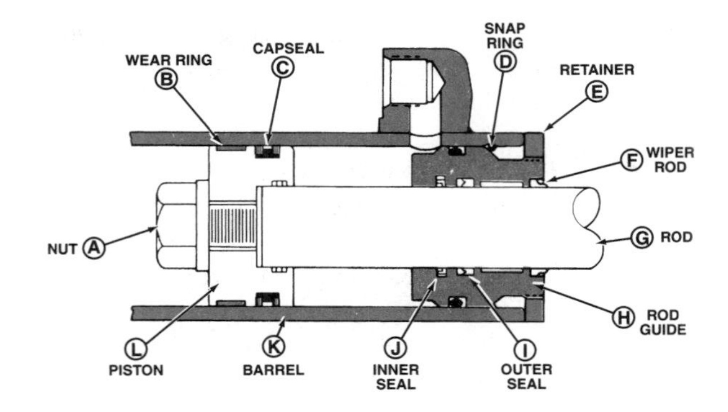

Fig. 14 - Typical Small Cylinder

The cylinder shown in Fig. 14 is typical of small cylinders used on modern machines.

The piston is retained to the rod with a nut. The outside of the piston has a wear ring to prevent metal to metal contact with the barrel. The cap seal prevents oil flow across the piston. It consists of an expander, which forces a teflon seal against the barrel wall. A backup ring on each side hold the seal in position.

To prevent leakage between the piston and rod, many cylinders will have a groove with a seal on the inside diameter of the piston.

The rod guide is held in position with a wire ring and a retainer nut. The retainer nut holds the rod guide against the ring. A locking shoulder on the rod guide prevents the ring from coming out of its groove.

The rod guide has a seal on the outside diameter. This is usually an O-ring or square packing in a groove.

The inner seal in the rod guide is a high-pressure seal. It is similar to the piston seal. An expander pushes a teflon seal against the rod. O-Rings and square packings are also used in this position.

The next seal is the backup or outer seal. It is normally an O-ring or packing. Many cylinders, especially older ones will not have this second seal. Some will place it between the wear ring and the wiper seal as shown in Fig. 15.

The wear ring prevents metal to metal contact with the rod.

The wiper seal is always a lip seal. It wipes the dirt from the rod as it enters the rod guide to prevent contamination and damage to the other seals.

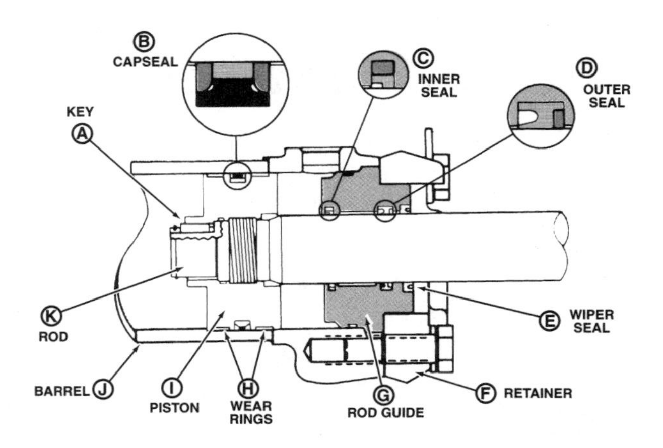

Fig. 15 - Typical Large Cylinder

The cylinder shown in Fig. 15 is more typical of larger cylinders. It has a threaded rod and piston with a key to prevent the piston from loosening. However, the threaded rod and nut shown in Fig. 14 is more commonly used.

The seals are almost identical to those used in Fig. 14 and serve the same purpose.

The rod guide uses a "bolt on" retainer to hold it in position. Another common design is a flanged rod guide, which is bolted to the barrel.

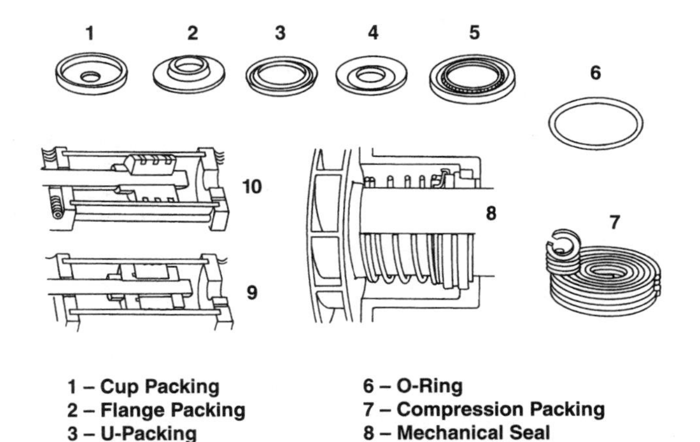

1 – Cup Packing
2 – Flange Packing
3 – U-Packing
4 – V-Packing
5 – Spring-Loaded Lip Seal
6 – O-Ring
7 – Compression Packing
8 – Mechanical Seal
9 – Non-Expanding Metallic Seal
10 – Cup Packing

Fig. 16 - Types of Seals

Many different types of seals are used in the cylinders. The pressures and temperature as well as whether the parts are moving or static determine selection. Fig. 16 shows some of these seals.

Chapter 12 of this manual covers seals in detail. Refer to this chapter for information on types and uses of seals, and correct maintenance.

IDENTIFYING SIZE OF CYLINDERS

The manufacturer sizes cylinders on mounted equipment for the job. However, with remote cylinders, there is a chance that the wrong cylinder may be used on a certain job.

Labels are used on most of these cylinders to identify the size. A typical label can be read as follows:

Label		Cylinder Diameter	
25	64	2 1/2 inches	64 millimeters
30	76	3 inches	76 millimeters
35	89	3 1/2 inches	89 millimeters

The numbers on the label give the diameter of the cylinder in tenths of inches and in millimeters. Merely place a decimal point between the first two digits and you have the size of the cylinder in inches. For example "25" = 2.5 inches. The "64" is the diameter in millimeters.

If you are not sure what size cylinder to use for a job, *check the operator's manual* for the machine or check the part's number in the parts catalog.

TEST YOURSELF

QUESTIONS

1. (True or false?) "Cylinders convert mechanical power to fluid power."

2. (Fill in the blanks.) "Piston-type cylinders give _______________ movement while vane-type cylinders give _______________ movement."

3. (Fill in the blanks.) "Cylinders which are ____________ acting give force in both directions. Cylinders which are _____________ acting give force in only one direction."

4. What fills the chambers on each side of the piston in a single-acting cylinder?

5. In a double-acting cylinder, how does a piston rod on one side only affect the working stroke?

6. How does a "hydraulic brake" or "cushion" work in a cylinder?

HYDRAULIC MOTORS

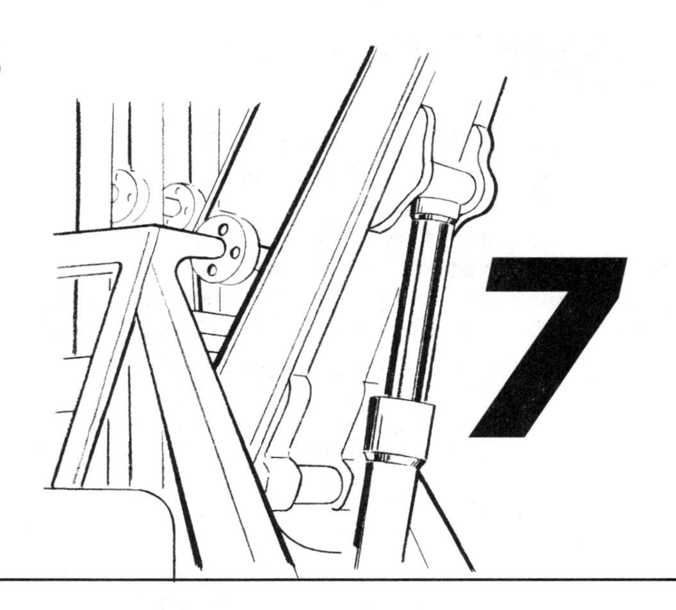

INTRODUCTION

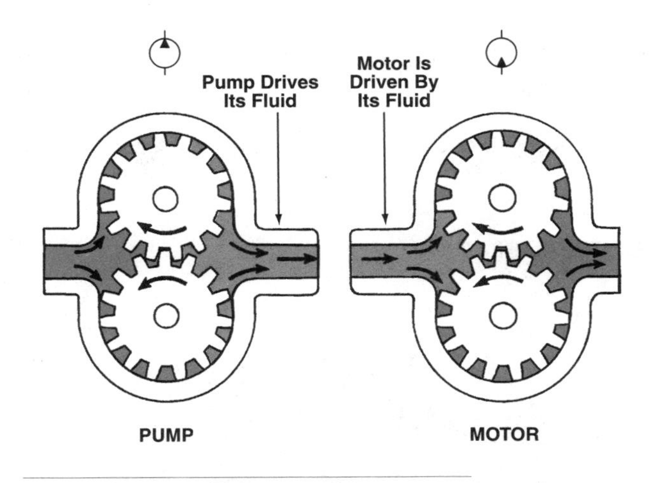

Fig. 1 — Hydraulic Pump and Motor Compared

A hydraulic motor works in reverse when compared to a pump (Fig. 1).

The pump drives fluid, while the motor is driven by fluid. Thus:

- **Pump - draws in fluid and forces it out, converting mechanical energy into fluid energy.**
- **Motor - fluid is forced in and exhausted out, converting fluid energy back to mechanical energy. (Fig.2)**

The pump and motor are often hydraulically coupled to provide a power drive. This is referred to as a hydrostatic drive system.

Like the cylinder (Chapter 6), the motor is referred to as an actuator. Unlike the cylinder, the motor is a rotary actuator which can rotate in a full circle (The vane-type cylinder is a limited rotary actuator).

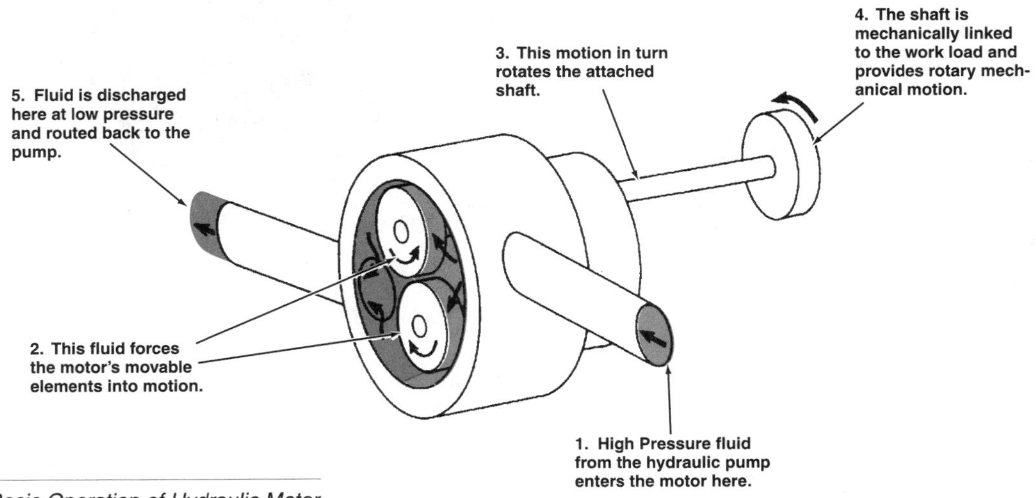

Fig. 2 — Basic Operation of Hydraulic Motor

DISPLACEMENT AND TORQUE OF MOTORS

The output force of a motor is called torque. This is a measure of the rotary (twisting) force of the motor drive shaft.

Torque is a measure of force x distance (measured in "pounds-foot" or Newton-meters). The inlet oil pressure pushing on the face of the gear teeth creates the force. Distance is the distance from the tooth face to the shaft center. This makes torque dependent on motor displacement and the inlet pressure.

Work is the torque x speed (measured in horsepower or kilowatts). Speed is determined by motor displacement and inlet flow.

COMPARING PUMP AND MOTOR DESIGN

The motor is designed much like the pump. Both use the same basic types - gear, vane, and piston. Often their parts are identical.

The internal sealing areas of both the pump and motor are the same. They both have the same close fitting parts and need pressurized endplates and vanes.

Motors, like pumps, can be **Fixed Displacement** or **Variable Displacement**. These possibilities will be covered in discussing each type of motor.

It is not normally possible to change a pump to a motor simply by forcing oil through it. The reasons will also be covered when each type of motor is discussed.

TYPES OF HYDRAULIC MOTORS

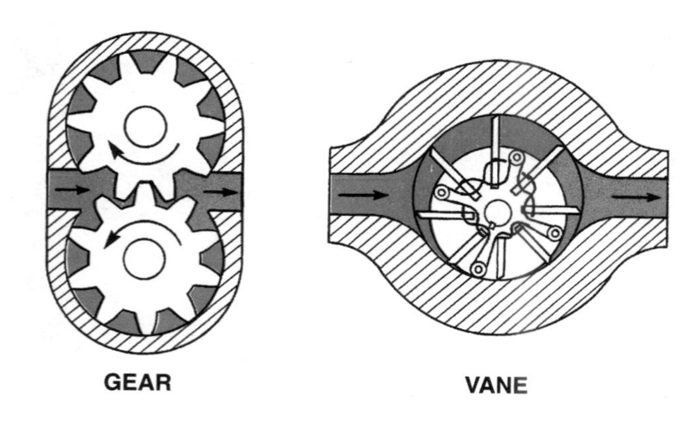

Fig. 3 — Types of Hydraulic Motors

There are three basic types of hydraulic motors (Fig. 3):

- **Gear Motors**
- **Vane Motors**
- **Piston Motors**

These are the same basic types as used in pumps (Chapter 4).

All three designs work on the rotary principle: a rotating unit inside the motor is moved by incoming fluid.

Let's discuss the operation of each type of motor.

GEAR MOTORS

Gear motors are widely used because they are simple and economical. Often they are used to drive small equipment in remote locations.

Usually small in size, gear motors are versatile and can be transferred from one use to another by using a universal mounting bracket and flexible hoses.

Gear motors are only fixed displacement and can be made to rotate in either direction.

They can be either **External Gear** or **Internal Gear Motors.**

Let's discuss the operation of each.

EXTERNAL GEAR MOTORS

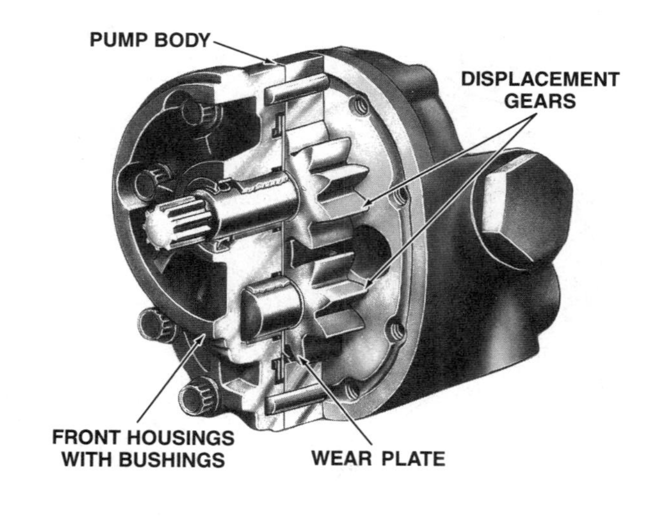

Fig. 4 — External Gear Motor

The external gear motor is a duplicate of the external gear pump. It has two equal-sized gears in mesh, sealed in a housing (Fig. 4).

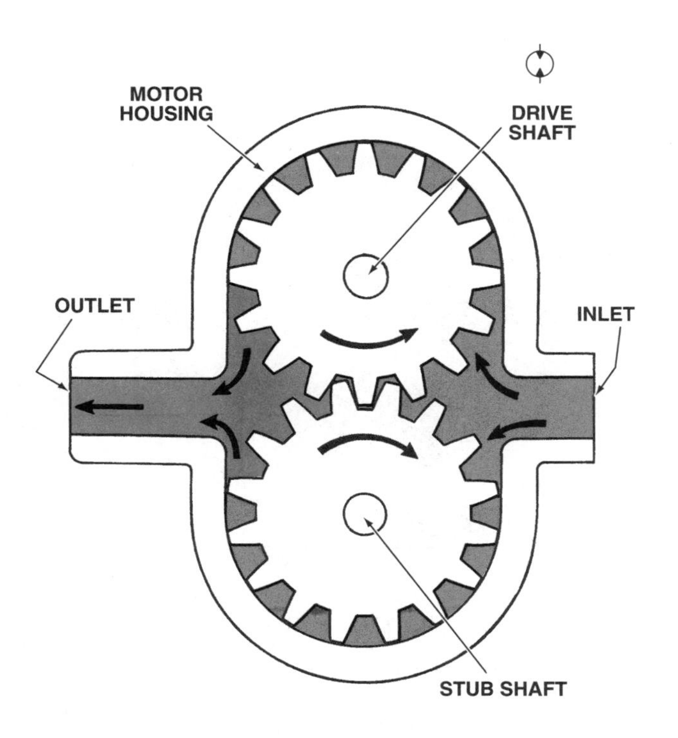

Fig. 5 — External Gear Motor in Operation

In operation, pressurized oil from the pump forces the motor gears to rotate away from the inlet port. This action rotates the motor shaft connected to the workload.

The force of the pressurized oil is expended as it travels between gear teeth and housing to the outlet port. Here it leaves the motor as low-pressure fluid and returns to the reservoir or the pump.

For high-pressure operations, the end plates need to be pressure compensated as they were in the pumps (Chapter 4).

A reversible motor requires check valves between the motor ports and the pressurized endcaps. This insures that the high-pressure port supplies oil to pressurize the endcap.

Internal Gear Motors

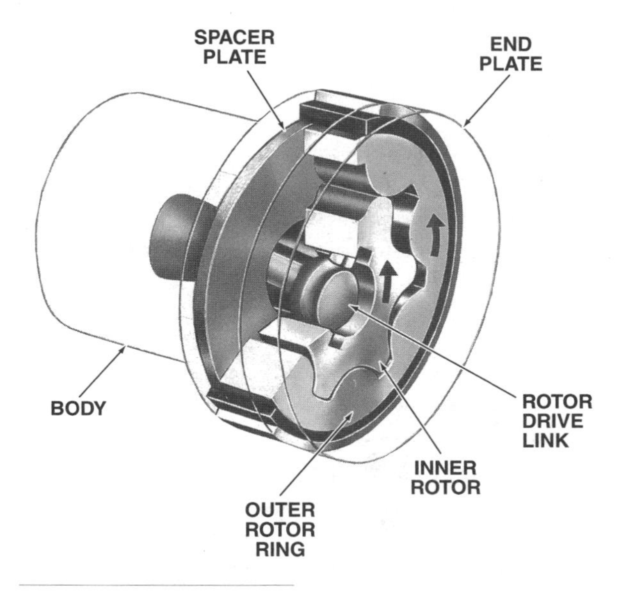

Fig. 6 — Internal Gear Motor

One popular internal gear motor is much like the rotor pump (Chapter 4). This motor is shown in Fig. 6.

Not actually gears, the moving parts are called the rotor and the rotor ring. The rotor is driven inside the rotor ring.

The rotor is mounted eccentric to the rotor ring. The ring has one more lobe than the rotor so that only one lobe is in full engagement with the outer ring at any one time. This allows the rotor's lobes to slide over the outer lobes, making a seal.

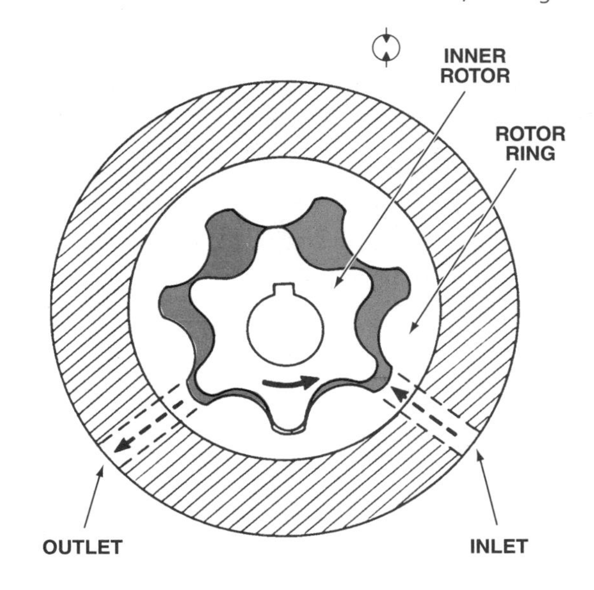

Fig. 7 — Internal Gear Motor in Operation

In operation, pressure fluid enters the motor, pushing on the rotor and rotor ring lobes, forcing both to rotate (Fig. 7). A seal is formed between the pressure and return cavities by the tooth tips of the rotor and rotor ring.

As the teeth reengage, oil is forced out the discharge port.

This motor is very compact. It is efficient only at low pressures. It is, however, used effectively as a follow-up motor in high-pressure power steering valves. In this application, the pressure is high but the difference between inlet and outlet pressure is very low.

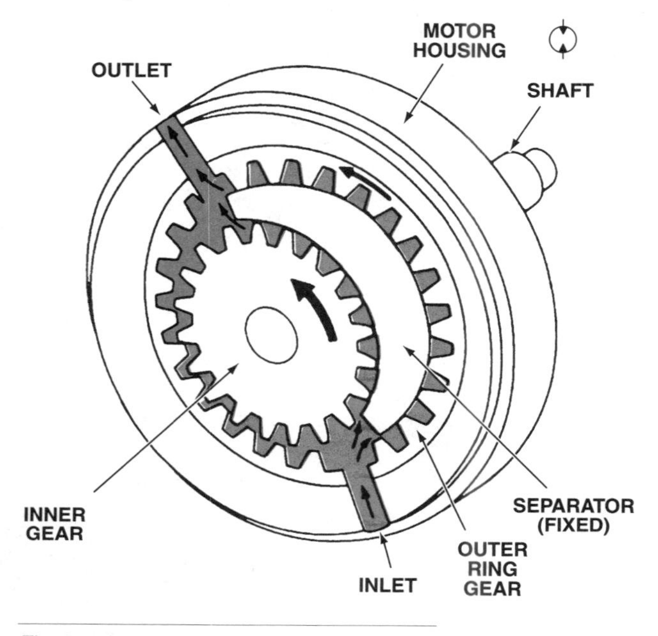

Fig. 8 — Internal Gear Motor with Separator

Fig. 8 shows another type of internal gear motor. Like the internal gear pump in Chapter 2, it uses a crescent-shaped separator between the inner and outer gears. Operation is the same as the lobe motor in Fig. 7. This motor also is used only in low-pressure applications because it is not practical to pressurize the end plates.

VANE MOTORS

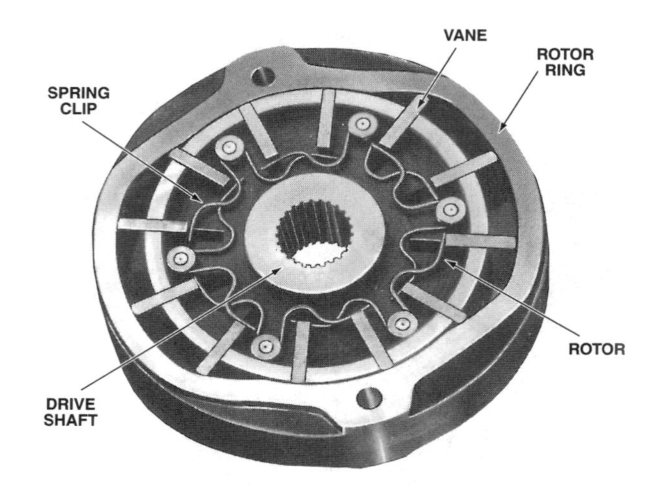

Fig. 9 — Vane Motor-Balanced Type

Vane motors, like the pumps, are available in the balanced and unbalanced type. Most of the vane motors on today's machines are the balanced type. Balanced motors have a longer service life because of the reduced bearing load.

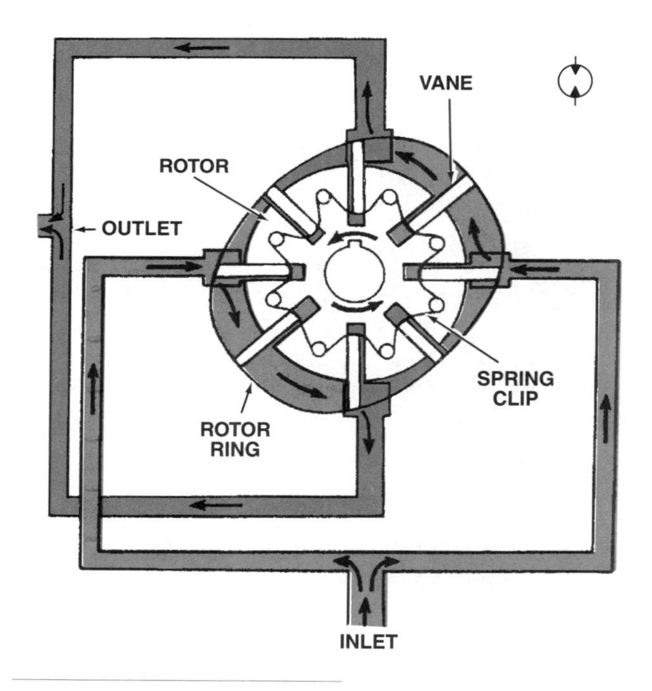

Fig. 10 — Vane Motor in Operation

As pressurized oil enters the motor inlet (Fig. 10), pressure will exert more force on the extended vane than the retracted one. This will move the rotor in a counterclockwise direction.

The vane pump relies on centrifugal force to hold the vanes in contact with the outer ring during startup. Because the motor is not turning at startup, it requires springs to hold the vanes against the ring (See Fig. 9 and 10).

Because the vane motor, like the pump, has passages to pressurize the bottom of the vanes and the end plates, it is very efficient and can be used in high-pressure applications (See Chapter 4).

PISTON MOTORS

Piston motors can be built with more precision and efficiency than the gear and vane motors. They are often favored in systems that have high speeds and/or high pressures. While they are more efficient, they are more expensive and require more careful maintenance.

Like its pump counterpart, the piston motor is available in two types:

- **Axial Piston**
- **Radial Piston**

On mobile systems, axial piston models are often used. The radial piston model is usually used where space is not limited and more power is needed.

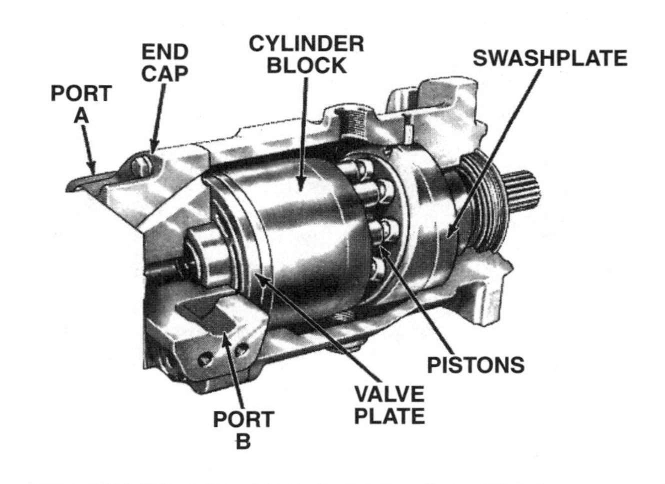

Fig. 11 — Axial Piston Motor - Fixed Displacement

Fig. 11 shows a fixed displacement version of an inline axial piston motor.

The motor shown is used as part of the hydrostatic drive on a self-propelled machine.

The end cap contains ports A and B, which accept pressurized oil from the pump to operate the motor. They also discharge low-pressure oil back to the pump.

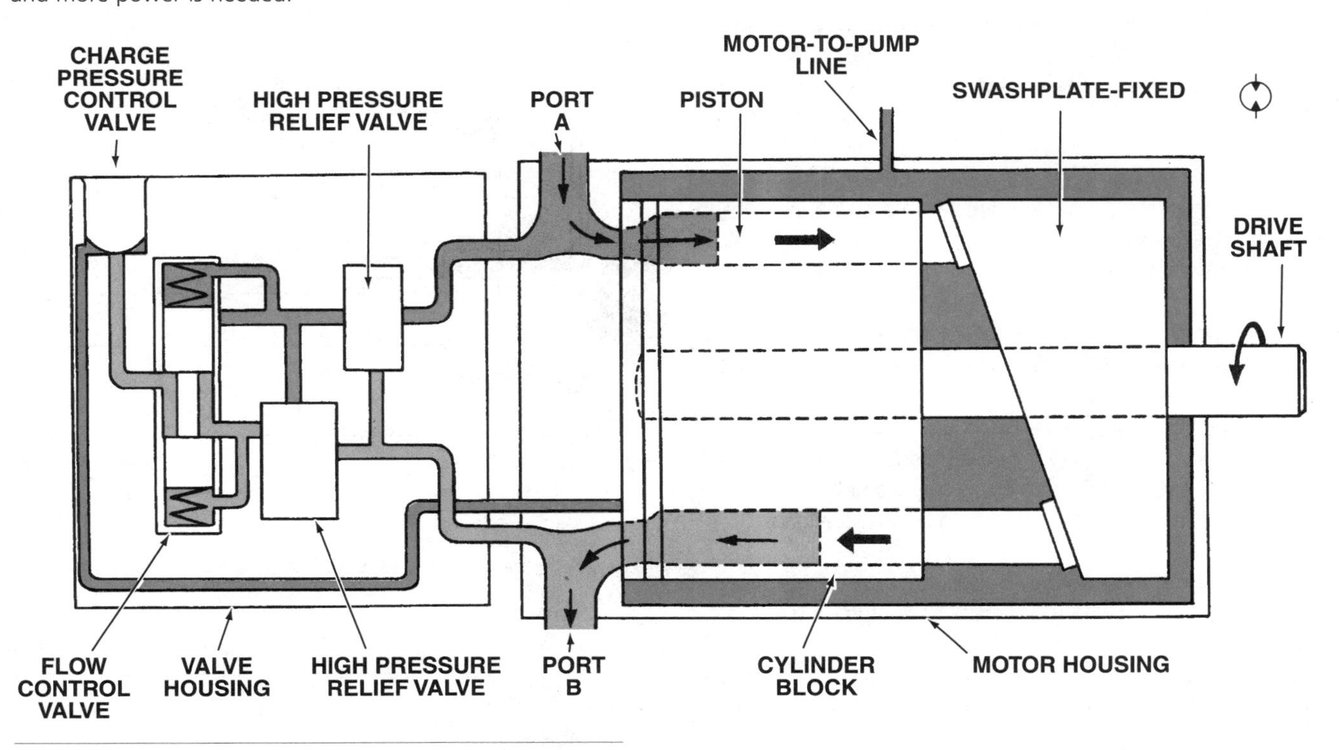

Fig. 12 — Axial Piston Motor in Operation - Fixed Displacement

The pistons work in bores in the revolving cylinder block and contact a fixed-angle swashplate as they revolve.

A fixed displacement motor is shown in Fig. 12. The cylinder block with pistons rotates in the housing and is attached to the output shaft.

In operation, high-pressure oil enters the pressure passage through port A. The pistons in the forward one half of the block are aligned with the pressure passage, forcing them down against the fixed angled swashplate. Because the swashplate is fixed, the piston slides down its angled face. This sliding action turns the cylinder block and shaft to rotate in a counterclockwise direction as viewed from the shaft end.

The piston will be aligned with the pressure passage until it is fully extended at the low point of the swashplate. As the cylinder block turns, other bores align with the pressure passage and are forced down the swashplate to continue the rotation.

During the second half of a pistons revolution, it will be aligned with the discharge passage and port B. As the piston is pushed into its bore by the swashplate, low pressure oil is discharged at port B.

To reverse the rotation, simply reverse the flow of oil, feeding pressure oil in at port B and returning it through port A. This can be done by connecting the motor to a pump which has reversible flow or to a three-position, four-way control valve.

When the oil flow is reversed, the pistons on the backside of the rotor are now pressurized and it will turn in the opposite direction.

When the motor is connected directly to a pump in a closed loop, it is necessary to protect the high and low pressure circuits from excessive pressures.

The valving shown in Fig. 12 consists of a flow divider valve, two high-pressure relief valves, and a charge pressure control valve.

There is a high-pressure relief valve for each circuit. When that circuit is the motor inlet, it will limit the maximum pressure. When pressure exceeds the valve setting, it opens and bypasses oil to the return circuit, protecting the system against overloads.

The flow divider valve moves back and forth in response to incoming pressure oil. It insures that the charge pressure valve is always exposed to the return pressure regardless of which circuit is pressurized. The charge pressure control valve is a low-pressure relief valve, which limits the maximum pressure in the return circuit.

VARIABLE DISPLACEMENT VERSION OF AXIAL PISTON MOTOR

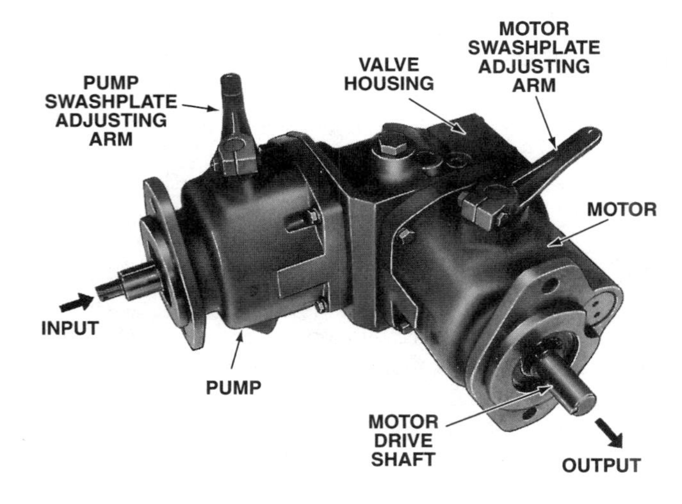

Fig. 13 — Axial Piston Pump and Motor Assembly

The assembly shown in Fig. 13 is an intregal pump and motor combination used to drive a self-propelled machine. This assembly has both a variable displacement pump and motor.

The pump and motor are connected by a housing which contains all oil passages, charge pump and relief valves.

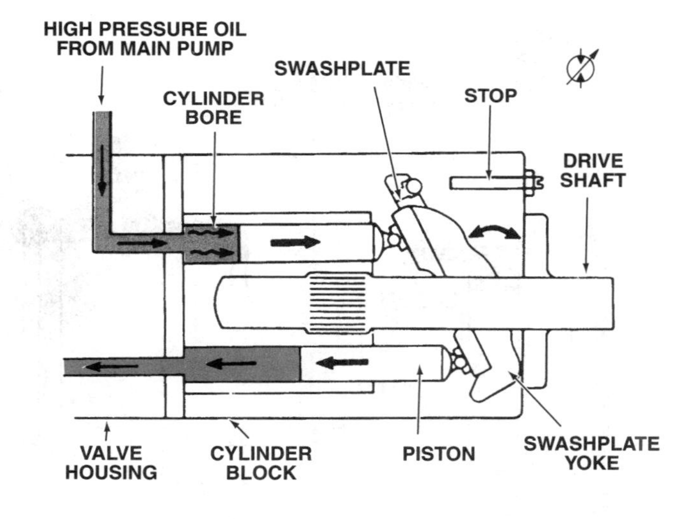

Fig. 14 — Operation of a Variable Displacement Motor

A variable displacement motor operation is shown in Fig. 14. This motor differs from the fixed displacement version described above only in that the swashplate angle is adjustable.

Operation is the same as for the fixed displacement model. High-pressure oil forces the pistons to slide down the face of the swashplate, rotating the cylinder block and drive shaft. Fluid at low pressure is expelled as the pistons are pushed back by the swashplate.

The displacement of oil in each cycle of the motor is determined by how far the piston must travel in its bore. This is controlled by the angle of the swashplate.

The displacement of the pump can be adjusted by changing the angle of the swashplate. This can be done with mechanical linkage but usually with hydraulic power as described in the pump section (Chapter 4).

An axial piston motor cannot operate unless there is some angle on the swashplate. A mechanical stop is used in this motor (Fig. 14) to establish a minimum angle for the swashplate.

A maximum angle of the swashplate will result in maximum displacement. This will result in the highest torque and the slowest speed. When the swashplate angle is at its minimum, the displacement is the least, resulting in the fastest speed and the lowest torque.

CAM LOBE MOTORS

The cam lobe motor is a radial piston motor. It consists of a cam, a carrier with pistons and a manifold.

The cam is attached to the motor housing and the carrier to the output shaft.

Pressurized oil is routed to the manifold. It is then fed through passages in the carrier to the piston bores.

Each passage in the manifold is timed to one cam ramp. There are 15 pressure oil and 15 return oil passages in the manifold.

The power stroke is developed when pressurized oil flows through the manifold and carrier oil passages, forcing the piston outward. As the piston follower is forced against the cam ramp, the carrier and shaft are forced to turn.

Return oil is routed from piston bores, through the carrier passages and into the manifold return passages.

The oil manifold is spring loaded against the piston carrier with a thin film of oil separating them.

In a 12-piston motor, as shown in Fig. 15, there are three sets of pistons doing the same thing at equal (120°) distances. This balances the forces on the carrier and axle assembly.

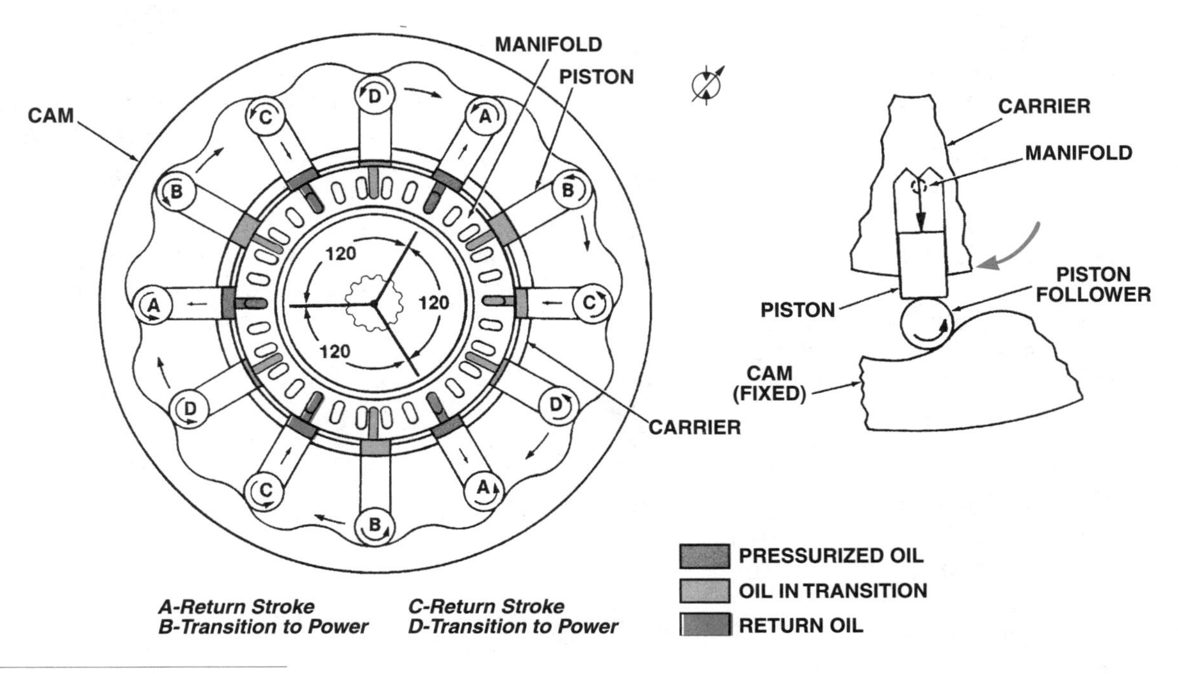

Fig. 15 — Cam Lobe Motor Operation

When the motor is engaged, one group of pistons (A) is in the power stroke. The group of pistons (C) push the return oil back through the manifold.

Pistons (D), at the top of their power stroke and (B), at the bottom of their return stroke are in transition between oil passages.

As one group ends its power stroke, another group has begun. This overlapping of strokes provides smooth power (torque) to the axle.

OPERATION OF THREE-SPEED MOTOR

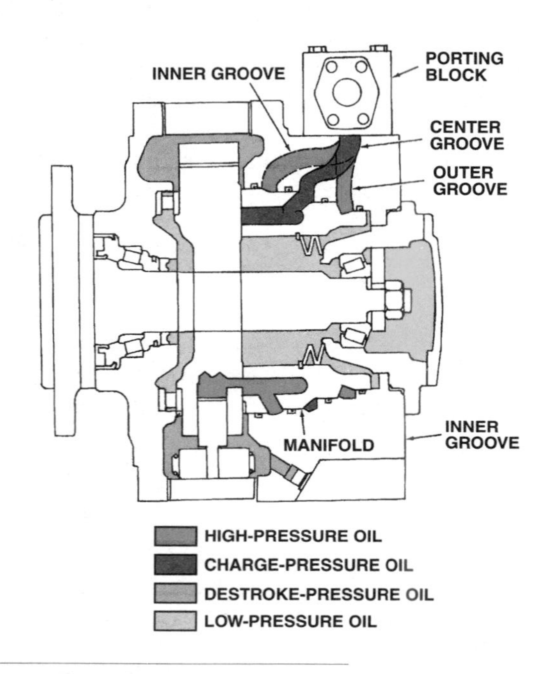

Fig. 16 — Three-Speed Cam Lobe Motor

The three-speed motor shown in Fig. 16 has three manifold grooves.

The inner groove is connected to 10 of the 15 pressure ports in the manifold. The outer groove is connected to the other 5 of 15 ports. The middle groove is connected to all of the return ports.

Seals separate the passages of the manifold. The one-speed motor described earlier would have only two passages - one for pressure and one for return.

The external control valve routes pressurized oil to the manifold ports depending on the speed selected.

Low Speed occurs when oil flow is routed to both the inner and outer grooves. High-pressure oil is directed to all 15 of the pressure ports in the manifold, providing the greatest displacement and torque and the slowest speed.

Middle Speed occurs when oil flow is routed only to the outer groove. High-pressure oil is routed to only 10 of the 15 pressure ports in the manifold.

The axle will turn one and one half revolutions using the same volume of oil required for one revolution in low speed. The motor will have more speed but less torque.

The pistons not involved are recirculating return oil from the return circuit via an external valve.

High speed occurs when pressurized oil is routed only to the inner groove. This groove feeds only five manifold passages that lead to the carrier pistons. High speed develops the lowest torque because there are 1/3 as many power strokes as compared to low speed. The axle will turn three revolutions using the same volume of oil required for one revolution in low speed.

Low-speed reverse is accomplished when high-pressure oil is routed to the center groove. This pressurizes all 15 of the ports, which in forward had been return ports. This provides reverse operation with maximum torque and minimum speed.

Middle reverse is accomplished when oil is routed to the 15 pressure ports through the center groove, but also routes high-pressure oil to the inner groove. This pressurizes the return for one third of the pistons and forces them to only recirculate high-pressure oil.

The axle will turn slightly less than one and a half revolutions using the same volume of oil as the low speed motor. This is because of the friction losses when the return is pressurized.

High-speed reverse is not possible because the friction loss of recirculating high-pressure oil by the majority of the pistons is greater than the functioning pistons can overcome.

OPERATION OF DESTROKE PUMP

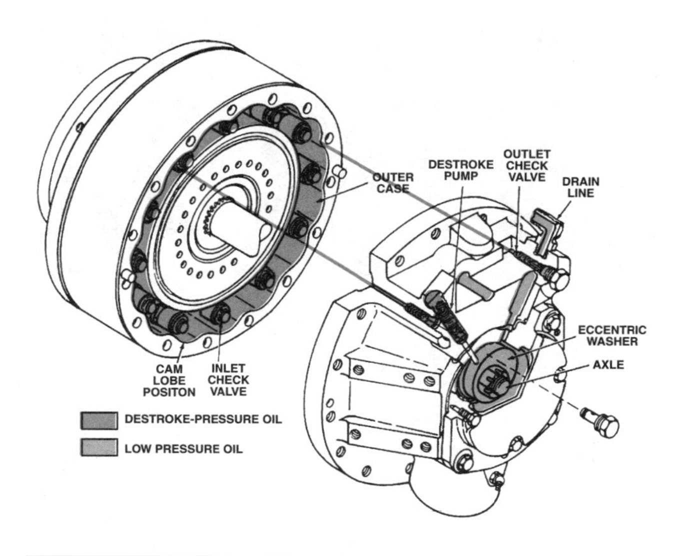

Fig. 17 — Destroke Pump Operation

Most cam lobe motors have a destroke pump (Fig. 17). The destroke pump pressurizes the outer case. When the motor is disengaged, the destroke pressure forces the pistons into the carrier so that they will not contact the cam lobes.

An eccentric washer on the axle activates the destroke pump once every revolution. The destroke pump draws low-pressure oil from the inner case. Oil moves through the inlet check valve to the outer case.

The outer check valve regulates the oil pressure in the outer case to 10 psi (70 kPa) over the pressure in the inner case.

ADVANTAGES OF HYDROSTATIC DRIVES

- *Variable speeds and torques*
- *Easy one-lever control*
- *Smooth shifting without "steps"*
- *Shifts "on-the-go"*
- *High torque available for starting up*
- *Flexible location—no drive lines*
- *Compact size for big power*
- *Eliminates clutches and gear trains*
- *Reduces shock loads*
- *Low maintenance and service*

SUMMARY OF MOTOR TYPES

This concludes our description of the three basic types of motors.

Before going into the application and efficiency of these motors, let's review some of the points we have just covered.

IN SUMMARY:

1. A hydraulic motor is the opposite of a pump: A pump drives fluid, while a motor is driven by fluid.
2. A pump converts mechanical energy into fluid energy, while a motor converts fluid energy back into mechanical energy.
3. The three basic types of motors are gear, vane, and piston. All three are rotary in operation.
4. A motor is often quite similar to a pump in appearance and construction.

We have covered only the basic types of motors. In actual use, there are many variations for many special needs.

HYDRAULIC MOTOR APPLICATION AND EFFICIENCY

The first part of this chapter has described the physical construction and operation of the basic types of hydraulic motors as used on modern farm and industrial equipment. Now we will discuss how the hydraulic motor is used, why it is used, and how the three types of motors are rated in regard to power output, efficiency, size, etc.

Again, because of the wide variety of motors and hydraulic systems, we will not attempt to prescribe a particular motor for a particular application. We can only describe in a general way the good and bad points of each type.

Fig. 18 — Hydrostatic Drive Propels This Combine

Fig. 20 — Large Motor Drives This Machine

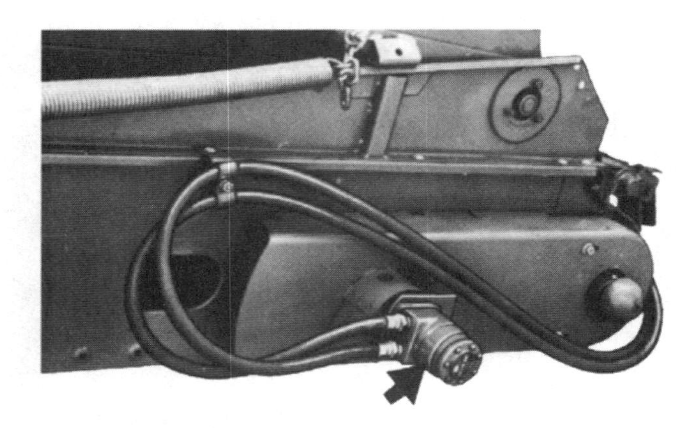

Fig. 19 — Small Motor Drives This Elevator

MOTOR SELECTION

To select a hydraulic motor, we must first know what it is expected to achieve. This means analyzing the system's requirements, then selecting a motor that best meets these needs.

The things that must be considered are:

- **The flow and pressure capabilities of the pump or circuit supplying the oil.**
- **The torque requirement of the load.**
- **The speed requirement of the load.**

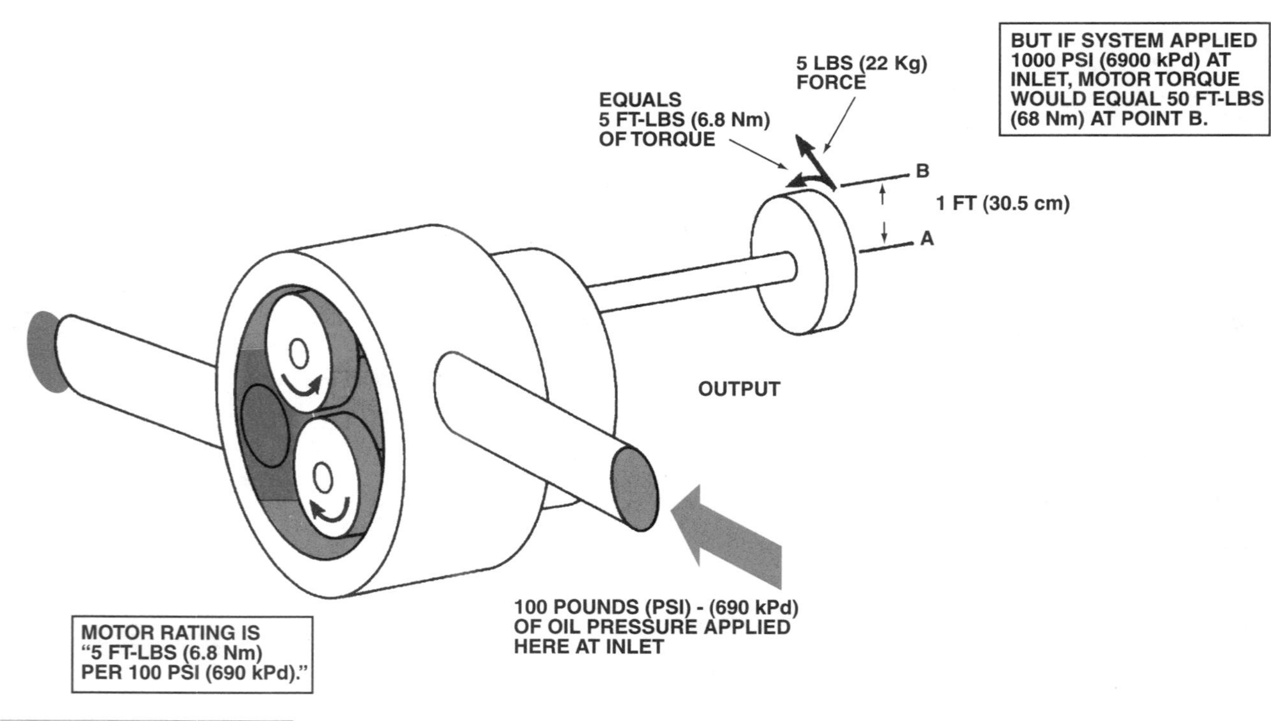

Fig. 21 — Motor Torque Rating

MOTOR TORQUE

The first consideration is to select a motor that will turn the load. Because torque is directly related to the input oil pressure, most motor torque ratings are given as torque per unit of input pressure.

Torque is expressed in the English system as "pounds-feet," which is one pound acting on the end of a one-foot long lever arm attached to the shaft. It is also expressed as "pound-inches" which is one pound acting on a one-inch lever arm.

The Metric system uses "newton-meters" as a torque unit. That is a force of one newton (0.1 kilogram) acting on the end of a lever arm which is one meter long.

One pound-foot equals 1.356 newton-meters (N-m)

One newton-meter equals 0.7376 pound-foot. (lb.-ft.)

Example: We'll use a motor rated "5 foot-pounds" (6.87 N-m) per 100 psi (690 kPa). This means that with 100 psi (690 kPa) of system pressure (see Fig. 21), the output torque would be a 5 pound (22.2 newtons) force one foot (0.305 meters) from the shaft center (point B). If the system pressure were 1000 psi (6890 kPa), the maximum torque would be 50 pounds (222 newtons) force at point B or 50 foot-pounds (68.7 N-m) of torque in the shaft.

In the actual selection of a motor, the process would be reversed. Example: We have a system pressure of 1500 psi (10,340 kPa), and our maximum load is 50 ft.-lb. (68.7 N-m) of torque. Our requirement will then be for a motor with a torque rating of at least 3.3 ft.-lb.(4.5 newton-meters) of torque per 100 psi (690 kPa) of input pressure.

You will note that torque is affected by oil pressure only. The volume of oil does not change torque.

Torque should always be calculated at maximum loads. The torque needed to start a load is always greater than the torque needed to maintain rotation, so the maximum load requirement should be used in selecting a motor.

When considering pressure, be certain that the motor can withstand the maximum system pressure.

MOTOR COMPARISON CHART

	GEAR MOTORS External	**GEAR MOTORS** Internal	**VANE MOTORS** (Balanced)	**PISTON (AXIAL) MOTORS** Fixed Displ.	**PISTON (AXIAL) MOTORS** Var. Displ.	**CAM LOBE MOTORS**
Physical Size	Small to Medium	Small to Medium	Small to Medium	Medium to Large	Medium to Large	Medium to Large
Average Weight to Power Ratio-lb/hp	0.9	0.9	1.0	1.4	3.2	5-13
Pressure Range (psi)	100-2000	100-2000	100-2500	100-5000+	100-5000+	100-5000+
Speed Range (rpm)	100-3000	100-5000	10-3000	10-3000	10-3000	1-220
Actual Torque (% of Theoretical)	80-85	80-85	85-95	90-95	90-9S	90-95
Starting Torque (% of Theoretical)	70-80	75-85	75-90	85-95	85-95	85-95
Momentarary Overload Torque (% of Acual)	110-120	115-130	120-140	120-140	120-140	130-150
Volumetric Efficiency (%)	80-90	85-90	85-90	95-98	95-98	95-100
Over-all Efficiency (%)	60-90	60-90	75-90	85-95	85-95	85-95
Estimated Bearing Life (hrs.) @ 1/2 load	5000-10000	5000-10000	7000-15000	15000-25000	15000-25000	20000-25000
Displacement	Fixed	Fixed	Fixed	Fixed	Variable	Variable in Steps
Reversibility	Possible	Possible	Possible	Very Good	Very Good	Good at Full Displacement
How Operates as a Pump	Good	Good	Good	Very Good	Very Good	Very Good
Estimated Bearing Life (hrs.) @ full load	2000-5000	2000-5000	3000-6000	7000-15000	7000-15000	3000-5000

NOTE: Remember that the values in this chart are not absolute. They may vary with each particulat model of motor.

SPEED

Having determined the torque requirements of the motor, you must then determine the displacement necessary to achieve the proper speed.

In choosing the displacement of a motor, the displacement and speed of the pump is needed.

The maximum recommended speed must be considered in motor selection.

With a given supply pump and motor, only the speed of the pump can affect the volume and, thus, the speed of the motor.

MOTOR HORSEPOWER

Horsepower is the measure of total work done by the motor. It is the combination of force x distance x speed. One horsepower being the work done in moving 550 pounds a distance of 1 foot in a time of 1 second (33,000 pounds, one foot in one minute). Expressed as torque, it is 550-lb. ft./sec.

The kilowatt is the metric unit of power. It is 0.7457 hp. Expressed as torque, one watt is 1 Nm/sec.

Motor horsepower is the most common method of rating hydraulic motors.

MOTOR EVALUATION

Now that we have given you some ideas as to how a motor is evaluated, let's see how the three basic motors compare with one another.

The chart below compares the three motors in a very general sense. As you will note there is no one pump which is superior in all categories. It is, therefore, necessary to evaluate the application when making a selection.

TEST YOURSELF

QUESTIONS

1. (Fill in the blanks.) "A hydraulic motor converts ___________force into ___________ force."
2. How is a motor different in operation from a pump?
3. What three types of motors are most popular on modern farm and industrial systems?
4. What is the "torque" of a motor?
5. (True or false?) "A pump can generally be used as a motor."
6. (Fill in the blanks) "The cam lobe motor described in the chapter has a ________________ (fixed or rotating) carrier and a __________________ (fixed or rotating) cam ring."

HYDRAULIC ACCUMULATORS

CAUTION: ACCUMULATORS STORE ENERGY. BEFORE WORKING ON A HYDRAULIC SYSTEM WITH AN ACCUMULATOR, YOU MUST RELIEVE ALL PRESSURE.

A spring is the simplest accumulator. When compressed, a spring becomes a source of potential energy. It can also be used to absorb shocks or to control the force on a load. Hydraulic accumulators work in much the same way. Basically, they are containers which store fluid under pressure.

USES OF HYDRAULIC ACCUMULATORS

Hydraulic Accumulators have four major uses (Fig. 1):

- **Store Energy**
- **Absorb Shocks**
- **Build Pressure Gradually**
- **Maintain Constant Pressure**

While most accumulators can do many of these things, they are usually placed in a system for only one.

Accumulators used to STORE ENERGY are often used to store oil under pressure for systems with fixed displacement pumps and closed-center valves. The accumulator stores pressurized oil during non use periods and feeds it back into the system to supplement the pump during periods of oil usage.

They are often used as a protection against failure of the oil supply. Example: power brakes on larger machines. If the system oil supply fails, the accumulator traps enough oil for several brake applications, thus providing emergency braking.

Accumulators which ABSORB SHOCKS take in excess oil during peak pressures and let it out again after the "surge" is past.

When a shock load occurs in the hydraulic system, there is an extra volume of oil with no where to go. Without an accumulator, that volume must be handled by line expansion and leakage. This can cause extremely high pressure spikes. With an accumulator, that volume of oil simply enters the accumulator. It compresses the gas slightly, resulting in an insignificant system pressure rise.

Closed-center systems with variable displacement pumps are subject to spike pressures as the pump goes out of stroke. When a valve is closed, the pump creates some flow as the swashplate is moved (axial piston) or as the pistons are pushed out (radial piston). The accumulator accommodates that flow with a minimum of pressure rise.

The accumulator also reduces vibrations and noise in the system.

Accumulators which BUILD PRESSURE GRADUALLY are used to "soften" the working stroke of a piston against a fixed load, as in a hydraulic press or the engagement of a hydraulic clutch. *See spring accumulators for more information.*

Accumulators which MAINTAIN CONSTANT PRESSURE are always weight-loaded types which place a fixed force on the oil in a closed circuit. Whether the volume of oil changes from leakage or from heat expansion or contraction, this accumulator keeps the same gravity pressure on the system. This type of accumulator has no practical application in mobile hydraulics.

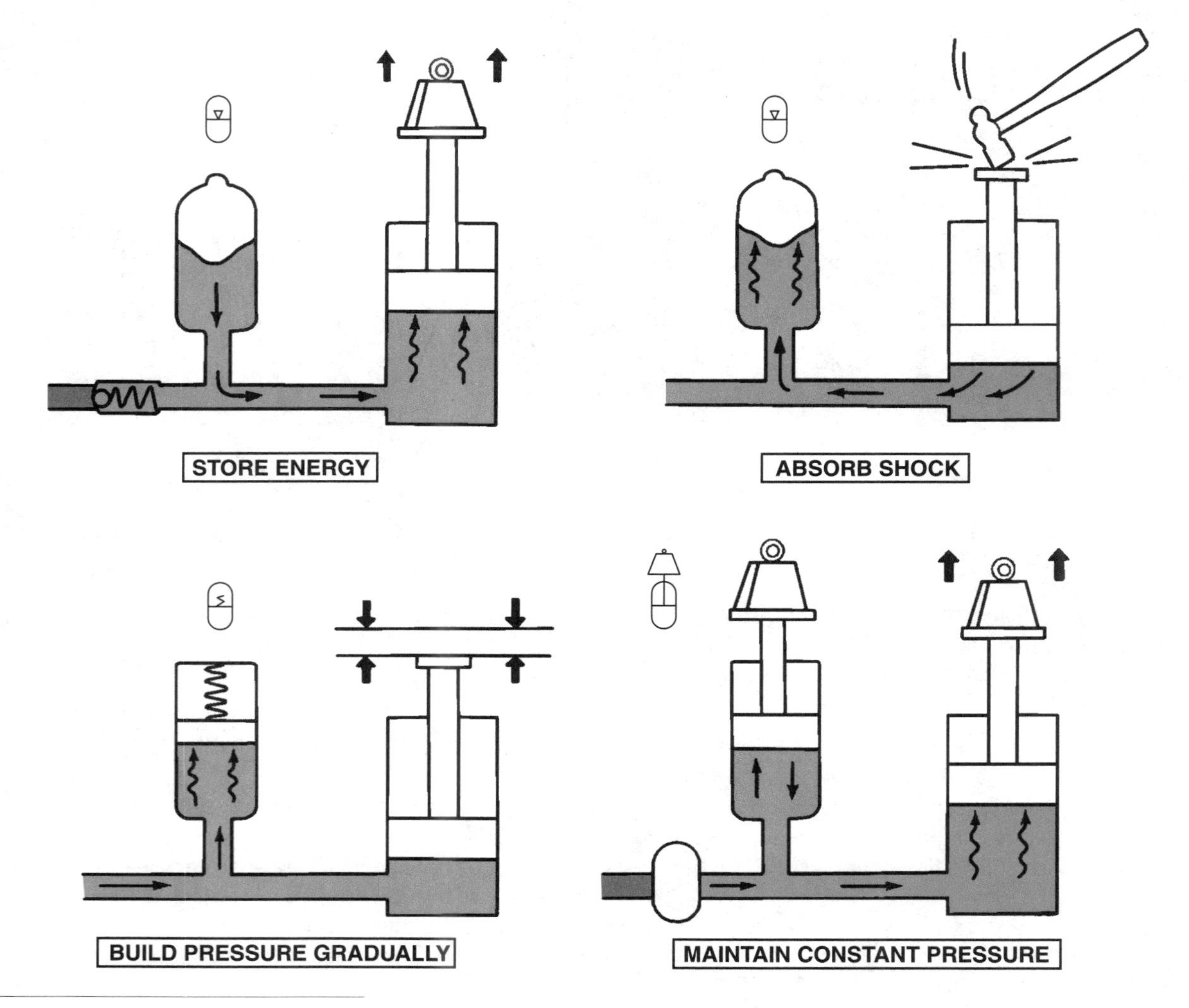

Fig. 1 — The four Uses of Accumulators

TYPES OF ACCUMULATORS

The major types of accumulators are:

- **Pneumatic (Gas-Loaded)**
- **Weight-Loaded**
- **Spring-Loaded**

PNEUMATIC ACCUMULATORS

We learned in Chapter 1 that fluids will not compress, but gases will. For this reason, many accumulators use inert gas as a way of "charging" a load of oil or of providing a "cushion" against shocks. An inert gas is a gas that will not explode.

"Pneumatic" means operated by compressed gas. In these accumulators, gas and oil occupy the same container. When the oil pressure rises, incoming oil compresses the gas. When oil pressure drops, the gas expands, forcing out oil.

In most cases, the gas is separated from the oil by a piston, a bladder, or a diaphragm. This prevents mixing of the gas and oil and keeps gas out of the hydraulic system.

Some accumulators for low-pressure or fairly static uses do not have a separator between the gas and oil, but their use is very limited in modern hydraulics. Also, there is no way to precharge these accumulators.

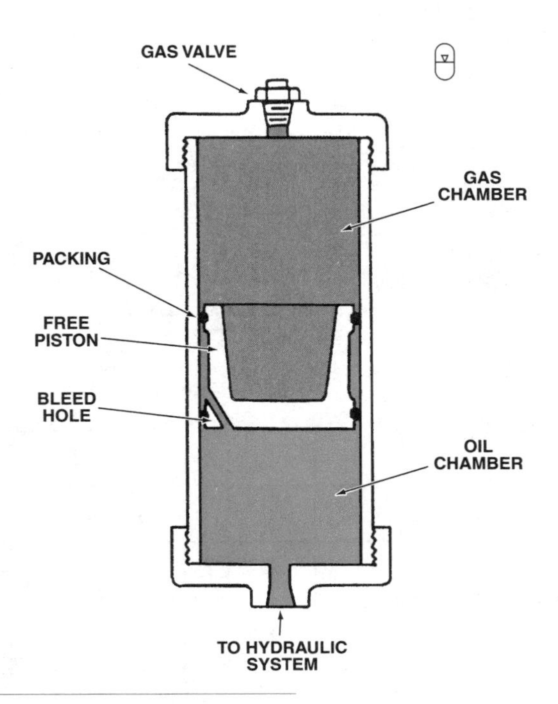

Fig. 2 — Piston-Type Accumulator

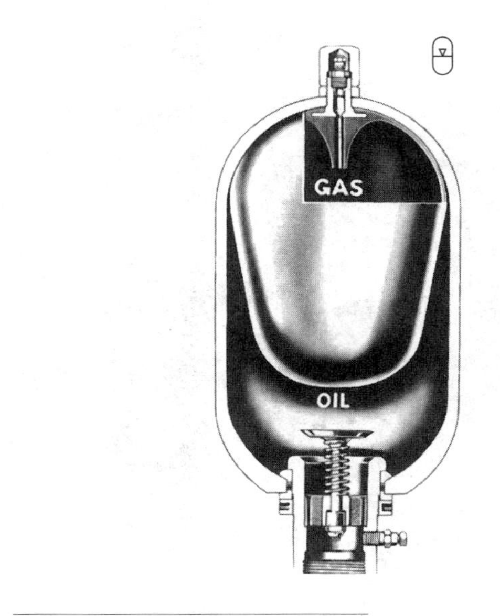

Fig. 3 — Bladder-Type Accumulator

A typical PISTON-TYPE ACCUMULATOR is shown in Fig. 2. It looks like a hydraulic cylinder minus the piston rod. A "free-floating" piston separates the gas from the oil.

The piston fits into a smooth bore and uses packings to separate the gas from the oil. It uses double packings to prevent cocking of the piston. A bleed hole is needed to make certain the top O-ring is lubricated.

The accumulator can be "precharged" with gas. This is done by filling the gas chamber to a desired pressure with an inert gas such as dry nitrogen.

Piston-type accumulators require careful service to prevent leakage. But they offer a high power output for their size and are very accurate in operation.

BLADDER-TYPE ACCUMULATORS are shown in Figs. 3 and 4. A flexible bag or bladder made of synthetic rubber contains the gas and separates it from the hydraulic oil. The bladder is molded to the gas charging stem located at the top of the accumulator.

Bladder-type accumulators can also be precharged.

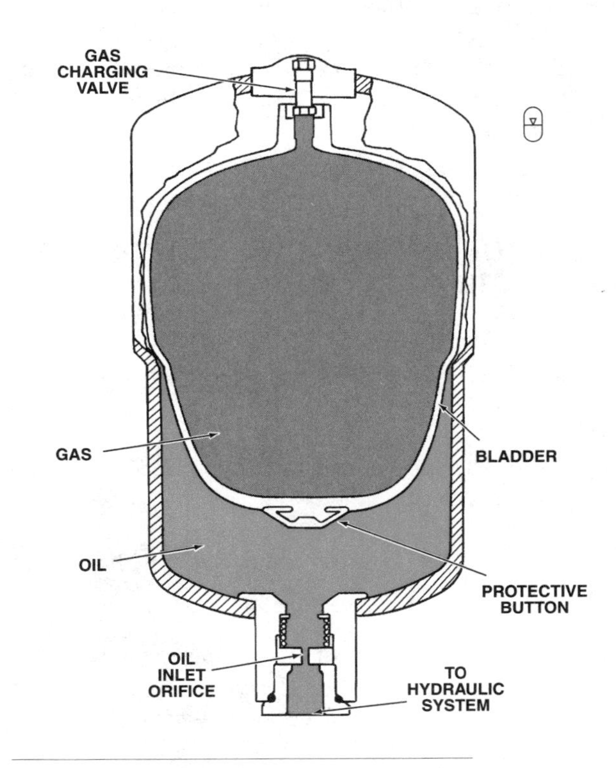

Fig. 4 — Bladder-Type Accumulator in Operation

When a bladder accumulator is precharged, protection is needed to prevent the bladder from being forced into the oil passage when the system pressure drops. To prevent that damage to the bladder, a valve is used in Fig. 3, a protective button in Fig. 4 and a screen in Fig. 5.

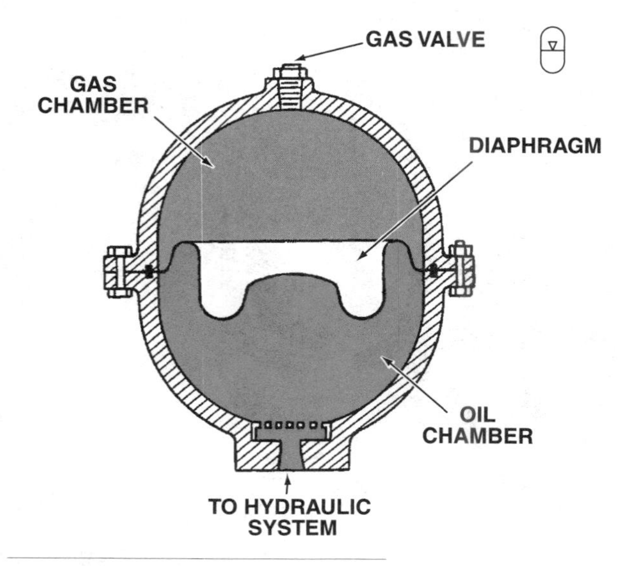

Fig. 5 — Diaphragm-Type Accumulators

DIAPHRAGM-TYPE ACCUMULATORS use a metal element to separate the gas from the oil. Molded to the element is a rubber diaphragm which flexes in response to pressure changes (Fig. 5). These accumulators are light in weight and are often used in aircraft systems.

EFFECTS OF DIFFERENT PRECHARGES ON PNEUMATIC ACCUMULATORS

The precharge is the gas pressure in the accumulator before any oil is admitted.

When oil enters the accumulator, the piston or bladder will move to maintain a gas pressure which is equal to the oil pressure.

When you precharge an accumulator with gas, how much pressure should you use?

It depends upon how you want the accumulator to work.

Fig. 6 shows how six different precharges effect the operation of an accumulator.

The gas precharges are shown with the red lines. They are 2000, 1000, 500, 300, 100 and O psi (13,790, 6890, 3450, 2070, 690 and 0 kPa) (see top line).

The accumulator has a 60 cubic inch (983 cubic centimeters) displacement for input oil (bottom line).

As system oil pressure rises (vertical scale at left), incoming oil compresses gas in the accumulator.

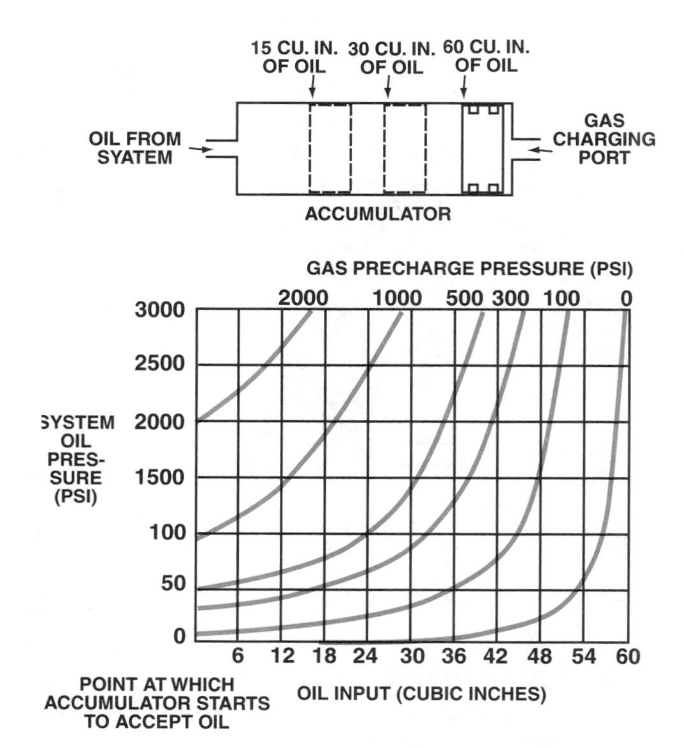

Fig. 6 — Effect of Different Precharges on a Pneumatic Accumulator

The accumulator will not accept oil until the hydraulic pressure exceeds the gas pressure. Therefore, the accumulator with the highest gas precharge (2000 psi)(13,790 kPa) starts to accept oil much later than the ones which have lower precharges.

The precharge effects how much oil the accumulator will accept at a given pressure.

The accumulator with little or no precharge will contain more oil at any given pressure. Its ability to furnish a quantity of high pressure oil to operate functions is very limited. Its limited ability to accept additional high pressure oil, makes it ineffective in dampening shock loads and pressure spikes.

Too high a preload can make the operating range of the accumulator too high for the hydraulic system.

In summary, the use of a pneumatic accumulator and its precharge depends upon the pressure and volume needs of the system. In other words, the operating pressures of the system, the volume of oil that is needed to supply functions, and the shocks loads to be dissipated, determine the size and precharge of the accumulator.

Precautions for Pneumatic Accumulators

Observe the following precautions when working on pneumatic accumulators. The correct procedures for service are given in detail later under "Servicing and Precharging Pneumatic Accumulators."

1. **⚠ CAUTION: NEVER FILL AN ACCUMULATOR WITH OXYGEN! An explosion could result if oil and oxygen mix under pressure.**
2. **⚠ CAUTION: NEVER FILL AN ACCUMULATOR WITH COMPRESSED AIR! An explosion could result if oil and oxygen mix under pressure.** When air is compressed, water vapor in the air condenses and can cause rust. This in turn may damage seals and ruin the accumulator.
3. **Always fill an accumulator with dry nitrogen.** This gas is inert and free of both water vapor and oxygen. This makes it harmless to parts and **it is safe to use.**
4. Never charge an accumulator to a pressure more than that recommended by the manufacturer. Read the label and observe the "working pressure."
5. Before removing an accumulator from a hydraulic system, release all hydraulic pressure.
6. Before you disassemble an accumulator, release both gas and hydraulic pressures.
7. When you disassemble an accumulator, make sure that dirt and abrasive material does not enter any of the openings.

SERVICING AND PRECHARGING PNEUMATIC ACCUMULATORS

CHECKING PRECHARGED ACCUMULATOR ON THE MACHINE

1. If you suspect external gas leaks, apply soapy water to the gas valve and seams on the tank at the "gas" end. If bubbles form, there is a leak.
2. If you suspect internal leaks, check for foaming oil in the system reservoir and/or no action of the accumulator. These signs usually mean a faulty bladder or piston seals inside the accumulator.
3. If the accumulator appears to be in good condition but is still slow or inactive, precharge it as necessary.

BEFORE REMOVING ACCUMULATOR FROM MACHINE

First be sure all hydraulic pressure is released. To do this, shut down the pump and cycle, a priority function in the accumulator hydraulic circuit, to relieve oil pressure (or open a bleed screw).

REMOVING ACCUMULATOR FROM MACHINE

After all hydraulic pressure has been released, remove the accumulator from the machine for service.

REPAIRING ACCUMULATOR

1. Before dismantling accumulator, release all gas pressure. Unscrew the gas valve very slowly. Install the charging valve first if necessary. Never release the gas by depressing the valve core, as the core might be ruptured.
2. Disassemble the accumulator on a clean bench area.
3. Check all parts for leaks or other damage.
4. Plug the openings with plastic plugs or clean towels as soon as parts are removed.
5. Check bladder or piston seals for damage and replace if necessary.
6. If gas valve cores are replaced, be sure to use the recommended types.
7. Carefully assemble the accumulator.

⚠ CAUTION: Incorrect charging procedures can be dangerous. Only charge the accumulator yourself if you have the know-how and equipment to do so safely. If in doubt, have it charged by a professional.

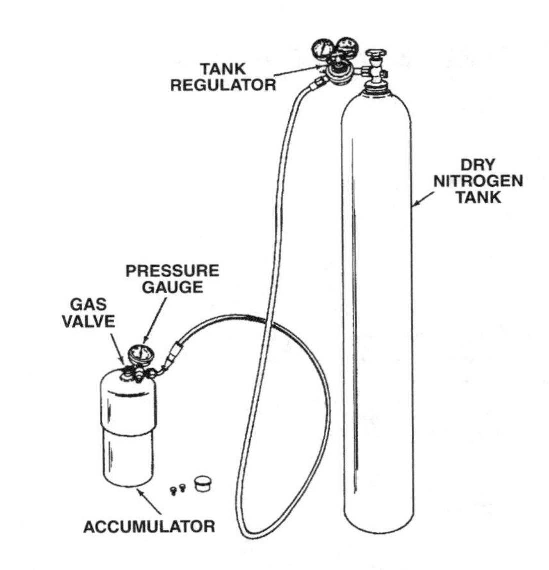

Fig. 7 — Precharging Accumulator (Pneumatic Type)

PRECHARGING ACCUMULATOR

Attach the hose from a Dry Nitrogen tank to the charging valve of the accumulator and open the accumulator charging valve (Fig. 7).

Open the valve on the regulator very slowly until pressure on the gauge is the same as that recommended by the manufacturer. Close the charging valve on the accumulator, then close the valve on the regulator. Remove the hose from the charging valve.

NOTE: When checking precharge on an accumulator installed on a machine, first release hydraulic pressure from the accumulator. Otherwise, you will be reading the hydraulic pressure rather than the gas precharge pressure.

INSTALLING ACCUMULATOR ON MACHINE

Attach accumulator to machine and connect all lines. Start machine and cycle a hydraulic function to bleed any air from the system. Then check the accumulator for proper action.

This accumulator uses a piston and cylinder, but weights on the piston do the job of loading or charging the oil. It is loaded by gravity.

Operation is very simple. The pressurized oil in the hydraulic circuit is pushed into the lower oil chamber. This raises the piston and weights. The accumulator is now charged, ready for work. It will provide oil and maintain a constant pressure in the system.

WEIGHT-LOADED ACCUMULATORS

The earliest form of accumulator is the weight-loaded type (Fig. 8).

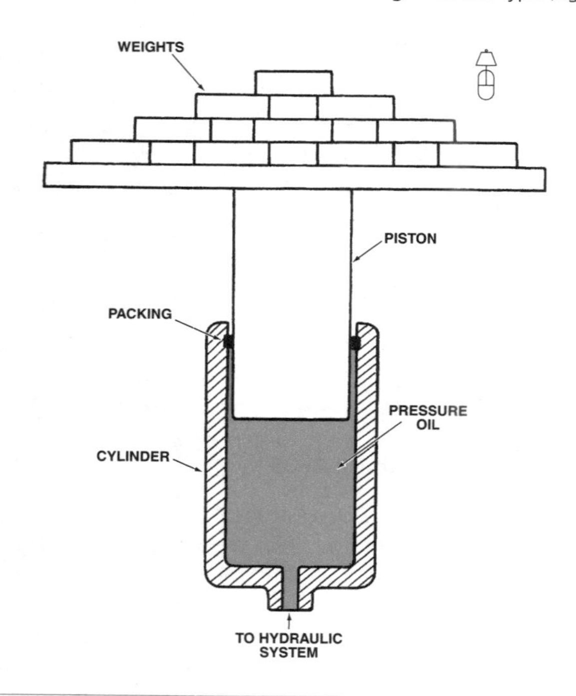

Fig. 8 — Weight-Loaded Accumulator

The advantage of the weight-loaded accumulator is that it can provide constant pressure.

The disadvantages is that they are the bulky and heavy and, therefore, have no applications in mobile machines.

SPRING-LOADED ACCUMULATORS

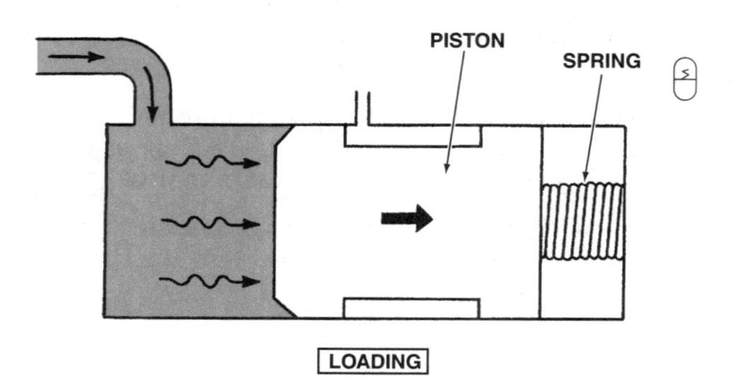

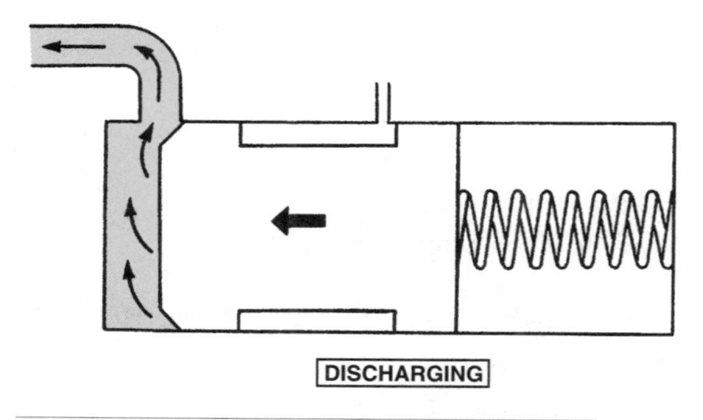

Fig. 9 — Spring-loaded Accumulator in Operation

This accumulator is very similar to the pneumatic-type except that springs do the loading.

In operation, pressurized oil loads the piston by compressing the spring (Fig. 9). When pressure drops, the spring forces oil into the system.

The accumulator shown in Fig. 9 is an internal type used to gradually build pressure in a hydraulic clutch of a transmission.

When the transmission is shifted, pressure drops and the accumulator discharges oil to fill the clutch piston. System oil is sent to the clutch and accumulator through an orifice. The clutch is already filled so oil entering through the orifice will charge the accumulator.

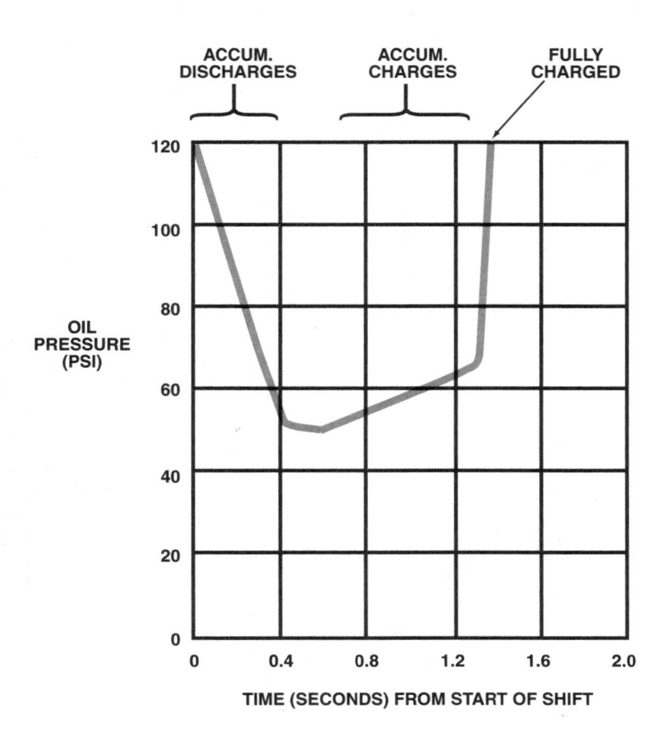

Fig. 10 — Operating Cycle of a Spring-Loaded Accumulator

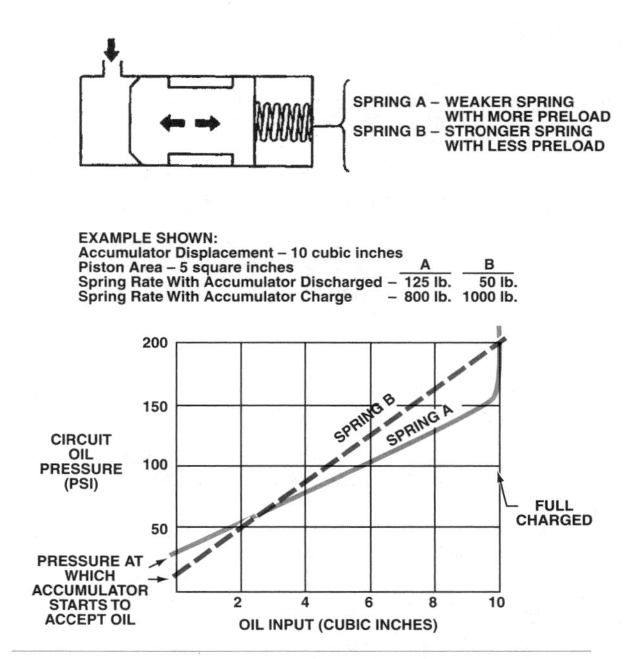

Fig. 11 — How Use of Two Different Springs Affects Accumulator Operation

Because the force of a spring increases as it compresses, the pressure in the system will gradually rise as the accumulator is filled. The time it takes to fill the accumulator (the time it takes to engage the clutch) is determined by the inlet orifice size.

The pressure builds gradually for a smooth engagement of the clutch.

In the graph shown in Fig. 10, note how pressure drops as the accumulator discharges, then builds slowly as it recharges, rising sharply to full pressure again when the accumulator reaches the end of its travel. The time lag during the charging cycle can be speeded up or slowed down by changing the orifice size which feeds the accumulator.

The operation of spring-loaded accumulators can be varied by changing: 1) the strength of the spring, 2) the length of the spring, 3) the preload on the spring, 4) the size of the piston or, 5) the length of the piston stroke.

Fig. 11 shows the effect of using a stronger spring with less preload (B) as opposed to a weaker spring (A).

Now we can see how important the correct spring is in the operation of these accumulators.

Be sure to use genuine parts and to follow the manufacturer's recommendations when repairing accumulators.

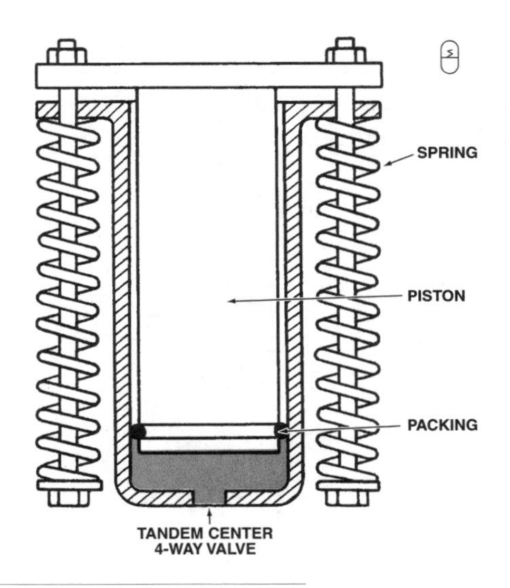

Fig. 12 — Spring-Loaded Accumulator

In another design shown in Fig. 12, the piston is retained by several spring-loaded rods. As oil enters the piston, pressure raises as the springs are compressed.

The springs are preset by tightening the adjusting nuts.

The advantage of spring-loaded accumulators is that they never have to be precharged or recharged.

The disadvantage is that these accumulators are too bulky when designed for high volume or high pressure systems. Therefore, they are practical only for low volume or low pressure uses.

TEST YOURSELF

QUESTIONS

1. (Fill in the blank) "The operating characteristics of a specific pneumatic accumulator is determined by its _______________."

2. Why is air not recommended for use in a pneumatic accumulator?

3. What should you do first before removing an accumulator from a machine?

4. What are the four major uses of accumulators?

HYDRAULIC FILTERS

WHY ARE FILTERS USED?

Did you ever stop to think that hydraulic fluids are lubricants for precision parts as well as a means of transmitting power?

Contaminated oil can score or completely freeze a precisely fitted valve spool. Dirty oil can ruin the close tolerance of finely finished surfaces, and a grain of sand in a tiny control orifice can put a whole machine out of operation. It's not hard to see that you have to keep the oil clean if you want a hydraulic system to operate without trouble.

Fig. 1 — Dust is a Major Source of Contamination

It's a constant battle because dirt is everywhere. Look at Fig. 1 and you see that the air surrounding a machine is a major source of contamination.

Another source of contaminants is the machine itself. As it works and wears normally, the machine generates burrs and chips of metal.

Measured in dollars and cents, it's a whole lot cheaper to buy a good filter, to maintain it properly, and to keep your oil clean than it is to replace a pump or a valve that is worn by contamination.

HOW FILTERS ARE USED

A FULL-FLOW system filters the entire supply of oil as it circulates in the hydraulic system. Filters in a full-flow system are usually located in the return line from the hydraulic functions. Additional filters, of course, may be located in front of or behind other hydraulic components if they are needed.

In contrast, a BYPASS filter system has its filter teed into a line so that only a small portion of the oil is diverted through the filter. The remainder of the oil goes unfiltered to the system or to the reservoir.

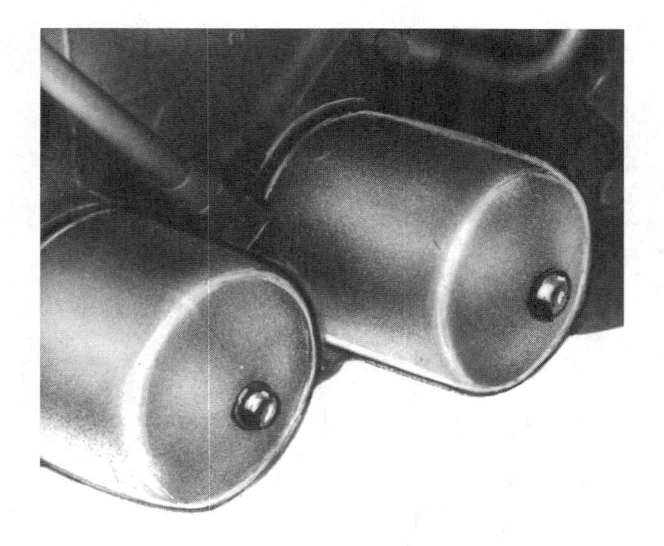

Fig. 2 — Tractor Mounted Hydraulic Filter

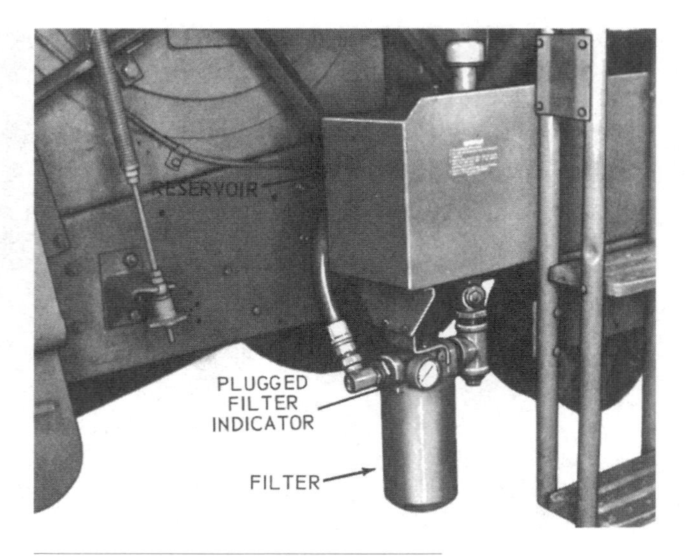

Fig. 4 — Remote Mounted Hydraulic Filter

The location of the filter in a hydraulic system will vary with the design of the machine. Fig. 2 shows a filter that is an integral part of the machine. Fig. 3 shows one mounted inside the reservoir and Fig. 4 shows a filter hooked into an outside line. Regardless of location, the one purpose of the filter is to keep the oil clean.

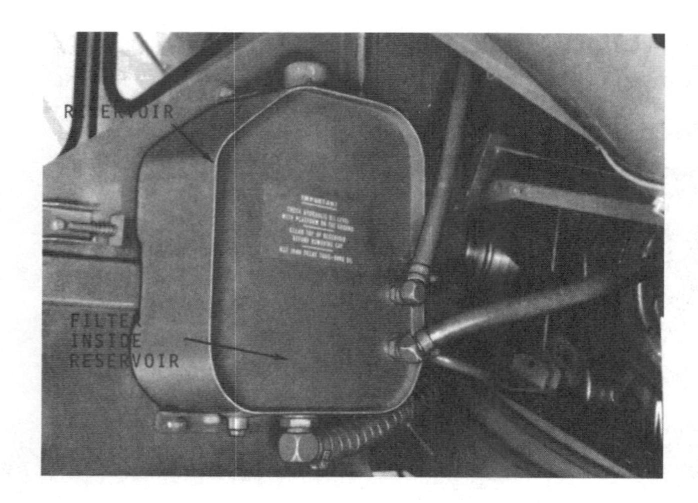

Fig. 3 — Reservoir Mounted Hydraulic Filter

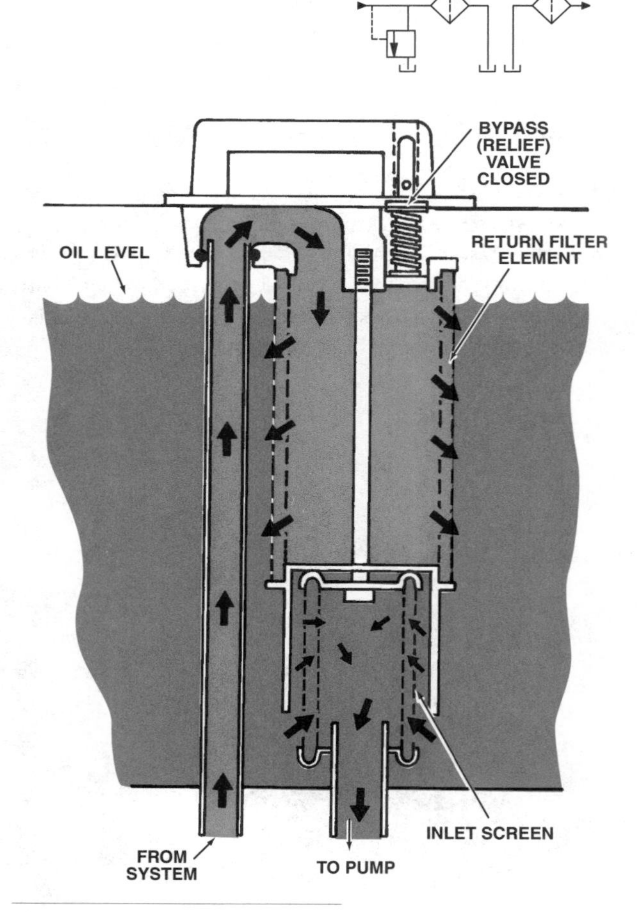

Fig. 5 — Full-Flow Hydraulic Filter

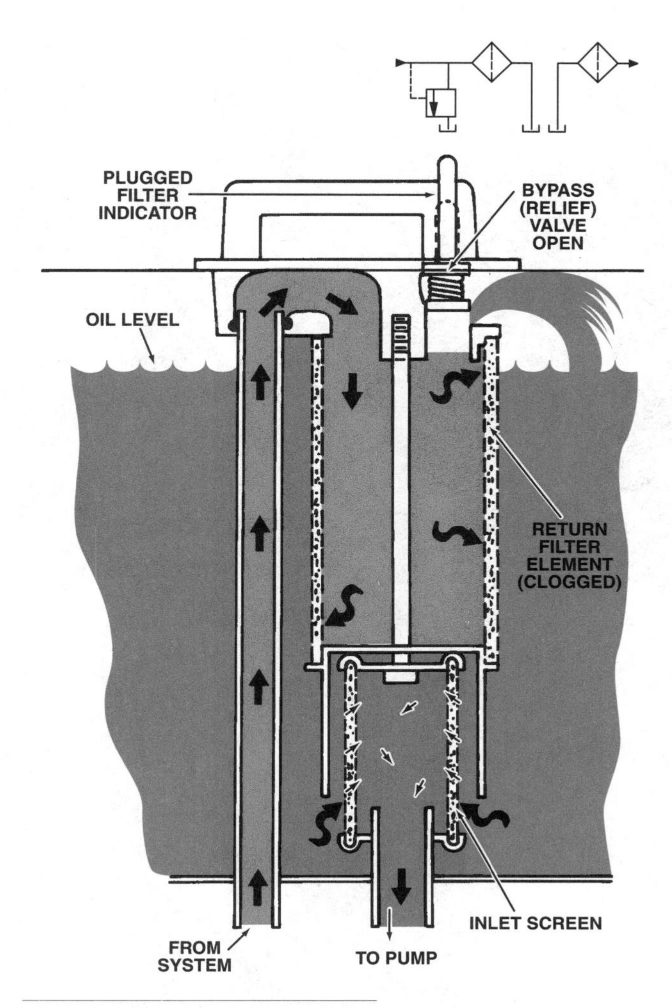

Fig. 6 — Hydraulic Filter Relief Valve

Filtration occurs as oil passes through the filter. Fig. 5 shows a full-flow system with a pump inlet screen and a return oil filter arranged in one package.

Oil from the reservoir goes through the inlet (suction) screen and flows to the pump. Because of the possibility of cavitation, it is a screen and not a filter. It is there to take out the largest particles that could cause catastrophic type failures to the pump and the system. This makes servicing of the filter important. Fine particles, which bypass the filter, can reenter the hydraulic system.

Oil returning from the control valve and cylinders passes through the return filter to the reservoir.

As oil passes through even a new filter, there is a slight pressure drop across the filter. As the filter gets dirtier, the pressure drop will increase. When the filter becomes completely plugged, no oil will flow.

As the pressure drop increases, the return pressure will rise. Pressure could build so high that it might break the filter, releasing all of its contamination to the system. It could also cause excessive backpressure on the hydraulic components.

This may also happen when very cold oil resists passing through the filter.

Because slightly dirty oil is better than no oil at all or than a ruptured filter, a bypass (relief) valve is usually incorporated into a filter assembly.

Fig. 6 shows such a bypass (relief) valve in operation. When the filter fills with particles, it will cause a restriction to returning oil. This builds a pressure that will act against the relief valve spring. When the restriction is enough to open the valve, unfiltered return oil will bypass the filter and flow into the reservoir (Don't confuse this "filter bypass" with the "bypass filter system" discussed earlier.).

This dirty oil entering the reservoir will then reenter the hydraulic system or could eventually plug the inlet screen. This makes the immediate servicing of plugged filters essential.

Indicators for plugged filters are shown in Figs. 4, 5, 6 and 7. They make it easy to see when filters require servicing.

The indicator shown in Fig. 4 is simply a pressure gauge that measures the pressure of the returning oil. This return pressure is an indication of the filter restriction. With a gauge, it is easy to tell that the filter is restricted well before it starts to bypass.

The indicator shown in Figs. 5 and 6 shows the position of the relief valve. The indicator is attached to the valve. Note that the valve must travel a distance from the fully closed position in Fig. 5 to the open position in Fig. 6. Because it takes less force to start compressing the spring than it does to fully compress it, the indicator will gradually protrude, as the filter becomes restricted. This gives ample warning of a plugged filter before it bypasses.

Fig 7 shows a filter warning light. The spring-loaded piston is pushed back as the filter restriction increases. When it moves far enough, it makes contact with an electrical terminal. This contact completes a ground circuit for a circuit board and/or warning light. The ground contact is made with less pressure than it takes to bypass the filter. This means the light will come on well before the filter bypasses, providing ample warning.

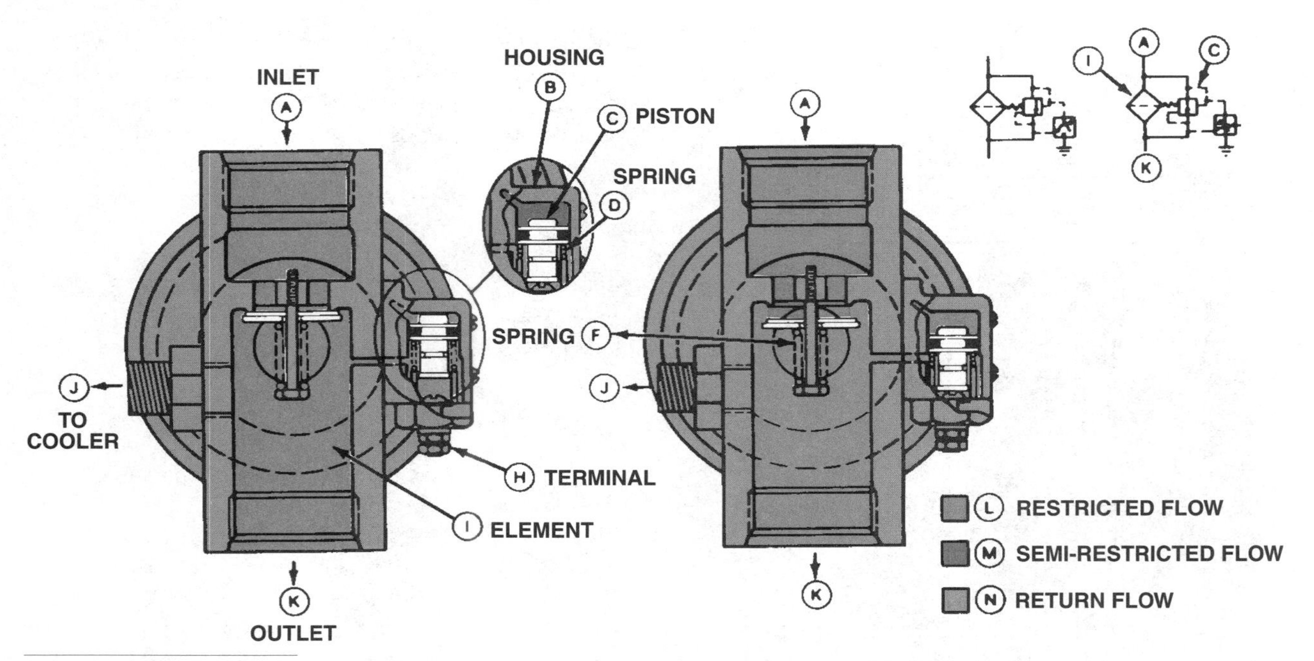

Fig. 7 — Filter Warning Light

The filter rating and weight of oil affect the resistance to flow through a filter. It is therefore important that the right filter and correct oil be used.

TYPES OF FILTERS

Now let's look at the type of filters used in a hydraulic system and just how much filtering they actually do.

Filters can be classified as either surface-type filters or depth-type filters depending on the way they remove dirt from hydraulic oil.

Fig. 9 — Wire Mesh Filter

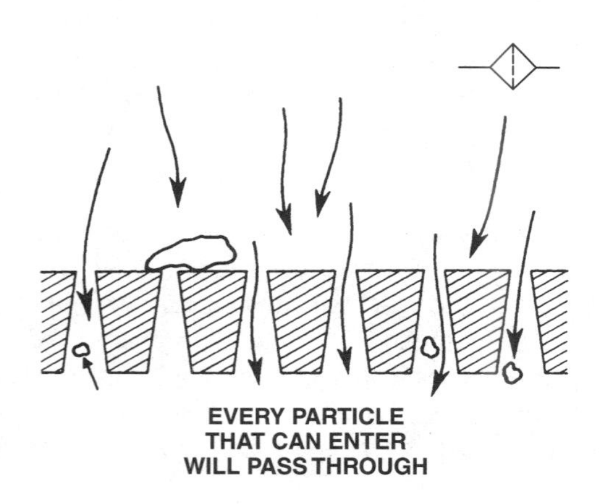

Fig. 8 — Edge-Type Filter

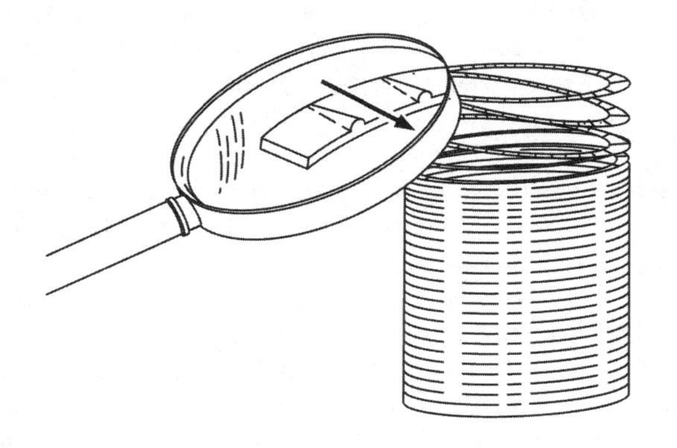

Fig. 10 — Metal Edge Filter

SURFACE FILTERS have a single surface that catches and removes dirt particles larger than the holes in the filter. Dirt is strained or sheared from the oil and stopped outside the filter as oil passes through the holes in a straight path (Fig. 8). Many of the large particles will fall to the bottom of the reservoir or filter container, but eventually enough particles will wedge in the holes of the filter to cause restriction or plugging. The filter must then be cleaned or replaced.

A surface filter may be made of fine wire mesh (Fig. 9). These filters are normally used as inlet (suction) screens.

The metal edge filter shown in Fig 10 consists of a metal ribbon wound edgewise to form a cylinder. This filter is very expensive but is cleanable and can be used in the high-pressure area in a system. Stacking metal or paper disks can also make edge filters.

DEPTH FILTERS, in contrast to the surface type, use a thicker filtering area. The filter material makes the oil move in many different directions before getting through. Material is trapped in the fibers as shown in Fig. 25

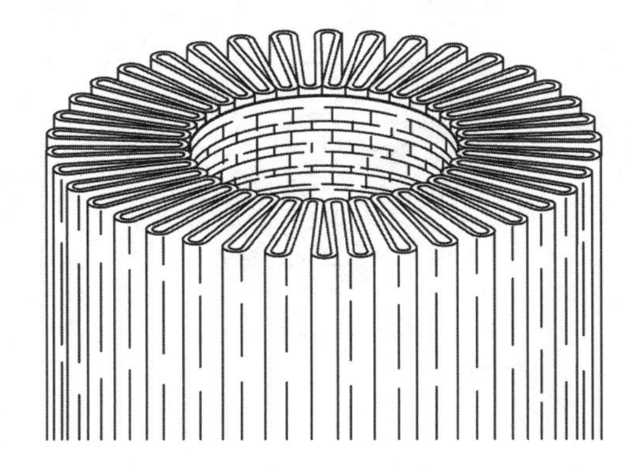

Fig. 11 — Pleated Paper Filter

The pleated paper filter shown in Fig. 11 is a depth-type filter. It is the most commonly used filter in hydraulic systems.

Depth filters can be classified as either **absorbent** or **adsorbent**, depending on the way they remove dirt.

Absorbent filters operate mechanically, by trapping the contamination in its porous material. That material can be cotton waste, wood pulp, wool yarn, paper, quartz or the current most popular, synthetic fiberglass.

Adsorbent filters operate the same way as absorbent filters but also are chemically treated to attract and remove contaminants. This filter may be made of charcoal, chemically treated paper or fuller's earth. It will remove contaminating particles, water-soluble impurities and, because of its chemical treatment, will also remove contamination caused by oil oxidation and deterioration. Adsorbent filters may also remove desirable additives from the oil and for this reason must be used only where specified.

FILTER RATING

In addition to the type, the filter rating is also important to a hydraulic system. It tells us the degree of filtration that the filter will provide. They are rated primarily by the size of particle they will pass.

The measurement used to determine the degree of filtration is a micron. A micron is 0.001 millimeters (approximately 0.00004 inch) (25,000 microns equal 1 inch)(10000 microns equal a centimeter).

There are two ways to rate a filter:

Micron Rating is a measure of the opening (hole) size of the filter. It would theoretically be an indication of the size particle the filter would pass.

Beta Rating is the measure of the tested ability of the filter to keep out certain sized particles. The tests are done in accordance with procedures established by the International Standards Organization (ISO).

Test ISO 4572 The Beta Rating (or beta ratio) is defined as the number of particles greater than a given size in the fluid upstream of the filter divided by the number of particles of the same size found downstream of the filter. Particle size is measured in micrometers. The chart in Fig. 13 shows the efficiency levels for various beta values.

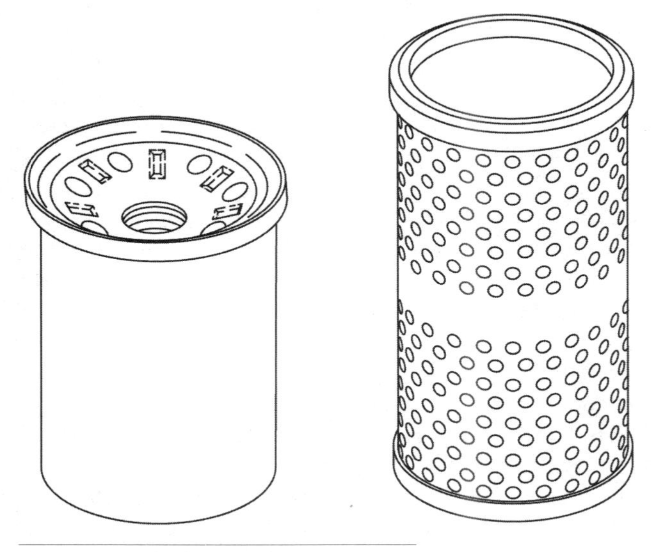

Fig. 12 — Screw-on and Cartridge Filters

Beta values and corresponding efficiency levels

ßx	=	1.0	=	0.0%	Efficient
ßx	=	1.2	=	16.67%	Efficient
ßx	=	1.5	=	33.33%	Efficient
ßx	=	2.0	=	50.00%	Efficient
ßx	=	3.0	=	66.67%	Efficient
ßx	=	4.0	=	75.00%	Efficient
ßx	=	16.0	=	93.75%	Efficient
ßx	=	20.0	=	95.00%	Efficient
ßx	=	50.0	=	98.00%	Efficient
ßx	=	75.0	=	98.67%	Efficient

Fig. 13 — Beta Rating vs. Efficiency

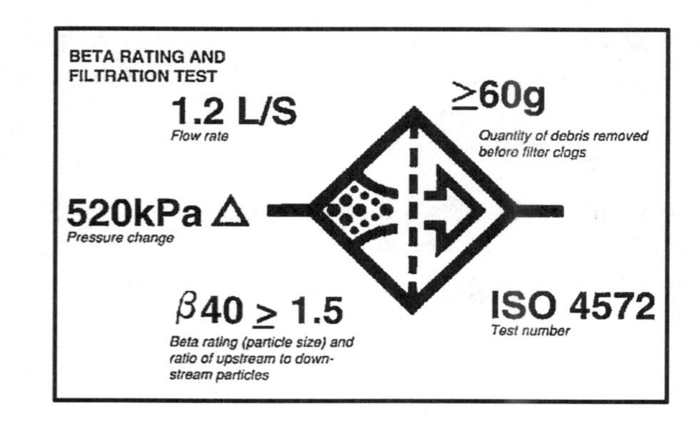

Fig. 14 — Beta Rating Decal

Decals showing the rating and other information are displayed on the filters or container (Fig. 12). They are the decals specified by ISO to be used on filters that meet the specifications of the ISO tests.

The arrow within the diamond (Fig. 14) indicates dirty fluid on the upstream side (left) passing through the filter (dotted line) and exiting on the downstream (right) side.

In this example, the figures at the lower left indicate that test results show the filter is trapping particles 40 micrometers or larger in size with 1.5 Beta rating or 33-1/3% efficiency. Example: For every 150 particles upstream, only 100 particles will remain downstream.

The flow, at upper left is the flow rate at which the test was run. It is 1.2 liters per second (19 gallons per minute).

The filter will collect 60 grams (2.1 ounces) of debris before it becomes clogged (upper right).

520 kPa (75.4 psi) is the drop in pressure which will occur across the filter in the clogged condition (left center).

While filters are rated by the size of dirt they can filter, there are many other tests, which go into the selection of filters for a specific application. They also test a used filter to test and rate the quality of a filter.

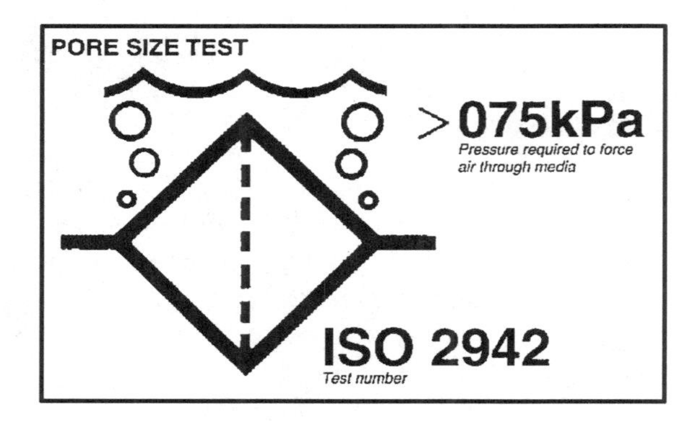

Fig. 15 — Pore Size Test Decal

Test ISO 2942 is the Pore Size Test that measures the amount of pressure required to force air through the media (Fig. 15). That determines the pore size ... the greater the pressure, the smaller the pores.

The bubbles and "water line" above the diamond symbolize the test. The test consists of submerging the filter in a fluid and forcing air through the media. The test results indicate more than 075 kPa (11 psi) were required.

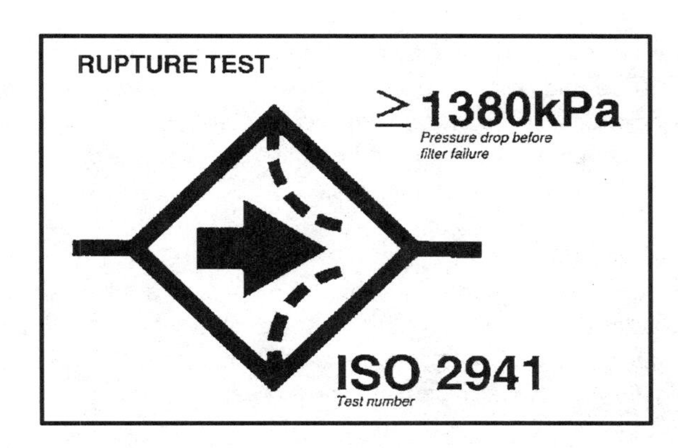

Fig. 16 — Rupture Test Decal

Test ISO 2941 Rupture Test is performed to determine the minimum pressure at which a clogged filter will rupture and contaminate the system (Fig. 16). The rupture test symbol is a diamond containing an arrow and a broken line.

The test results indicate 1380 kPa (200 psi) or more was the minimum pressure (or pressure drop) it incurred before rupture occurred.

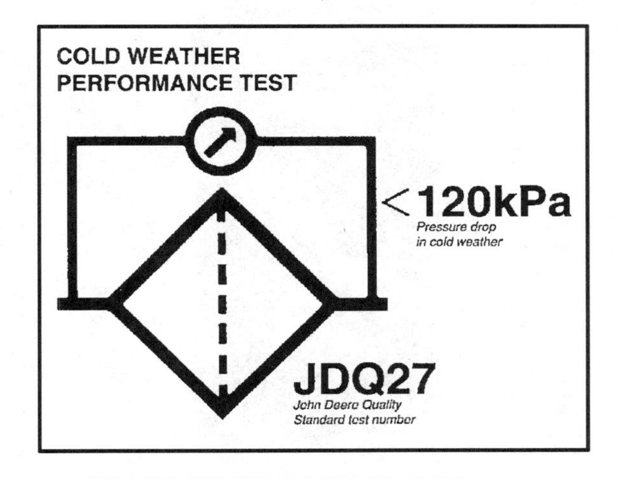

Fig. 17 — Cold Weather Performance Test Decal

ISO has established the decal but has not yet published the specifications for the **Cold Weather Performance Test** (Fig. 17). Some filter and machine manufacturers have established their own tests.

It is a test for cold weather performance. The test is performed by running a thick fluid through a clean element and measuring the drop in pressure on the upstream side.

The test results (Fig. 17) indicate that there will be less than 120 kPa (17.4 psi) pressure drop across the filter during cold weather operation.

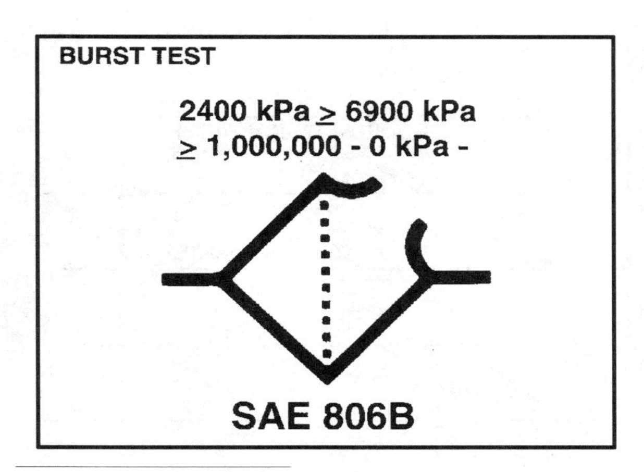

Fig. 18 — Burst Test Decal

SAE Test 806 Burst Test (specifications not yet published by ISO) is a test for external leakage of the filter. Two separate tests are performed to meet the requirements.

The Pressure Test - the filter needs to be pressurized to 6900 kPa(1000 psi) without leaking.

The Fatigue Test - Pressure is run from 0 to 2400 kPa (350 psi). Filter must not leak after 1,000,000 cycles.

Test ISO 4572 is an **Initial Restriction Test**. It measures the oil pressure loss on a new clean element. If the initial oil restriction is great enough, the filter may not be very effective until the oil warms up.

Test Procedure: A new clean element is checked with low viscosity oil (hot oil conditions) and checked with high viscosity oil (cold oil conditions). Pressure loss from filter inlet to outlet is measured.

Test ISO 2943 Material Compatibility Test checks a filter's material compatibility to survive fluid application.

Test Procedure: First, the element is conditioned for 72 hours in oil at a temperature 15°C above maximum fluid operating temperature. The filter is then pressurized in a bursts/collapse test to determine if its material is compatible with the oil.

Test ISO 3724 Flow Fatigue Test measures a filter's ability to withstand differential oil pressures created by variable flow rates.

As you can see there is a lot of things considered when selecting a filter for an application. ***It is important, therefore, that the manufacturer's recommendations be followed when replacing a filter.***

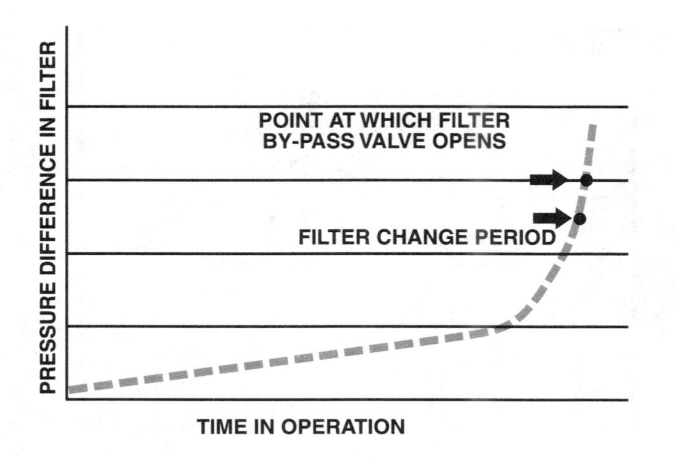

Fig. 19 — Life of a Filter Element

As a filter is used, the particles accumulate in the filter. This gradually reduces the size of the filter pores. The pressure difference between the inside and outside of the filter (pressure drop) rises as the filter pores become clogged.

Fig. 19 shows the rise in pressure as the filter becomes clogged over a period of time. Note how sharply the pressure rises near the end of filter life. At this point, the filter stops being effective and should be cleaned or replaced.

CONTAMINATION

What is contamination and how does it get into a hydraulic system?

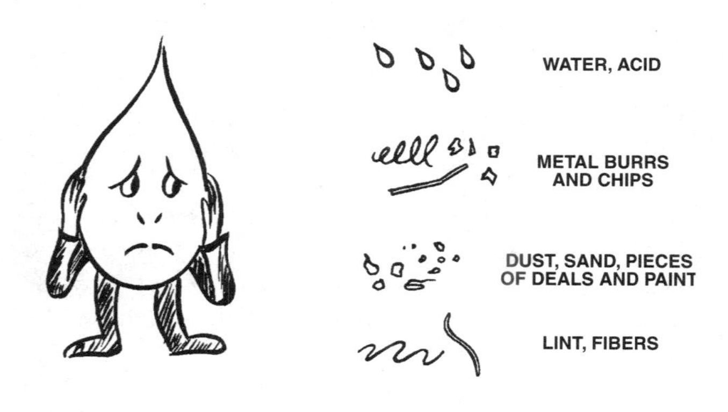

Fig. 20 — Contamination

Liquids, metallic particles, non-metallic particles and fibers are all forms of material that can contaminate the oil (Fig. 20). This material can come from both inside and outside the hydraulic system.

The air is a prime source of contamination. It may contain moisture and particles from the atmosphere, as well as road or field dust. These contaminants can enter through breathers and filler pipes, past seals and gaskets, or when the system is opened for repair or maintenance.

The machine itself is a significant source of contamination. During break-in, particularly, bits of metal and other abrasive particles will contaminate the oil. These particles will generate more contamination as fragments of paint, pieces of seals and gaskets, and metallic particles cause wear.

This is why manufactures will use a special short time break-in filter or recommend that the filter be changed after a short break-in period.

Hydraulic oil can be contaminated during maintenance and service if unclean containers, dirty oil, or dirty and linty wiping cloths are used.

IMPORTANT: Always cap or plug open lines or connectors to reduce the possibility of contamination.

Fig. 21 — Mishandled Filter

Mishandling of filters can cause damage such as bending the container to restrict flow, tearing the element or breaking the element from the end piece to by-pass oil, etc.

The oil itself is another source of contamination. As oil works in a system, sludge and acids form because of a chemical reaction to water, air, heat and pressure.

All of these contaminants can have a serious effect on the efficiency of the hydraulic system.

Fig. 22 — Completely Plugged Filter

Water, even in small amounts, rusts polished metal surfaces, causes wear and plugs filters. It also combines with other contaminants in the oil to form sludge and acids.

Sludge coats moving parts and plugs orifices and filters (Fig. 22). Plugged filters can cause increased circulation of dirty oil when bypassed or may even starve the hydraulic pump.

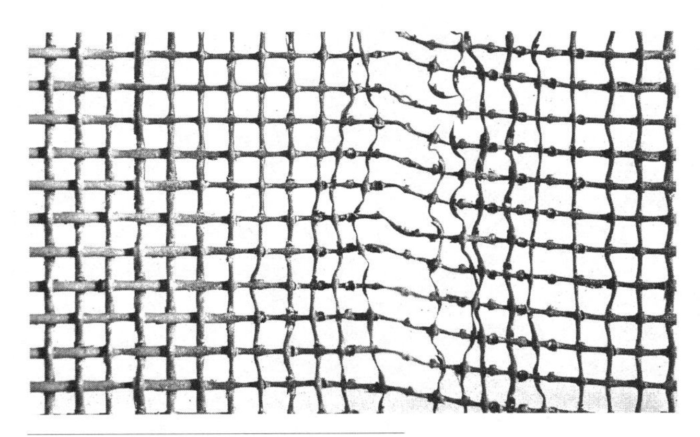

Fig. 23 — Chemically Corroded Filter

Acid in the oil pits metal surfaces causing them to be rough and create more contamination. It can also deteriorate part as shown in Fig. 23.

Sludge and acids are not normally filtered out so it is important that oil be changed when this type of contamination occurs. It is also important that the oil be changed in accordance with the machine operator's manual.

Metallic and n/on-metallic particles circulating in the oil cause damage that is usually quite apparent. Large particles get caught on the edges of moving parts and wear off sharp metering edges in valves (Fig. 24).

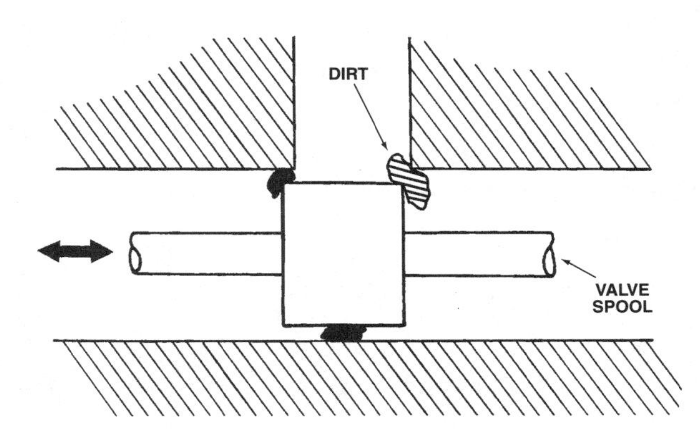

Fig. 24 — Dirt will Wear Valve Spool

Smaller particles become trapped between closely fitted parts, causing them to stick or freeze completely. Other particles embed themselves in softer metals to grind away moving parts and seals, causing internal leakage and loss of efficiency.

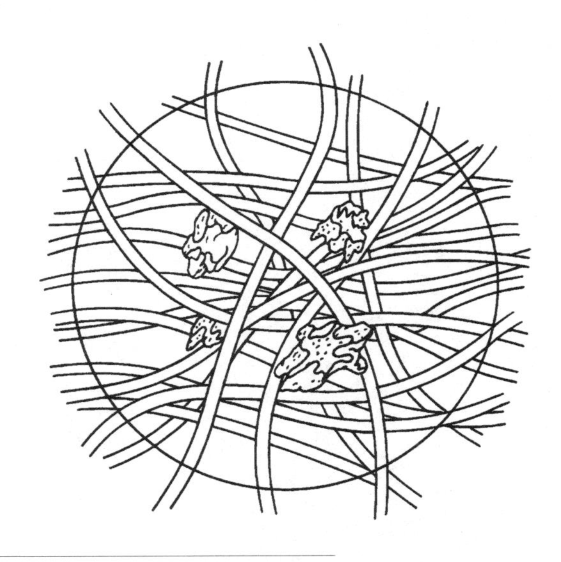

Fig. 25 — Fibers Can Plug an Orifice

Tiny fibers or lint from wiping cloths or even from your clothes may mat together in tiny pores and orifices (Fig. 25). The fibers themselves do little harm but the solid particles that they trap can plug and wear hydraulic components.

The important thing to remember about contamination is that the damage it causes is not constant. Each dirt particle is an abrasive "seed" that produces more contamination, which finally results in permanent damage to the machine.

TEST YOURSELF

QUESTIONS

1. Compare full-flow and bypass filtering in a hydraulic system.

2. What does a filter relief valve do?

3. What are the two methods of rating a filter?

4. Of the two rating methods, which one is based on performance?

5. Name at least four performance tests that are done to qualify filter quality.

RESERVOIRS AND OIL COOLERS

INTRODUCTION

This chapter covers the parts of the hydraulic system that store and cool the oil. While less complex than the other parts of the system, these components are vital to the operation of the system.

RESERVOIRS

Every hydraulic system must have a reservoir. The reservoir not only stores the oil, it also helps keep the oil clean, free of air, and cool.

CAPACITY OF RESERVOIRS

A reservoir should be compact, yet large enough to:

- **Hold all the oil and not overflow when all the cylinders in the system are retracted. If it is a sealed reservoir, additional air space is needed.**
- **Maintain the oil level enough above the suction line so that air is not drawn into the pump intake when all the cylinders are extended.**
- **Dissipate heat during normal operation (See "Oil Coolers," also in this chapter).**
- **Allow air and foreign matter to separate from the oil.**

FEATURES OF RESERVOIRS

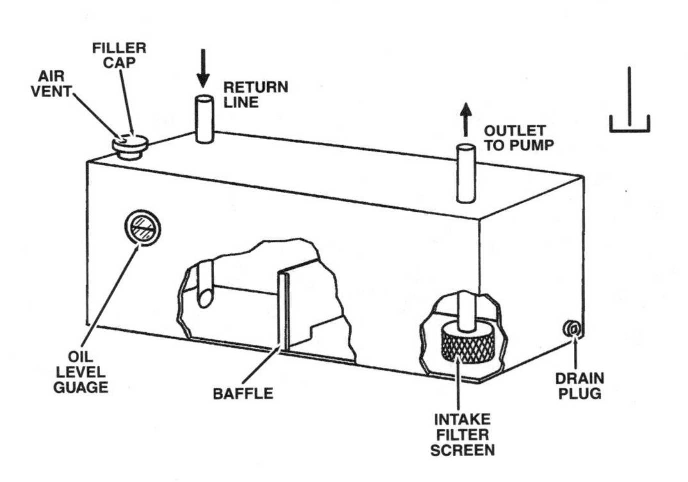

Fig. 1 — Reservoir

To serve its purpose, the reservoir must have several features (Fig.1).

1. The FILLER CAP should be air tight when closed, but may contain an AIR VENT which filters air entering the reservoir. The air vent filter must be kept clean to prevent oil contamination. Most modern systems, however, are designed with a sealed reservoir which also requires a sealed filler cap.

   The sealed reservoir normally uses a radiator cap. This cap will allow air to exit the reservoir when the pressure reaches the pressure value of the cap. It also has a low-pressure intake valve to allow air into the reservoir when a vacuum occurs.

In normal operation, the reservoir will exhaust a small amount of air as it is used and warmed up in the morning. During the day, the air pressure will simply raise and lower slightly as the oil level fluctuates. Contaminated air is not entering the system as the machine is working.

When the system cools at day's end, a small amount of relatively clean air will be drawn into the reservoir.

Some sealed reservoirs use a pipe plug or cap which do not allow breathing. As the system warms and as the cylinders are activated, the air simply compresses.

2. The OIL LEVEL GAUGE shown in Fig. 1 gives the level of oil in the reservoir without opening it. Dipsticks are still widely used, however.
3. The BAFFLE helps to separate return oil from directly entering the pump inlet. It slows the circulation of return oil giving it time to settle. In many modern systems, the effect of the baffle is achieved by placement of the lines and filters.
4. The OUTLET and RETURN LINES are designed to enter the reservoir so they create the least amount of turbulence. If the return line is above the oil level, turbulence and aeration will result.

   NOTE: Be careful when placing extra returns from auxiliary equipment in the reservoir. If not placed correctly, they can cause foaming of return oil.
5. The INTAKE FILTER is usually a screen. It needs to be well below the oil level at all times to prevent air from entering the system. The system return oil filter may also be installed in the reservoir
6. The DRAIN PLUG allows all oil to be drained from the reservoir. Some drain plugs are magnetic to hold metallic particles and prevent their circulation in the system.

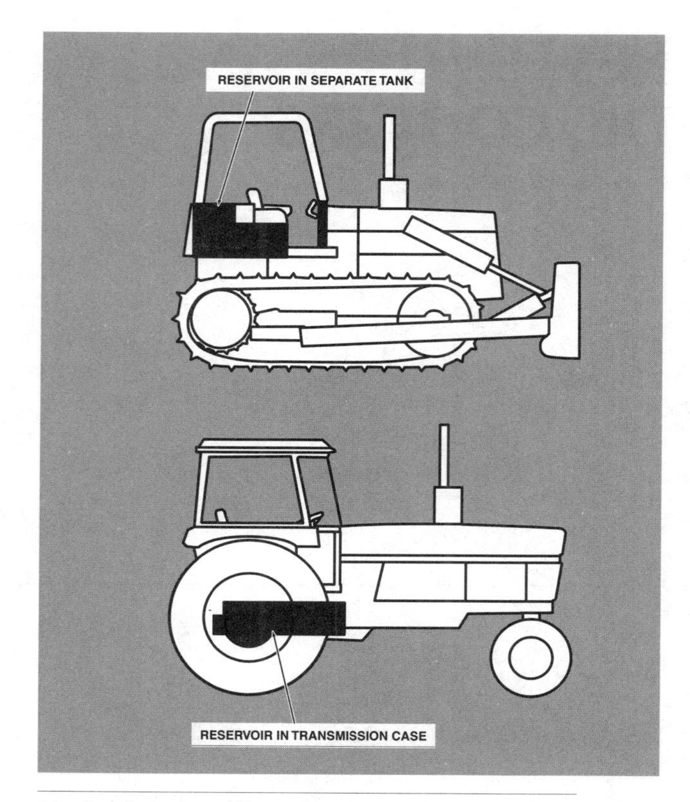

Fig. 2 — Location of Reservoirs on Two Types of Tractors

On modern farm and industrial machines, the reservoir must be compact and light. The bulldozer shown in Fig. 2 has a separate tank while the wheel tractor uses its transmission case as a reservoir. The location of the reservoir depends upon the design of the machine, the available space and the size of the reservoir needed.

OIL COOLERS

The hydraulic system creates a lot of heat. It is cooled by radiation at the valves, cylinders, lines and the reservoir.

On some of the modern hydraulic systems, additional cooling is required. This is why oil coolers are becoming more common on modern equipment.

Two types of oil coolers are used:

- **Air-to-Oil Coolers**
- **Water-to-Oil Coolers**

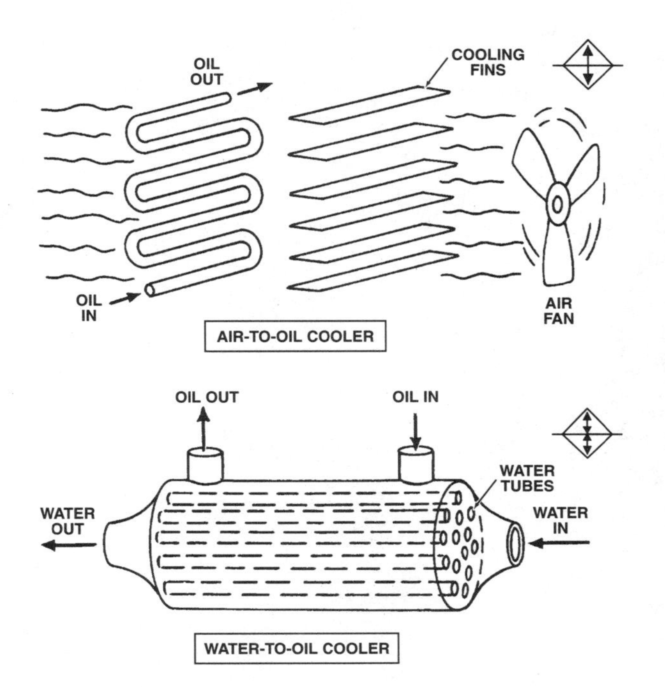

Fig. 3 — Oil Coolers

Fig. 3 compares the operation of the two types of coolers.

AIR-TO-OIL COOLERS use moving air to dissipate heat from the oil. On mobile machines, the engine cooling system (radiator) fan normally supplies the air blast (see Fig. 4).

Coolers can be mounted in front of the radiator "front mounted", as shown. They can also be "side by side" where a separate radiator is mounted next to the engine radiator.

The cooler has fins that direct the air over long coils of oil tubes, which expose more oil to the air. The cooler may also have a top and bottom tank to store oil.

Excessive pressures can damage most coolers. It is, therefore, necessary to provide relief or bypass protection in case of a plugged cooler or when operating with cold oil. That protection is often located in valving at the source of the cooler oil. On others, a bypass valve will be located at the cooler.

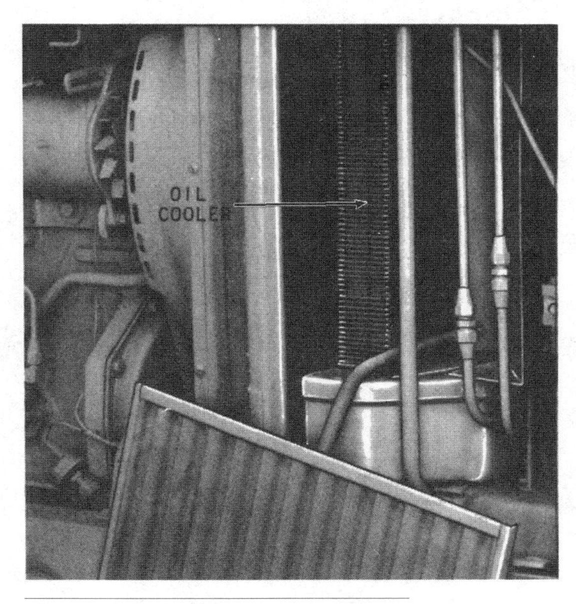

Fig. 4 — Location of a Typical Oil Cooler

WATER-TO-OIL COOLERS use moving water to carry off heat from the oil. The water flows through many tubes and the oil circulates around the cooling tubes as shown in Fig. 3. On mobile machines, the engine coolant is usually circulated through the cooler.

The cooler is usually a separate unit, connected to the system with hoses, as shown. Sometimes the oil tubes are placed in the engine radiator lower tank.

Another less common type of water-to-oil cooler uses the evaporation of water to cool oil. Water is sprayed over coils of oil tubes while forced air is blown in from the bottom. Part of the water evaporates, cooling the remaining water, which in turn draws heat from the oil in the tubes. This cooler is not as compact and is not practical for mobile equipment.

TEST YOURSELF

QUESTIONS

1. Name two functions of the reservoir besides storing oil.

2. What is the purpose of a pressurized reservoir?

3. What two mediums are used for cooling in most oil coolers?

LINES AND COUPLERS

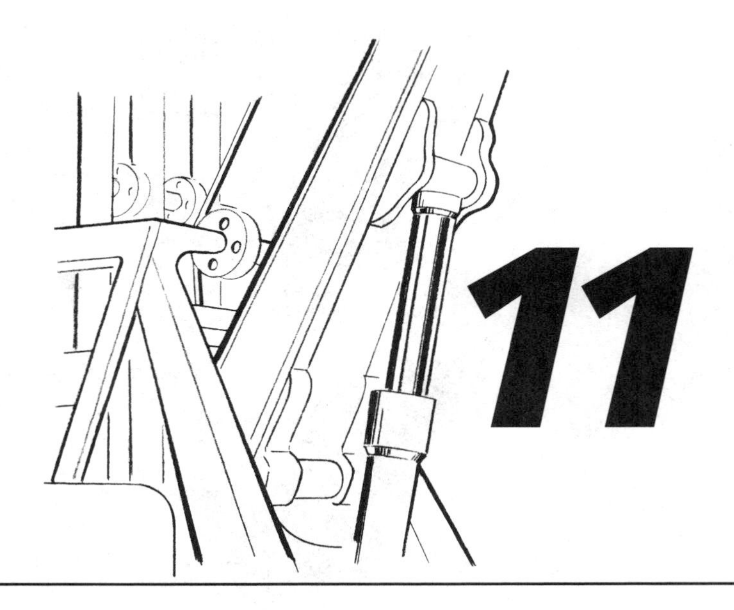

INTRODUCTION

The components in a hydraulic system must be connected together with lines and couplers (Fig. 1). While less complex than some other parts, lines and couplers are essential because they conduct fluid from one hydraulic component to another. A clear understanding of how they work and what they are designed to do will help avoid costly failures. When we refer to lines we include both rubber hoses and metal tubes. Couplers are often called fittings or connectors.

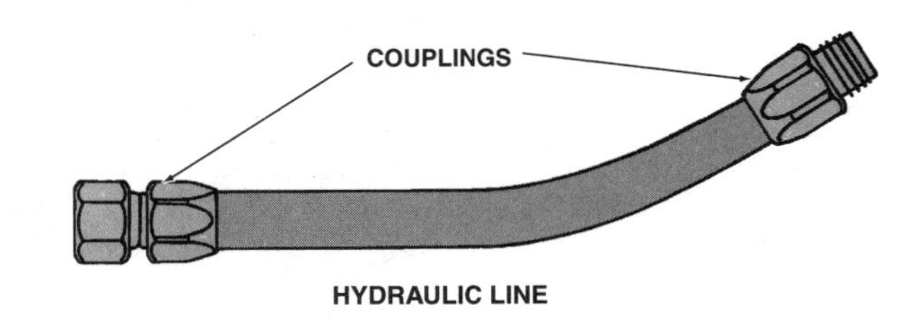

Fig. 1 — Lines and Couplers

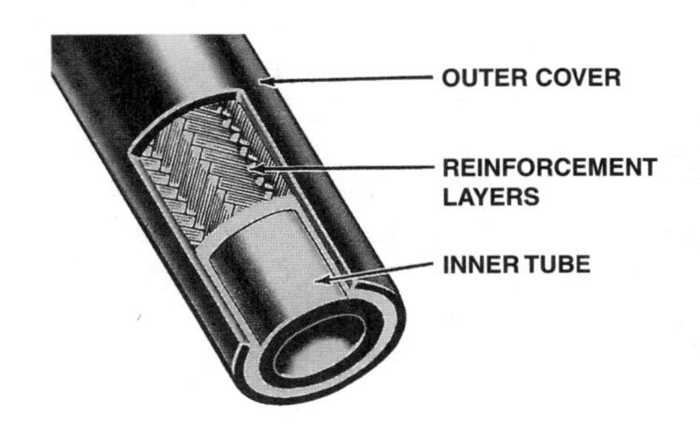

Fig. 2 — Hose Construction

HOSES

Hoses provide flexible oil conduits between hydraulic components. Hoses permit some movement of the components and absorb vibration and pressure surges. Hose walls have many layers (Fig. 2).

HOSE SELECTION

Always replace damaged hoses with the kind of hose specified in the technical manual or parts catalog. Hydraulic hose construction standards usually follow specifications suggested by the S.A.E. (Society of Automotive Engineers). Some of the specified tolerances are:

- **Inside diameter**
- **Outside diameter**
- **Maximum operating pressure**
- **Minimum burst pressure**
- **Minimum bend radius**
- **Temperature limits**

Hose size is determined by the amount of flow needed. If a hose is too small, it will restrict flow, slow operation, and generate heat. If a hose is too large, hydraulic pressure may rupture the hose because of the greater inner wall area.

Hose pressure rating is based on **operating pressure** and **burst pressure**. The hose must be strong enough to withstand normal operating pressure plus pressure surges. Hydraulic fluid temperature is also critical in hose selection. **High temperature hoses** are specified if hydraulic fluid temperature is exceptionally high. Heat resistant covers may be used to protect hose from heat.

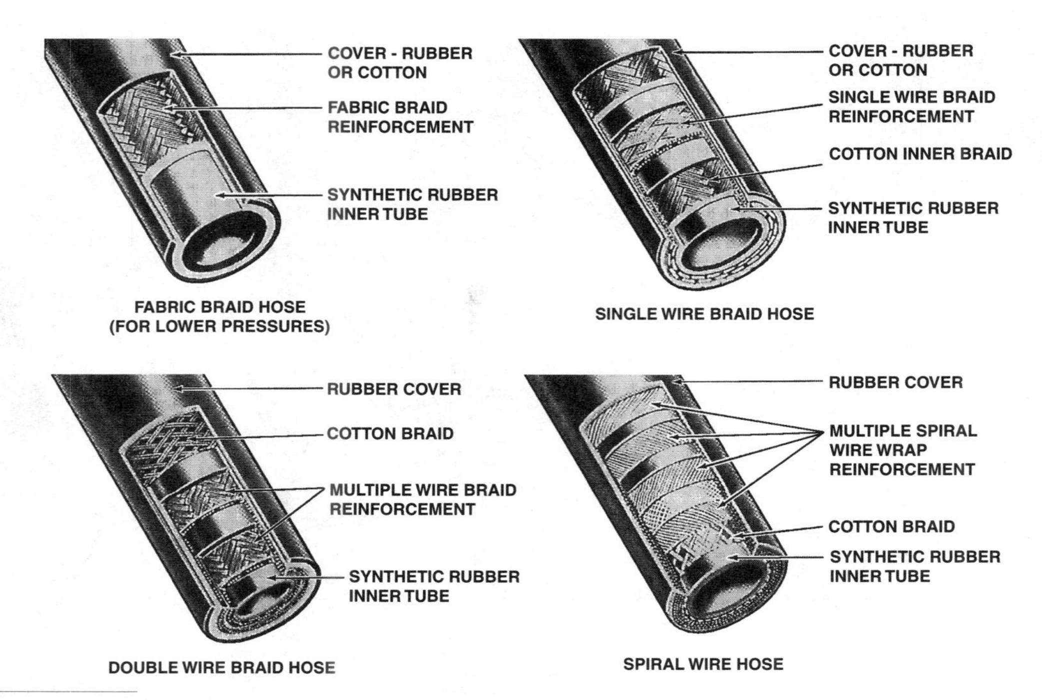

Fig. 3 — Four Kinds of Hose

Hoses are specified by equipment manufacturers. Substituting a different kind of hose is extremely risky. Even a stronger hose is no guarantee of safety and economy.

There are four common kinds of hoses (Fig 3):

- **Fabric braid**
- **Single wire braid**
- **Double wire braid**
- **Spiral braid**

FABRIC BRAID HOSE

Construction	*Uses*
Inner Tube: Black synthetic rubber Reinforcement: Woven fiber reinforced with spiral wire to prevent collapse. Cover: Synthetic rubber, oil- and abrasion resistant.	Lines for Petroleum base hydraulic oils, gasoline or fuel oil. In suction lines or in low-pressure return lines. Temperature Range: -40 to 250°F (-40 to 121 °C) Vacuum: 30 In. Hg.

Construction	*Uses*
Inner Tube: Black synthetic rubber, oil-resistant Reinforcement: One-fiber braid Cover: Black synthetic rubber, oil- and abrasion- resistant.	Hydraulic oil return lines only, or general purpose fuel oil, gasoline, water, anti-freeze mixtures, air, and other chemicals. Temperature Range: -40 to 250°F (-40 to 121 °C)
Inner Tube: Black synthetic rubber, oil- resistant. Reinforcement: Two fiber braids. Cover: Black synthetic rubber, oil- and abrasion- resistant.	Hydraulic oil return lines only, or general purpose fuel oil, gasoline, water, anti-freeze mixtures, air, and other chemicals. Temperature Range: -40 to 250°F (-40 to121°C)

IMPORTANT: Fabric braid (low pressure) hoses are NOT RECOMMENDED FOR PRESSURE LINE USE IN HYDRAULICS. Therefore, they will not be included in the chart on selecting hoses which follow these charts.

DOUBLE WIRE BRAID (SAE J517 100R2) USE WHEN SIZE AND PRESSURE ARE WITHIN THESE LIMITS

1/4 in.	6.35 mm	5000 psi	35 475 kPA
3/8 in.	9.53 mm	4000 psi	27 580 kPa
1/2 in.	12.70 mm	3500 psi	24 133 kPa
5/8 in.	15.88 mm	2750 psi	18 961 kPa
3/4 in.	19.05 mm	2250 psi	15 514 kPa
1 in.	25.40 mm	2000 psi	13 790 kPa
1 1/4 in.	31.75 mm	1625 psi	11 204 kPa
1 1/2 in.	38.10 mm	1250 psi	8619 kPa

SINGLE WIRE BRAID (SAE J517 100 R1) USE WHEN SIZE AND PRESSURE ARE WITHIN THESE LIMITS

1/4 in.	6.35 mm	2750 psi	18 961 kPA
3/8 in.	9.53 mm	2250 psi	15 513 kPa
1/2 in.	12.70 mm	2000 psi	13 790 kPa
5/8 in.	15.88 mm	1500 psi	10 342 kPa
3/4 in.	19.05 mm	1250 psi	8618 kPa
1 in.	25.40 mm	1000 psi	6895 kPa
1 1/4 in.	31.75 mm	625 psi	4309 kPa
1 1/2 in.	38.10 mm	500 psi	3448 kPa

FOUR-SPIRAL STEEL WIRE REINFORCED HOSE (SAE J 100R9) USE WHEN SIZE AND PRESSURE ARE WITHIN THESE LIMITS

3/8 in.	9.53 mm	4500 psi	31 026 kPa
1/2 in.	12.70 mm	4000 psi	27 580 kPa
3/4 in.	19.05 mm	3000 psi	20 685 kPa
1 in.	25.40 mm	3000 psi	20 685 kPa
1 1/4 in.	31.75 mm	2500 psi	17 237 kPa
1 1/2 in.	38.10 mm	2000 psi	13 790 kPa
2 in.	50.80 mm	2000 psi	13 790 kPa

SINGLE WIRE BRAID HOSE

Construction	*Uses*
Inner Tube: Black synthetic rubber. Reinforcement: Two fiber braids. Cover: Synthetic rubber, oil- and abrasion resistant.	Hydraulic oil lines, fuel oil, anti-freeze solutions, or water Lines. Temperature Range: -40 to 250°F (-40 to121°C)
Inner Tube: Black synthetic rubber, oil-resistant. Reinforcement: One braid of high tensile steel wire. Cover: Black synthetic rubber oil- and abrasion-resistant.	Hydraulic oil lines, fuel oil, gasoline or water. Temperature Range: -40 to 250°F (-40 to121°C)

DOUBLE WIRE BRAID HOSE

Construction	*Uses*
Inner Tube: Black synthetic rubber oil-resistant. Reinforcement: Two braids or more of high tensile steel wire. Cover: Black synthetic rubber, oil- and abrasion-resistant.	High-pressure hydraulic oil lines, fuel oil, gasoline or water lines.
Inner Tube: Black, synthetic rubber. Reinforcement: Two braids or more of high tensile steel wire. Cover: Synthetic rubber oil- and abrasion-resistant green color.	Hydraulic lines using phosphate ester base fluids. (Should not be used with petroleum oils.) Temperature Range: -40 to 200°F (-40 to 93°C)

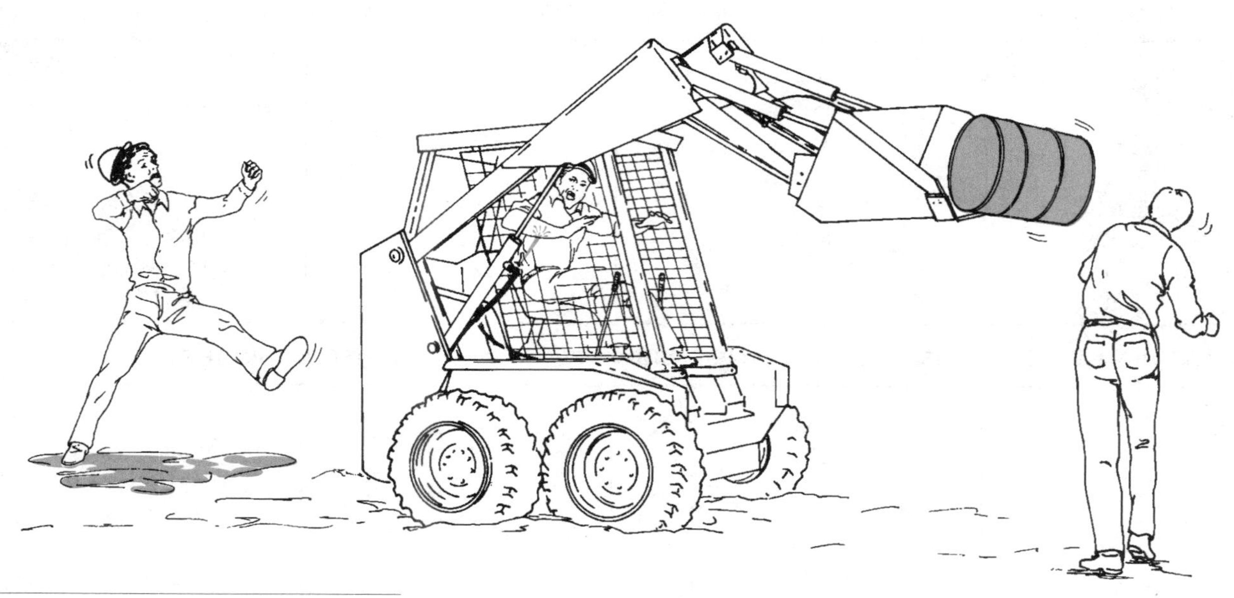

Fig. 4 — Personal Injury Mechanical Damage

SPIRAL WIRE HOSES

Construction	*Uses*
Inner Tube: Black synthetic rubber oil resistant.	Very high-pressure hydraulic oil lines or fuel oil lines.
Reinforcement: Multiple spiral of high tensile steel wire and one fiber braid.	Temperature Range: -40 to 200°F (-40 to 93°C)
Cover: Black synthetic rubber, oil- and abrasion-resistant.	

IMPORTANT: Spiral wire hose is recommended in high flexing applications. The wire strands of the spiral wrap do not rub and wear as much as the braided strands do.

As the hose size increases, the pressure limit lowers.

HOSE FAILURE

Hydraulic hose failures are dangerous in several ways.

Personal Injury

When a pressure hose bursts, equipment fails or retracts (Fig. 4).

Pressurized hydraulic fluid bursting out of a ruptured hose or fitting can penetrate skin and cause serious injury requiring surgery to remove fluid and infection.

Leaking hydraulic fluid is slippery enough to cause a bad fall.

Mechanical Damage

If a hydraulic line bursts and loses fluid, it can start a chain reaction of damage (Fig. 5). Parts like pumps, motors, and seals that depend on the fluid for lubrication will also fail because pieces of the failed part work through the system and damage them. The best thing to do if a line bursts is shut off the machine immediately. Operation without enough fluid will increase the system temperature. Heat will reduce the strength of seals and hoses and can affect the proper sealing of metal parts because of abnormal thermal expansion.

Fig. 5 — Mechanical Damage

TROUBLESHOOTING HOSES

Hose troubleshooting is mainly inspection and replacement.

Carefully check for:

- **Cracks and splits**
- **Pinhole leaks**
- **Incorrect hose length**
- **Improper installation**
- **Wrong type of hose**
- **Wrong fitting**

Cracks and Splits

Cracks and splits in the outer cover of a hose are common after extended use, and do not always mean the hose has failed. Leaks, however, are failures. Their source must be found and corrected immediately. Leaking fluid reduces efficiency and deprives the system of proper lubrication. It will soon overheat and damage itself.

Leaks

Leaks reduce efficiency and lead to heat and wear problems. Find leaks and replace the leaking hose immediately.

CAUTION: When looking for small leaks, use cardboard (Fig. 7). Oil injected under your skin will cause serious injury.

Fig. 6 — Avoid Fluid Being Injected Into the Body

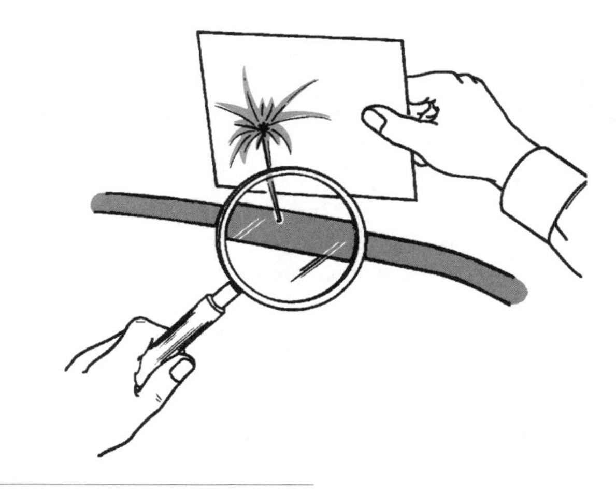

Fig. 7 — Find Small Leaks Safely

AVOID HIGH-PRESSURE FLUIDS

CAUTION: Escaping fluid under pressure can penetrate the skin causing serious injury. Avoid the hazard by relieving pressure before disconnecting hydraulic or other lines. Tighten all connections before applying pressure. Search for leaks with a piece of cardboard. Protect hands and body from high pressure fluids.

If an accident occurs, see a doctor immediately. Any fluid injected into the skin must be surgically removed within a few hours or gangrene may result. Doctors unfamiliar with this type of injury should reference a knowledgeable medical source.

Length

Hoses must be the correct length (Fig. 8). If too short, stretching will cause failure. If too long, the hose will vibrate and flex until it fails. Loops should never be allowed.

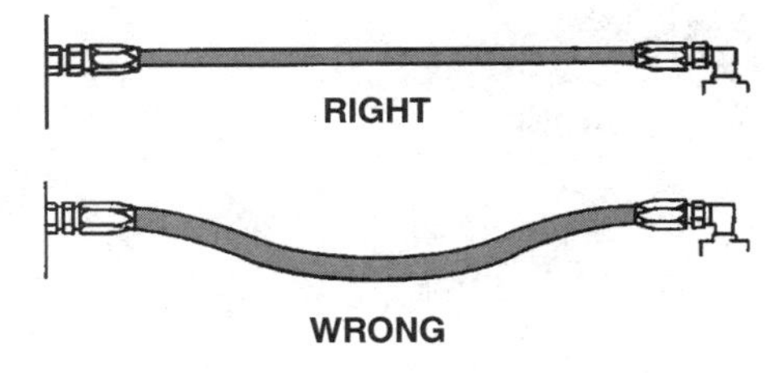

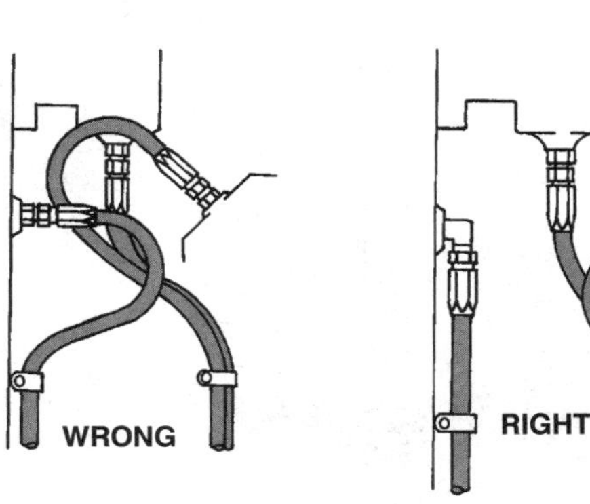

Fig. 8 — Don't Stretch or Loop Hoses

Installation

Sometimes a hose is surrounded by a sleeve or has a metal or plastic protector around it. This protection is for abrasion, fire protection or operator safety in case of a leak in the hose. Install protective covering if originally used .

Twisted hoses (Fig. 9) can cause restricted flow. Twisting is usually caused by tightening couplers wrong. Allow as much free motion as possible, and hold the hose to prevent twisting as you tighten. Always turn fittings on by hand until you are sure they are not cross threaded.

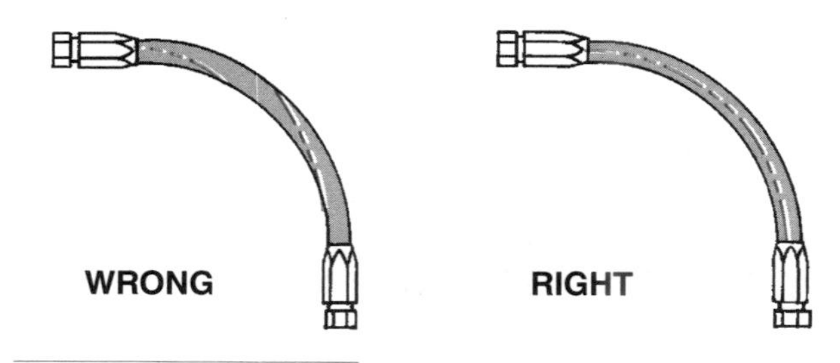

Fig. 9 — Don't Twist Hoses

Improper routing is the most common cause of premature hose failure. There is often one correct path for the hose, but many wrong paths. The results of incorrect routing are abrasion (Fig. 10), twisting, and kinking (Fig. 11) and burning (Fig. 12).

Keep hoses away from heat, especially exhaust heat. If hoses must be close to heat, use brackets, covers, and heat shields to protect or hold them away from the most intense heat.

When removing hoses that are routed in peculiar paths, lay a wire in the path before removing the hose. The wire will guide your installation.

Do not move clamps (Fig. 13) to a new location when you replace a hose.

Be sure all original brackets, covers, and heat shields are back in place if you replace a hose. If you see an unprotected hose in a hot spot, check for improper routing or a missing clamp, bracket, cover, or heat shield.

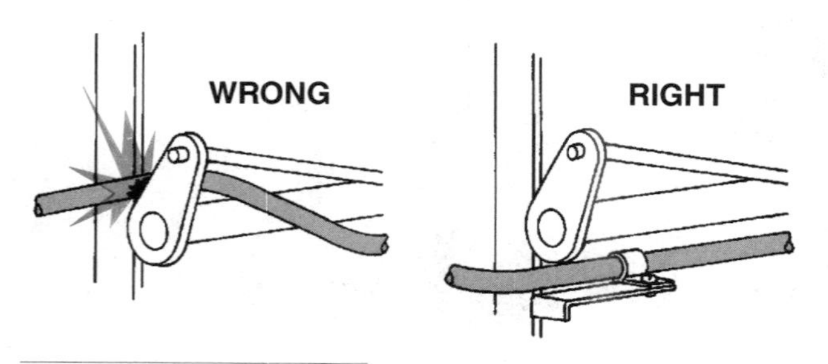

Fig. 10 — Don't Pinch Hoses

Hose Type

The wrong type of hose is never economical. Even hose that is heavier than specified may not do the job, it may be too stiff.

Fitting Type

Do not make do by trying to adapt fittings to hoses. Use the correct fitting. Adapted fittings tend to leak and cause routing problems.

HOSE SUMMARY

If a defective hose is found, replace it immediately. Here is a summary of the basic rules for correct hose installation (Fig. 14).

- **Taut Hose — Allow some slack. Hoses that are stretched will fail sooner than correctly installed hoses.**
- **Loops — Used angled couplers to prevent long loops. The correct coupler makes installation neater and reduces the length of hose required.**
- **Twists — Twists cause couplers to loosen and hoses to weaken as twisted hoses flex during operation.**
- **Rubbing — Use clamps or brackets to hold hoses away from sharp edges and moving parts. Hose guards of wire spring or flat armor may be necessary.**
- **Heat — Route hydraulic hoses away from hot spots, such as exhaust pipes and exhaust manifolds. If a hose cannot be moved away from heat, insulate the hose with shielding.**
- **Sharp Bends — The minimum bend radius depends on the hose construction, diameter, and the pressure. At low pressure, the diameter can be smaller than the same hose under higher pressure. Reroute the hose or use angle fittings rather than make sharp bends. Allow slack for freedom of movements, but don't allow kinks and loops.**

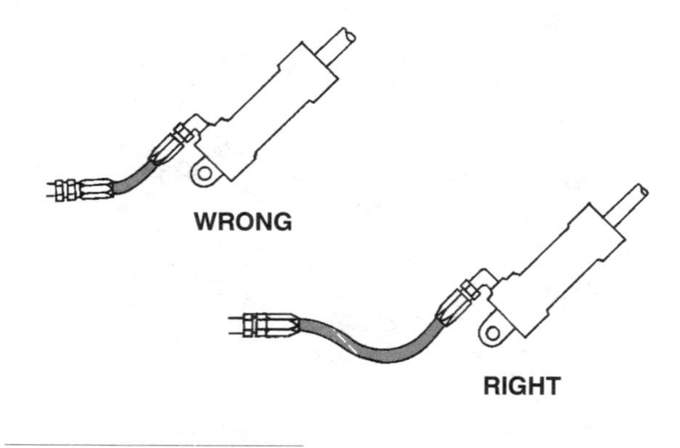

Fig. 11 — Don't Kink Hoses

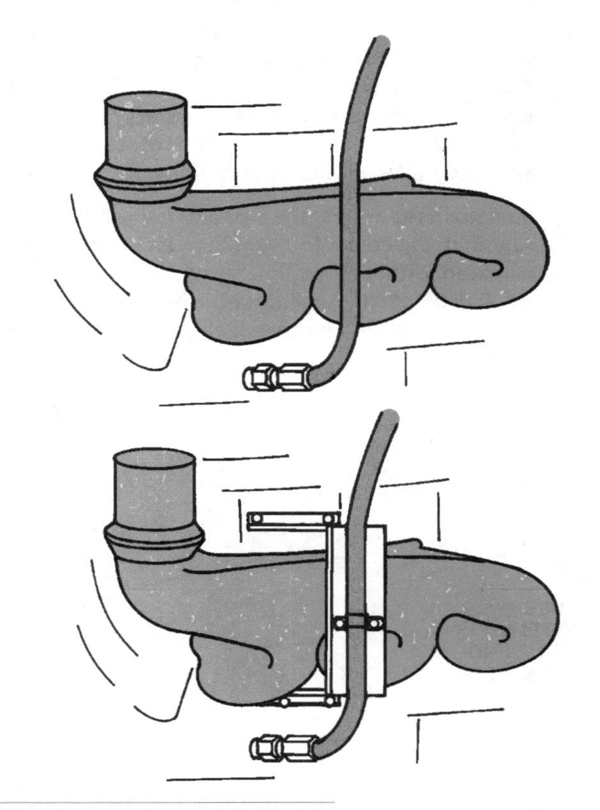

Fig. 12 — Protect Hoses from Heat

Fig. 13 — Replace Every Clamp

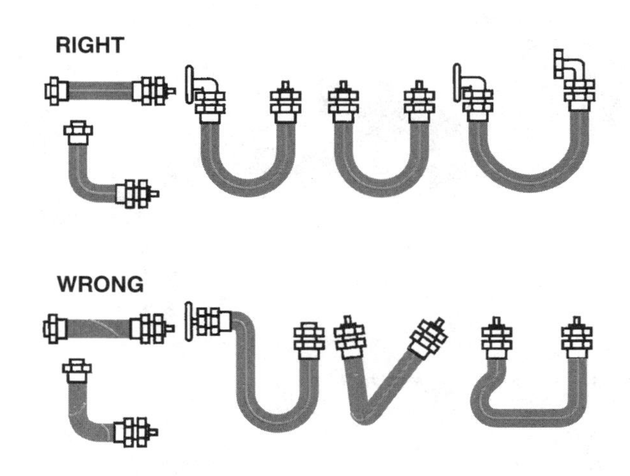

Fig. 14 — Hose Installation Guide

TUBES

Tubing is used where flexibility is not needed. Correctly sized, shaped, and installed, tubing adds to reliability.

TUBE SELECTION

Just as with hoses, tubing must have the correct specifications for inside and outside diameter, maximum operating pressure, minimum burst pressure, minimum bend, and temperature capacity.

The kind of material used in tubing construction is important.

Copper

Copper tubing is only used for low pressure, moderate temperature, and low vibration jobs. Copper tubing becomes brittle and breaks easily when flared or heated.

Aluminum

Aluminum tubing is easily bent and flared; however, it should only be used in low pressure applications.

Plastic

Nylon is suitable for use in low pressure, moderate temperature hydraulic systems. Nylon tubing will melt at high temperature or break in cold temperature.

Steel

Cold drawn steel tubing makes the strongest hydraulic lines. There are two types: seamless and electric welded. Steel can take high pressure and temperature extremes, but is harder to form and bend than copper, aluminum, and plastic.

Fig. 15 — Correct and Incorrect Bends

Fig. 16 — Be Careful with Hydraulics

TROUBLESHOOTING TUBING

Rigid lines rarely fail. But, a failure can cause the same serious personal injury and damage to the equipment as failed hoses. Troubleshooting tubes is mainly inspection and replacement.

Tubing should not be free to vibrate. Vibration can loosen clamps and brackets. Check for loose clamps and brackets. Tighten or replace if necessary.

Dents, flattened sections, kinks, and sharp bends restrict fluid flow (Fig. 15). Replace if necessary.

Check for wet spots which could be caused by small cracks or pinhole leaks. Look for pinhole leaks with cardboard, not your hand (Fig. 16).

AVOID HIGH-PRESSURE FLUIDS

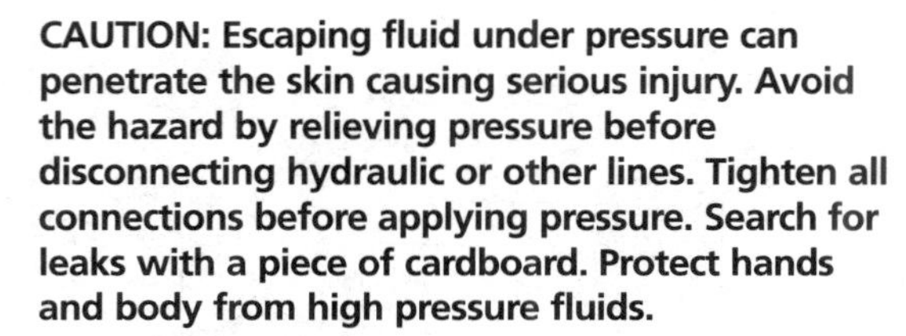

CAUTION: Escaping fluid under pressure can penetrate the skin causing serious injury. Avoid the hazard by relieving pressure before disconnecting hydraulic or other lines. Tighten all connections before applying pressure. Search for leaks with a piece of cardboard. Protect hands and body from high pressure fluids.

If an accident occurs, see a doctor immediately. Any fluid injected into the skin must be surgically removed within a few hours or gangrene may result. Doctors unfamiliar with this type of injury should reference a knowledgeable medical source.

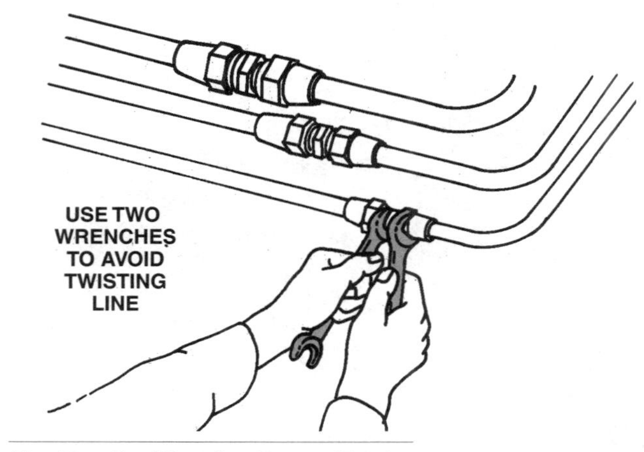

Fig. 17 — Two Wrenches Prevent Twisting

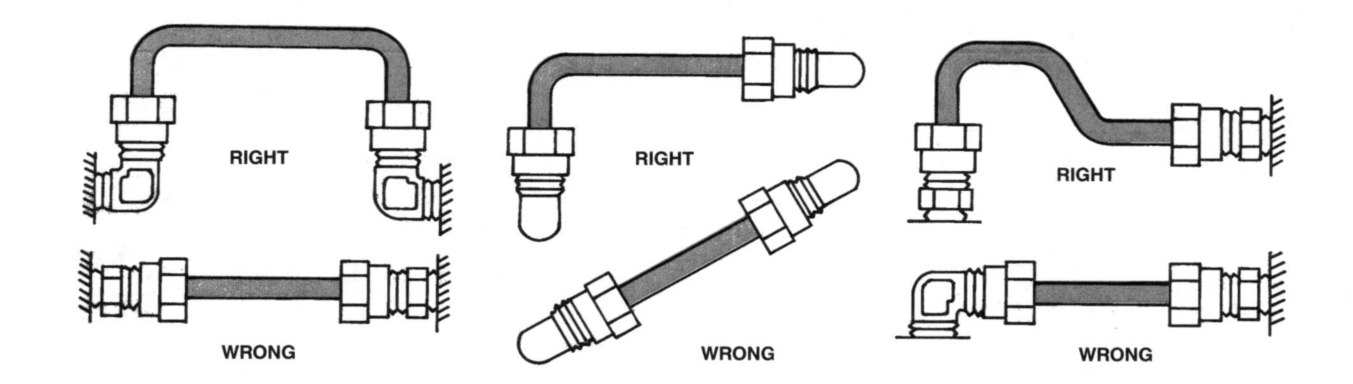

Fig. 19 — Tube Routing Guide

HOW TO TIGHTEN FLARE NUTS

Line Size (Outside Diameter)	Flare Nut Size (Across Flats)	Tightness ft-lb	Recommended Turns of Tightness (After Finger Tightening)	
			Original Assembly	Re-assembly
3/16 in.	7/16 in.	10	1/3 Turn	1/6 Turn
1/4 in.	9/16 in.	10	1/4 Turn	1/12 Turn
5/16 in.	5/8 in.	10-15	1/4 Turn	1/6 Turn
3/8 in.	11/16 in.	20	1/4 Turn	1/6 Turn
1/2 in.	7/8 in.	30-40	1/6 to 1/4 Turn	1/12 Turn
5/8 in.	1 in.	80-110	1/4 Turn	1/6 Turn
3/4 in.	1 1/4 in.	100-120	1/4 Turn	1/6 Turn
		(N.m)		
5mm	11mm	(13.5)	1/3 Turn	1/6 Turn
6mm	14mm	(13.5)	1/4 Turn	1/12 Turn
8mm	16mm	(13.5-20)	1/4 Turn	1/6 Turn
9.5mm	17.5mm	(27)	1/4 Turn	1/6 Turn
13mm	22mm	(41-54)	1/6 to1/4 Turn	1/12 Turn
6mm	25mm	(108-149)	1/4 Turn	1/6 Turn
9mm	32mm	(136-163)	1/4 Turn	1/6 Turn

Fig. 18 — Tighten Flare Nuts Correctly

REMOVE AND INSTALL

If defective tubing is found, replace it immediately. Always install new tubing of the same size, type, and shape as the original. Be sure the new tubing has the correct fittings and is lined up perfectly at the attachment points.

Start fittings gently by hand to make sure they don't cross thread.

Use two wrenches to tighten or loosen fittings to avoid twisting (Fig. 17).

Use a tightening chart as a guide when tightening tubing flare nuts (Fig. 18). A chart is usually printed in the technical manual.

The most important rule for tightening tube fittings is: ***Tighten only until snug. Do not overtighten. More damage has been done to tube fittings by overtightening than from any other cause.***

MAKING REPLACEMENT TUBING

- The fewer the couplings the better.
- Bends in replacement tubing should be the same as the original.
- Replacement tubing should follow the original path.
- Use the fewest bends, the simplest bends, and bends that follow the largest radius possible.
- Keep lines from protruding and avoid interference with other components and the operator. Do not run tubing across access doors, controls, or moving parts. Shields should be used to protect tubing from moving parts. Install all shields that are removed during work.
- Avoid straight-line hookups wherever possible. Straight tubing is difficult to remove and may not allow for expansion and contraction. Never install short, straight sections (Fig. 19).

- Support long sections of tubing with brackets and clamps. Several lines can sometimes be clamped together. Properly supported lines are easier to maintain and are neater.
- Do not route replacement tubes through bulkheads without bulkhead connectors or grommets. Grommets will protect tubing from abrasion. Bulkhead connectors will permit easier removal.
- Use quality bending tools and follow the tool manufacturer's instructions.
- Don't make flattened, kinked, or wrinkled bends. Bends must be accurate and smooth so flow will not be restricted. A general rule is the radius of a bend should be more than 3 to 5 times the tubing diameter. Thick walled tubing requires the larger radius.

COUPLERS

Couplers join hoses to components, change hose size, connect lines, and route hoses through bulk heads (Fig. 20).

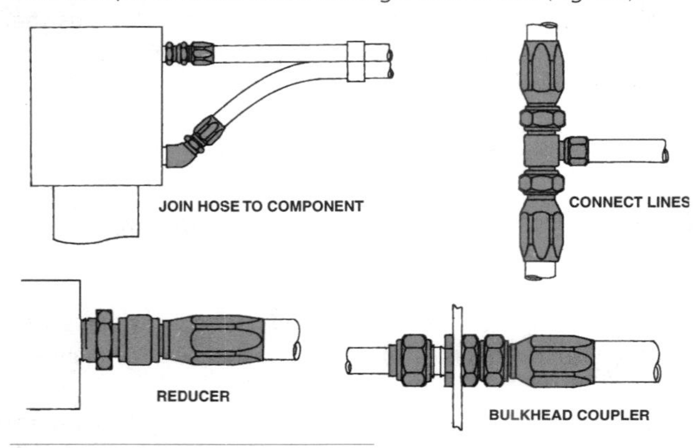

Fig. 20 — Couplers Do These Jobs

HOSE COUPLERS

The fittings attached to the ends of hoses are made of aluminum, brass, stainless steel, plastic, or steel. Steel fittings are usually used because they can take the most heat and pressure. The fittings must provide a strong, long lasting seal. High pressure and medium pressure fittings look much the same, but manufacturers mark the high pressure fittings with a notch (Fig. 21) to avoid confusion.

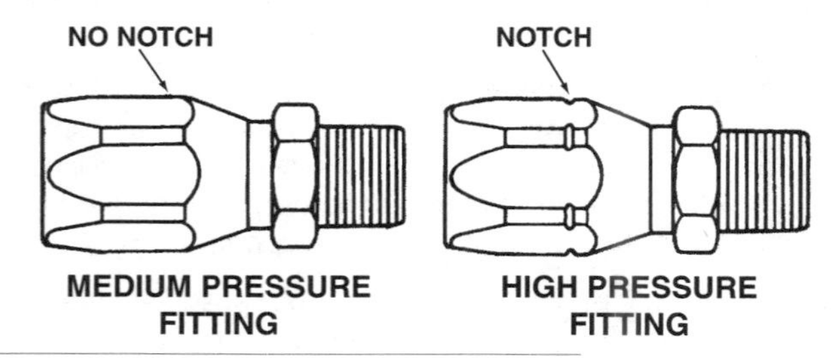

Fig. 21 — High Pressure Fittings Are Notched

There are permanent and reusable fittings (Fig. 22).

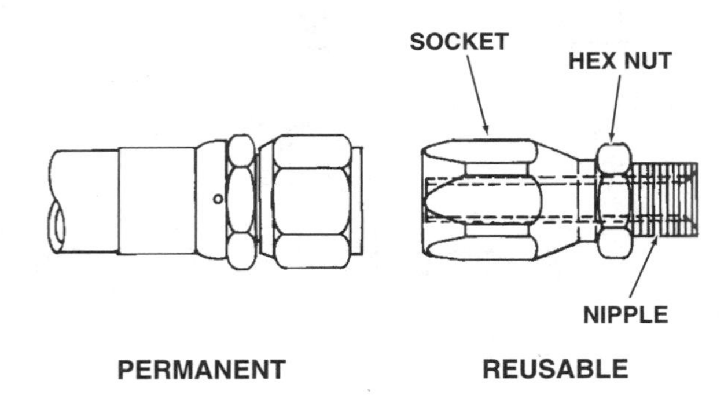

Fig. 22 — Permanent and Reusable Couplers

Permanent Hose Fittings

Permanent hose fittings are discarded with the hose. They are either crimped or swedged onto the hose. Some dealers have crimping machines which can make up hose assemblies using permanent fittings and stock hose cut to length.

Reusable Hose Fittings

Reusable hose fittings are either pushed on, screwed on, or clamped onto the hose. When the hose wears out, the fittings can be removed and used on a new hose cut from stock. Most reusable fittings can be converted to another thread type by changing the nipple in the socket. As might be expected, reusable fittings are slightly more expensive.

Skive, no-skive and clamp type fittings are common designs for reusable hose couplings (Fig. 23).

SKIVE fittings require the hose cover removed (skived) from the end before installation. The fittings are then screwed on the hose. This type of fitting is used for hoses with thick covers.

NO-SKIVE fittings are also screwed on the hose ends but the hose cover remains in place. The fitting is used on hoses with thin covers.

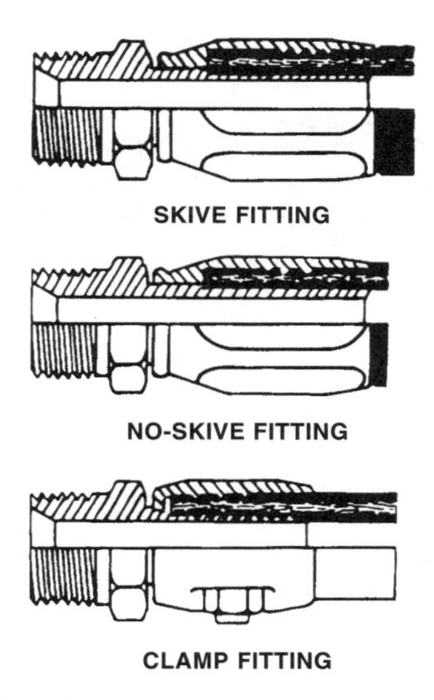

Fig. 23 — Types of Reusable Hose Couplers

CLAMP fittings are similar to permanent style fittings but instead of swedging or crimping, the fittings are held on the hose end using nuts and bolts. The bolts pull the fittings together causing the fitting barbs to bite into the hose.

If hose and fittings are matched ***incorrectly***, then the results can be pinhole leaks, ruptures, heat build ups, pressure drops, cavitation, and other failures.

There are six standard sealing methods for hose couplers to components and other lines (Fig. 24):

- **Pipe thread**
- **Dry seal for cone seat**
- **O-ring seal**
- **Split flange O-ring seal**
- **O-ring face seal**

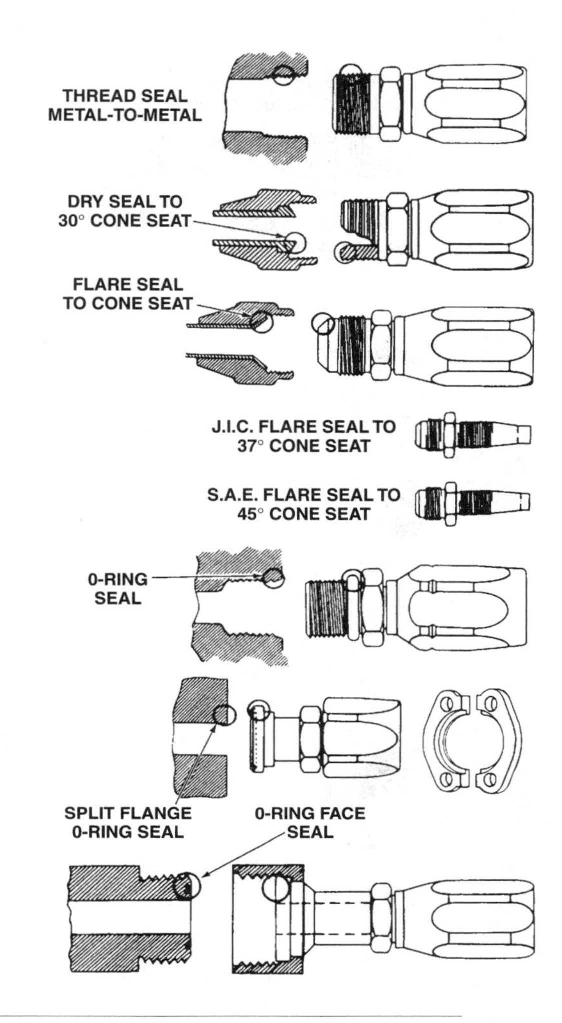

Fig. 24 — Six Sealing Methods for Hose Couplers

Installing Reusable Fittings

Reusable fittings can be removed and installed with hand tools and a vise (Fig. 25).

Low pressure, reusable fittings may use a hose clamp.

Medium pressure, reusable fittings (Fig. 26) are made specifically for single wire braid hose. DO NOT USE MEDIUM PRESSURE FITTINGS WITH THE WRONG HOSE.

High pressure, reusable fittings are made for multiple wire braid hose. DO NOT USE ANY OTHER HOSE WITH HIGH PRESSURE FITTINGS. High pressure fittings are marked with a notch around the hose socket. Install high pressure fittings carefully (Fig. 27).

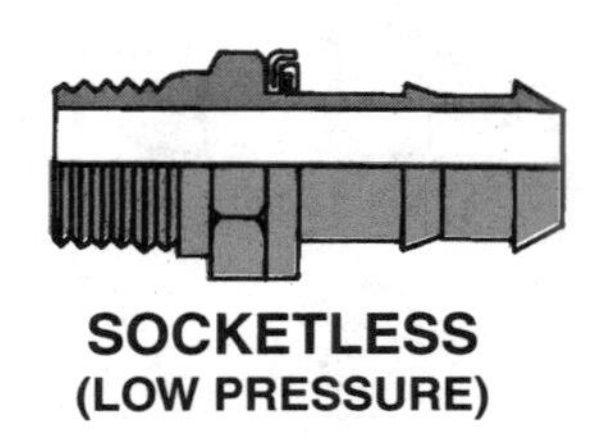

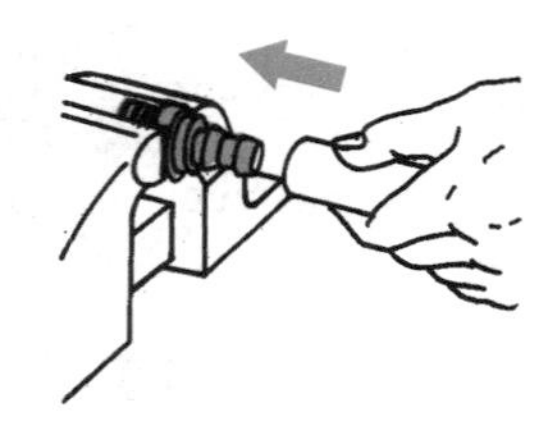

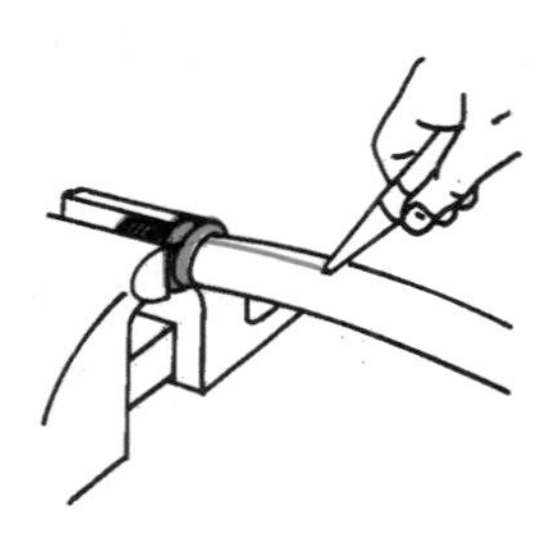

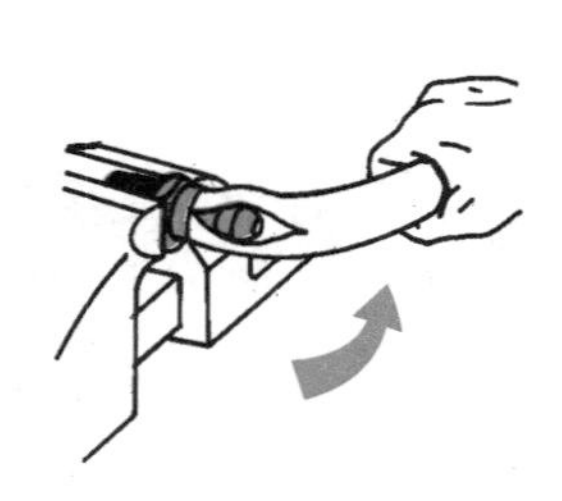

Fig. 25 — Low Pressure, Reusable Hose Couplers

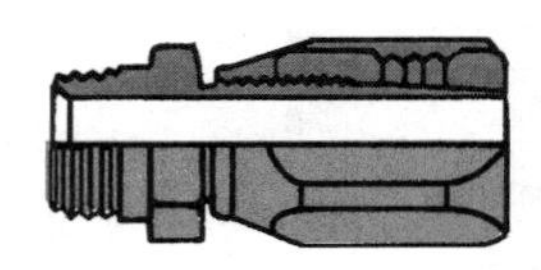

SOCKET AND NIPPLE
(MEDIUM PRESSURE, SINGLE WIRE BRAID HOSE)

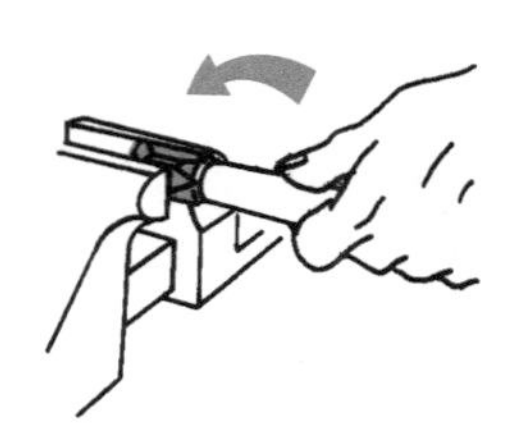

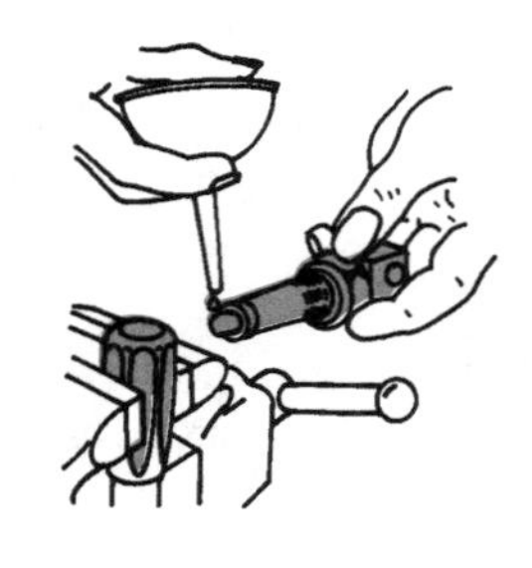

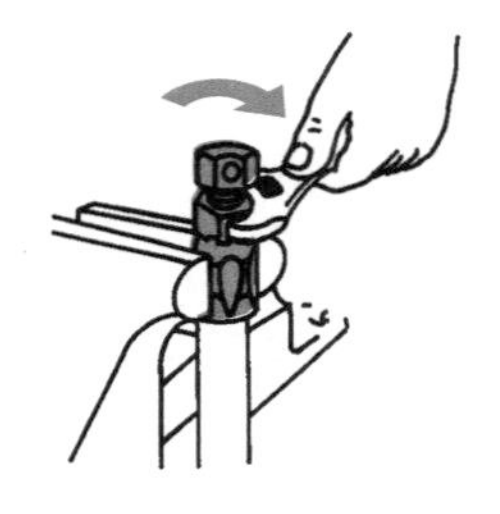

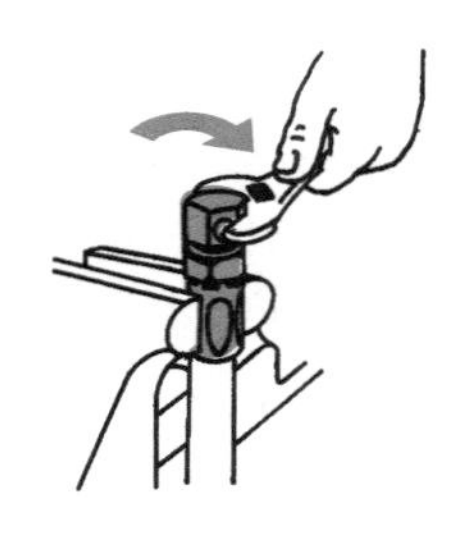

Fig. 26 — Medium Pressure, Reusable Hose Couplers

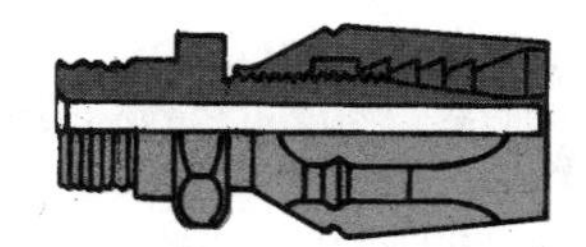

SOCKET AND NIPPLE
(HIGH PRESSURE - NOTCH ON SOCKET)
(MULTIPLE WIRE BRAID HOSE)

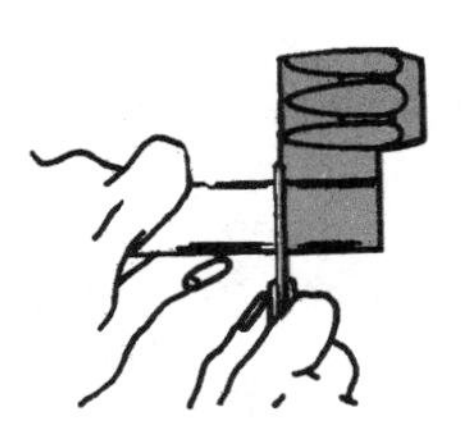

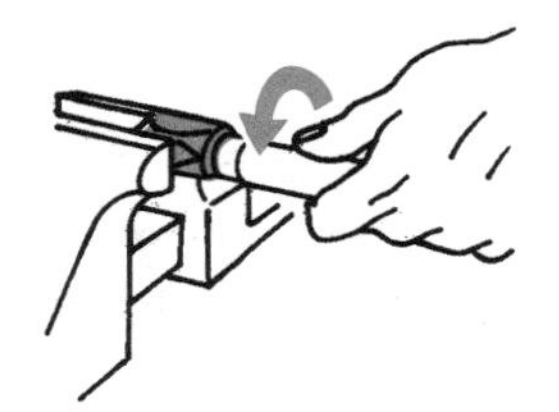

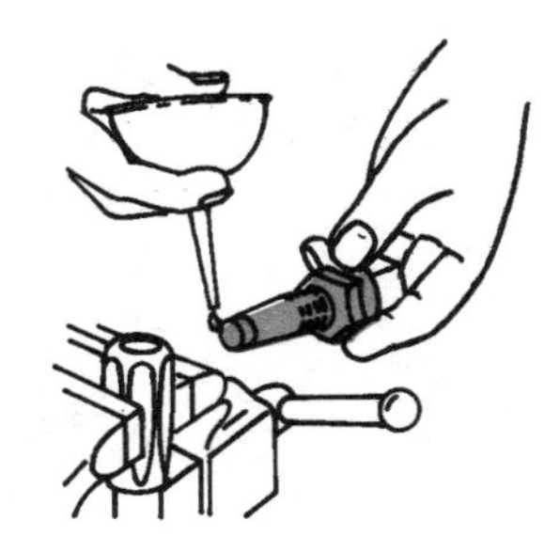

Fig. 27 — High Pressure, Reusable Hose Couplers

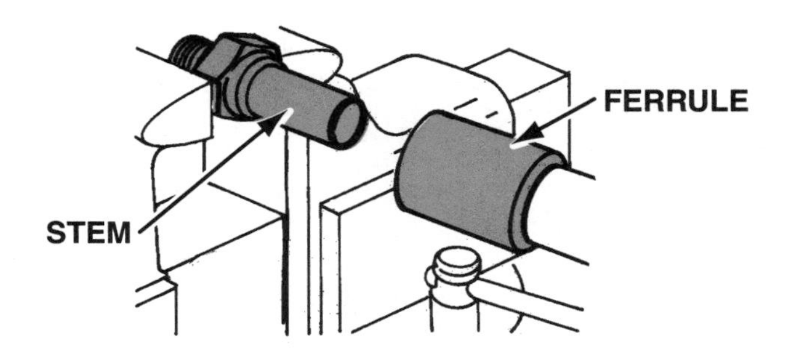

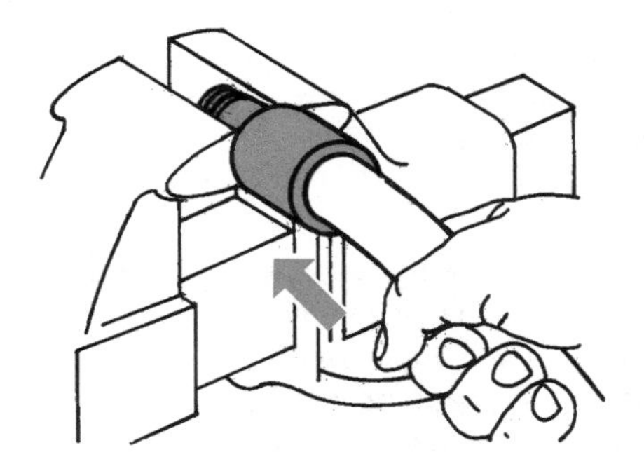

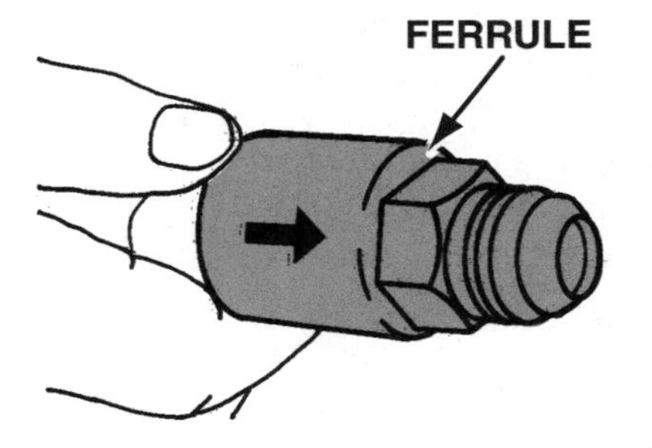

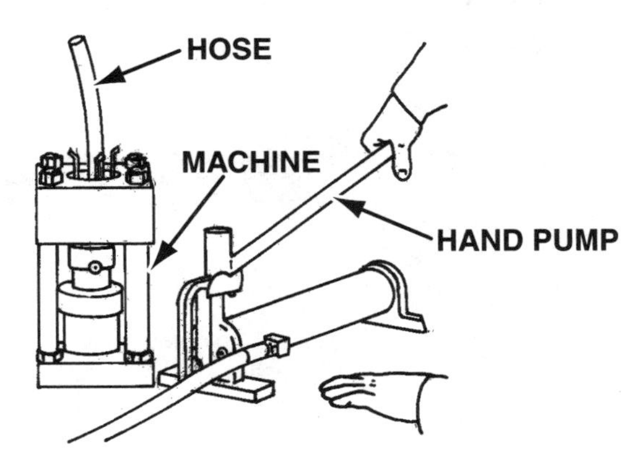

Fig. 28 — Permanent Hose Couplers

Installing Permanent Fittings

Use a crimping machine (Fig. 28) to crimp a permanent fitting onto a hose. Some crimping tools will automatically adjust for small differences in hose size. However, the correct hose should be used.

Crimping machines are powered by air or hydraulic pressure and the actual crimping is done by die fingers powered by a cylinder. The die fingers are adjusted for the hose fitting size. Depth stop must be adjusted to suit the length of the fitting.

TUBE FITTINGS

Most tube fittings are joined to the tube so the fitting can be tightened without turning the tube. The few fittings that require turning the tube should be replaced by installing the complete tube and fitting. The two most common types of tube fittings are FLARED and FLARELESS.

Flared Fittings

Flared fittings are used with thin walled tubing that is easily shaped. They seal with a metal to metal contact (Fig. 29). The flared end of the tube is squeezed between the flare fitting and a mating part as the fitting is tightened onto the threads of the mating part. The flare angle and the threads are important because sealing is not possible if the flare is not captured and squeezed squarely. A 37 degree flare is the accepted standard for high pressure. Fittings for use with 37 degree flare are always made of steel. Fittings with a 45 degree flare are used for low pressure and are usually brass.

Flared fitting designs are:

- **Three Piece Flare**
- **Two Piece Flare**
- **Inverted or Double Flare**
- **Self Flaring**

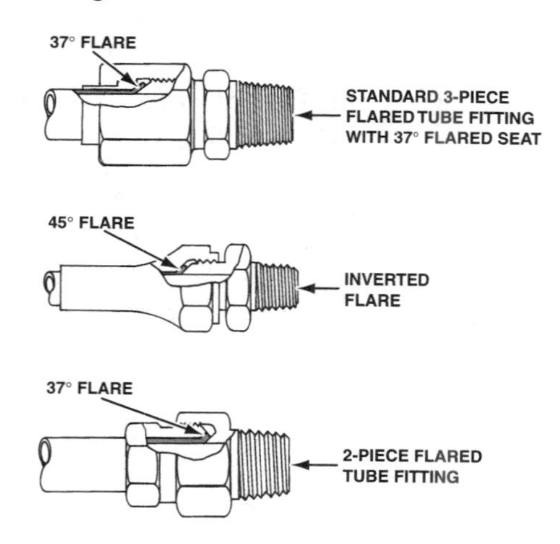

Fig. 29 — Flared Tube Couplers

THREE PIECE FLARED FITTING

The standard three piece flared fitting has a body, a sleeve, and a nut which fits over the tube. The free floating sleeve provides clearance between the nut and the tube, aligns the fitting, and is a lock washer for the tightened assembly. Advantages include the locking action of the sleeve and the fact that the flared tube is not rotated, and damaged, during assembly.

TWO PIECE FLARED FITTING

Two-piece flared fittings have no sleeves but use a tapered nut to align and seal the flared end of the tube. This fitting has some disadvantages. When this fitting is tightened, friction may cause the flare to bind and cause a leak. The same friction may also twist the tubing as it is tightened and damage it.

INVERTED OR DOUBLE FLARE FITTINGS

Inverted flare fittings or double flare fittings have a 45 degree flare inside the body and **should not be used for farm or industrial application**. This type is used primarily in auto industries.

SELF FLARING FITTING

Self flaring fittings are made with a wedge sleeve. When a nut is tightened, the wedge presses against the tube end and a female part of the fitting to form the flare. This fitting is strong and resists vibration with a minimum amount of tightening.

Flareless Fittings

The advantage of flareless fittings (Fig. 30) is, they do not require a special tool to flare the tubing. There are many sizes of flareless fittings and except for ferrules, the fittings may be reused. The most common types of flareless fittings are:

- **Ferrule**
- **Compression**
- **O-Ring**
- **Inverted Flareless**

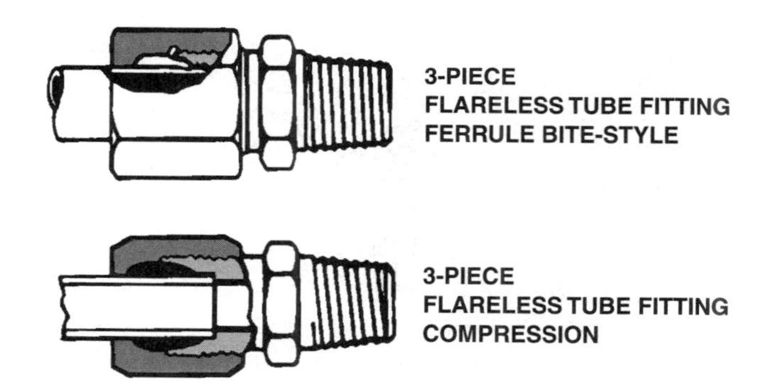

Fig. 30 — Examples of Flareless Fittings

FERRULE FITTING

Ferrule fittings have a body, ferrule, and nut. The ferrule is made in a wedge shape and is compressed against (into) the tubing by tightening the body and nut. The shape of the body ferrule and nut seals these parts when they are tightened together. At the same time, a cutting edge on the ferrule is forced into the tubing wall to keep the tube from pulling out of the fitting. The shape of the ferrule and the mating surfaces of the body and nut may differ with the manufacturer.

Flareless fittings require less workmanship than flare fittings but are very critical when tightening. The joint will leak if not tightened enough or if tightened too much. If over-tightened, the ferrule and tube must be replaced. Follow the manufacturer's assembly instructions closely. As a rule of thumb, after the joint is assembled and finger-tightened, tighten only 1/6 to 1/3 of a turn with a wrench.

COMPRESSION FITTING

Compression fittings use a similar design as ferrule fittings. Instead of biting ferrule, a soft material is used which compresses around the tube as the nut is tightened. This type of ferrule is usually replaceable.

With this type of fitting, the tubing is more likely to be pulled out of the fitting compared to the bite-style. Some fittings have a metal collar between the nut and ferrule to ensure electrical continuity through the fitting.

O-RING FITTING

O-Ring fittings have the advantage of a replaceable sealing element. The O-ring can be replaced. Some movement can be tolerated and exact length is not critical. O-Ring fittings cannot be used for high pressure. However, they are often used for oil pick up tubes and return tubes.

INVERTED FLARELESS FITTING

Inverted flareless fittings use the ferrule sealing method, but the ferrule is part of the fitting body. The advantage of this type of fitting is that it only uses two parts and one of the possible surfaces that could leak (between ferrule and body) is eliminated.

O-RING STYLE FITTINGS

O-Ring connectors have become very popular for their sealing ability. They are also easily repaired if a leak develops. We will discuss three types of connectors:

- **O-Ring Boss Connectors**
- **Four Bolt Flange Connectors**
- **O-Ring Face Seal Connectors**

O-Ring Boss Connector

The O-ring boss connector is also referred to as a straight thread fitting. It is classified in two styles, fixed-hex and adjustable nut. The adjustable nut fitting allows the position of the fitting to be located where desired and then sealed/locked in place (Fig. 31).

If the fittings leak:

— the O-ring was damaged during assembly

— O-ring has lost its flexibility from use

— the O-ring machined seat (boss) may be scored or damaged

— O-ring material is not compatible with system fluid and/or temperature

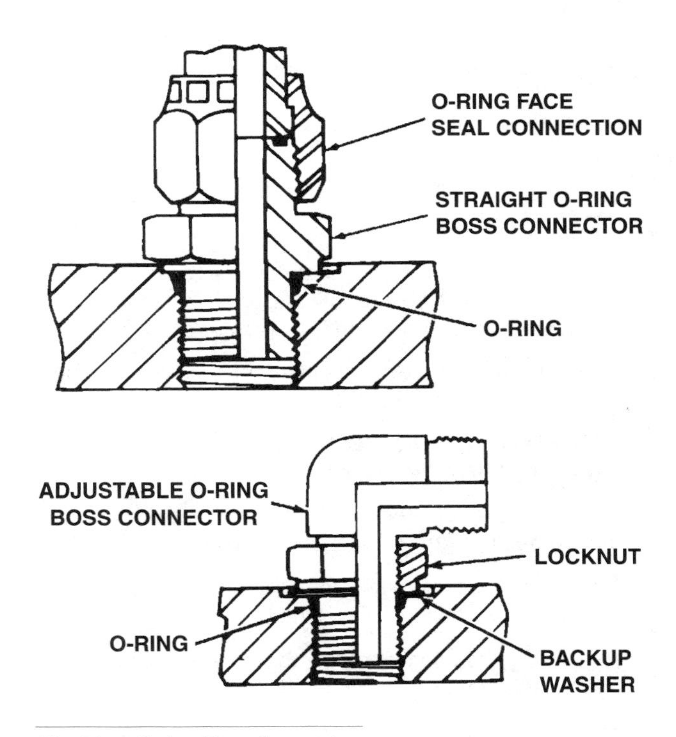

Fig. 31 — O-ring Boss Connectors

O-RING BOSS TORQUE VALUES

Pressure Applications Below 4000 psi (27,580 kPa)

Dash Size	Thread Size	lb-ft	N•m
-3	3/8-24 UNF	6	8
-4	7/16-20 UNF	9	12
-5	1/2-20 UNF	12	16
-6	9/16-18 UNF	18	24
-8	3/4-16 UNF	34	46
-10	7/8-14 UNF	46	42
-12	1-1/16-12 UNF	75	102
-14	1-3/16-12 UNF	96	122
-16	1-5/16-12 UNF	105	142
-20	1-5/8-12 UNF	140	190
-24	1-7/8-12 UNF	160	217

Fig. 32 — O-ring Boss Connectors Suggested Torques

INSTALL O-RING BOSS CONNECTOR

Follow these procedures closely for a leak-free connection:

1. Inspect O-ring boss seat for dirt or defects.
2. Lubricate O-ring with the fluid conducted or petroleum jelly. Use plastic or metal thimble or electrical tape over threads and slide O-ring onto undercut area of fitting.
3. For the fixed-hex fitting, install and tighten fitting to specification as shown in Fig. 32.

   IMPORTANT: DO NOT let hoses twist when tightening fittings. Use two wrenches.
4. For adjustable nut fittings, turn the nut and washer fully toward bend of fitting (counterclockwise).
5. Turn fitting (clockwise) into the port until washer contacts face of the boss and O-ring is in boss.
6. Turn fitting (counterclockwise) to desired position up to one turn maximum.
7. Hold the fitting with a wrench and tighten the nut to specification as shown in Fig. 32.

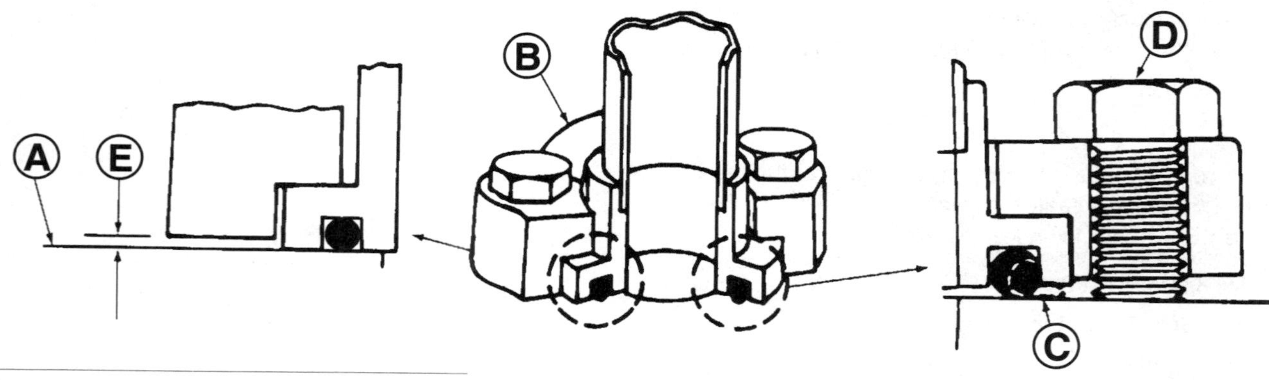

Fig. 33 — Four Bolt Flange Connection Installation

Fig. 34 — O-ring Face Seal Connector

FLAT FACE O-RING SEAL FITTING TORQUE

Dash Size	Thread Size	Swivel Nut lb-ft	Swivel Nut N•m
-4	9/16-18	12	16
-6	11/16-16	18	24
-6	13/16-16	37	50
-10	1-14	51	69
-12	1-3/16-12	75	102
-14	1-3/16-12	75	102
-16	1-7/16-12	105	142
-20	1-11/16-12	140	190
-24	2-12	160	217

Fig. 35 — O-ring Face Seal Connector Suggested Torques

Four Bolt Flange Connector

The four bolt flange connectors are widely used in industrial machine application, mainly because of the large diameter oil pipes required. Another advantage is the relative ease with which they can be installed.

There are actually two flanges involved.

- A flange welded or brazed to the pipe or tube. This flange has an O-ring groove cut into its face.
- A second flange fits over the welded or brazed flange to hold it against a flat mating surface using four cap screws. This compresses the O-ring or packing to seal the joint. This flange can be a one- or two piece (split) design.

INSTALL FOUR BOLT FLANGE CONNECTOR

Refer to Fig. 32.

1. Clean sealing surfaces (A). Inspect. Scratches cause leaks. Roughness causes seal wear. Out-of-flat causes seal extrusion. If the defects cannot be polished out, replace the component.
2. Install the O-ring (and backup washer if required) into the groove using petroleum jelly to hold it in place.
3. Split flange: Loosely assemble split flange (B) halves. Make sure the split is centrally located and perpendicular to the port. Hand tighten the cap screws to hold parts in place. Do not pinch O-ring (C).
4. Single piece flange: Place the hydraulic line in the center of the flange and install the cap screws. The flange must be centrally located on the port. Hand tighten the cap screws to hold the flange in place. Do not pinch the O-ring.
5. Tighten one cap screw, then tighten the diagonally opposite cap screw. Tighten the two remaining cap screws.

DO NOT use air wrenches. DO NOT tighten one cap screw fully before tightening the others. DO NOT over-tighten.

Cap screws must be tightened evenly!

DO NOT over-tighten which can cause bent bolts (D, Fig. 33) and flanges. There should be 0.010 to 0.030 inch (0.3 to 0.8 mm) clearance (E) between sealing surface and mounting flange.

THREAD SIZE (INCH)			
DASH SIZE	NOMINAL TUBE O.D.	O-RING BOSS	O-RING FACE SEAL
-3	0.188	3/8-24	
-4	0.250	7/16-20	9/16-18
-5	0.312	1/2-20	
-6	0.375	9/16-18	11/16-16
-8	0.500	3/4-16	13/16-16
-10	0.625	7/8-14	1-14
-12	0.750	1-1/16-12	1-3/16-12
-14	0.875	1-3/16-12	1-3/16-12
-16	1.000	1-5/16-12	1-7/16-12
-20	1.250	1-5/8-12	1-11/16-12
-24	1.500	1-7/8-12	2-12
-32	2.000	2-1/2-12	

Fig. 36 — O-ring Face Seal/O-ring Boss Connector Sizes

O-ring Face Seal Connector

O-ring face seal connection offers the best leakage control available today. The male connector has straight threads and O-ring in the face. The female swivel side has straight threads and a flat machined face. The seal occurs by compressing the O-ring against a flat surface. Proper O-ring size is important; it is not the same sizes as O-ring boss or four bolt flange O-rings (Fig. 34).

INSTALL O-RING FACE SEAL CONNECTORS

1. Inspect the fitting sealing surfaces. They must be free of dirt of defects.
2. Inspect the O-ring. It must be free of damage or defects.
3. Lubricate O-rings and install into groove using petroleum jelly to hold in place.
4. Push O-ring into the groove with plenty of petroleum jelly so O-ring is not displaced during assembly.
5. Position angle fittings and push joint together while tightening by hand. This will keep the O-ring in place.
6. Tighten fitting or nut torque value shown in chart (Fig. 35) per the dash size stamped on the fitting.

The table in Figure 36 compares thread sizes for O-ring boss and O-ring face seal connectors.

IMPORTANT: DO NOT let hoses twist when tightening the fittings. Use two wrenches.

QUICK DISCONNECT COUPLERS

Quick disconnect couplers are used where conductors must be connected or disconnected frequently. They are self-sealing devices and do the work of the two shutoff valves and a coupler.

These couplers are fast and easy to use and keep fluid loss at a minimum. More importantly, there is no need to drain or bleed the system every time a hookup is made. However, dust plugs must be inserted in the coupler ports when the oil lines are disconnected. This will prevent contamination of the hydraulic system.

There are five basic types of quick couplers (Fig. 37):

- **Poppet**
- **Sleeve and Poppet**
- **Ball Bearing**
- **Straight Through**
- **Rotating Ball**

Poppet Coupler

Poppet couplers have a self-sealing poppet in each coupler half. When closed, the poppets seal in oil. When connected, the poppets push each other off their seats to allow oil flow. When disconnected, the poppets close again by spring action before the two halves release their seal. The coupler halves are locked in place by a ring of balls which are held into a ring groove in the inserted coupler half by a spring loaded outer sleeve.

Sleeve and Poppet Coupler

Sleeve and poppet couplers have a self-sealing poppet in one half and a tubular valve and sleeve in the other. The sleeve is inserted first and gives an added margin of sealing against oil loss or air entry.

Ball Bearing Coupler

Ball bearing couplers utilize a spring loaded ball bearing for a metal-to-metal sealing valve. This type is not recommended for vacuum systems.

Straight Through Coupler

Straight through couplers allow full flow whether they are connected or not. This type of coupler is used when full flow is required and a separate shut-off valve is installed in the line.

Rotating Ball Coupler

Rotating ball couplers utilize a quarter-turn ball valve in each half. Coupling is disconnected by manually turning each ball valve handle. An interlock prevents the coupling from being separated when the valves are open. This type of coupling is used for low pressure, high volume applications.

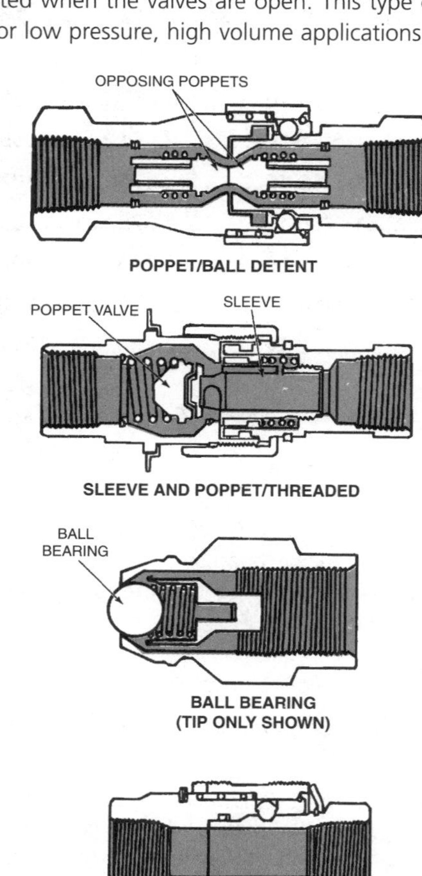

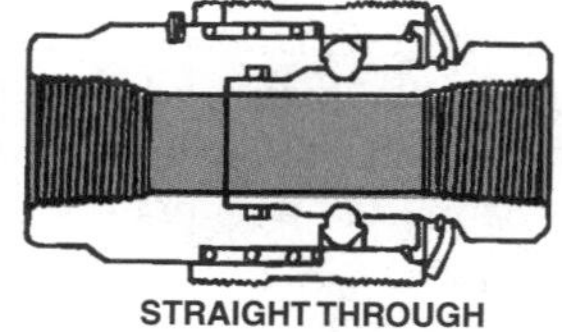

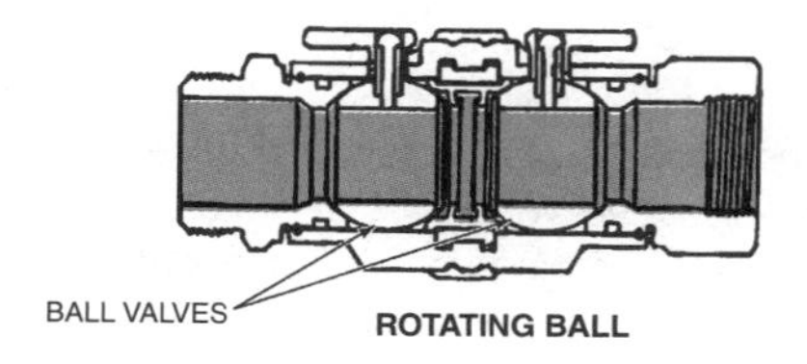

Fig. 37 — Quick Disconnect Couplers

QUICK DISCONNECT LATCHES

There are three types of latches for quick disconnect couplers:

- **Threaded**
- **Ball Detent**
- **Bar/Pin Detent**

Threaded Latch

The threaded latch (Fig. 37) uses the mechanical advantage of threads to connect and disconnect under pressure. This type of latch gives greatest holding power.

Ball Detent Latch

The ball detent latch (Fig. 37) uses a series of balls that fit into a groove of the male half to hold the coupler together. It can also be used as an emergency breakaway when the hose is pulled.

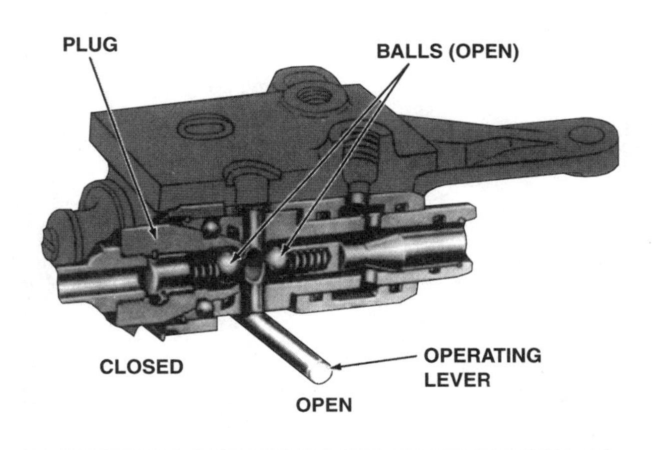

Fig. 38 — Ball Bearing Quick Disconnect with Lever Actuated Valve

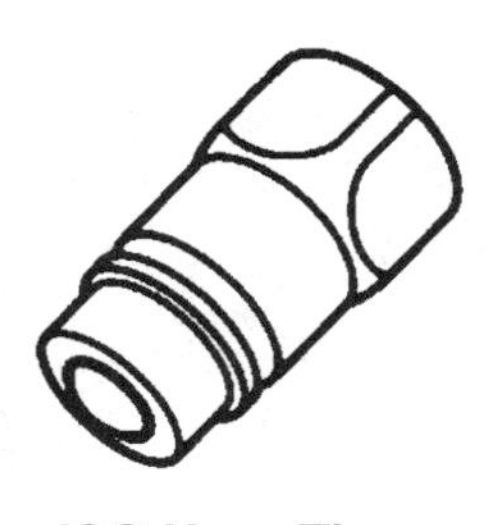

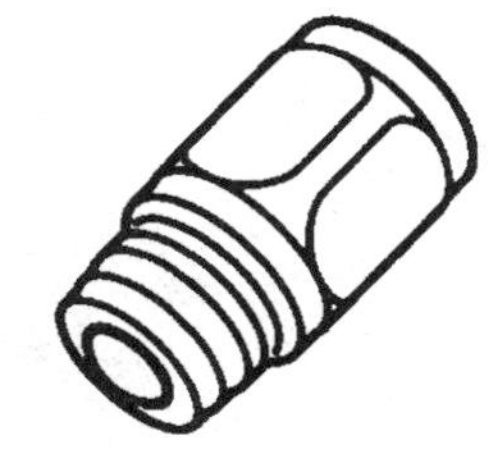

Fig. 39 — Hose Tip Styles

Bar/Pin Detent Latch

Bar/Pin Detent latch is similar to the ball detent. It utilizes two bars or pins that fit into the groove of the male part of the coupler. This type is usually used for air pressure applications less than 300 psi.

Some couplers have special applications such as the one in Fig. 38. The male plug side is pushed into a valve body. The lever opens and closes the valve balls in both the plug and body, controlling oil flow. When the coupler is disconnected by pulling the connected line and plug, the lever immediately rotates to close the valve balls automatically without loss of oil.

USE CORRECT HOSE TIPS

It is important that the hydraulic line connectors are compatible otherwise couplers can separate under pressure causing injury or damage.

New machines have connectors that conform to ISO and SAE standards. The hose tips on older equipment may be of an older and different design (Fig. 39). If so, the tips must be changed to the newer style so the equipment can be hooked up safely.

Also inspect coupler tips for wear or damage. Worn tips can cause leaks and flow problems. This can lead to erratic implement operation.

TEST YOURSELF

QUESTIONS

1. (True or false) "The best routing for tubing between two points is a straight line."
2. (True or false) "Hoses should be installed with a slight twist to keep the fittings tight."
3. What are the two types of tubes approved for use on high pressure hydraulic systems?
4. When a new hose is installed what should be done with brackets and clamps removed during hose removal?
5. What are skive fittings?
6. What are two flare angles mentioned and where is each used?
7. Are quick couplers dust proof?

HYDRAULIC SEALS

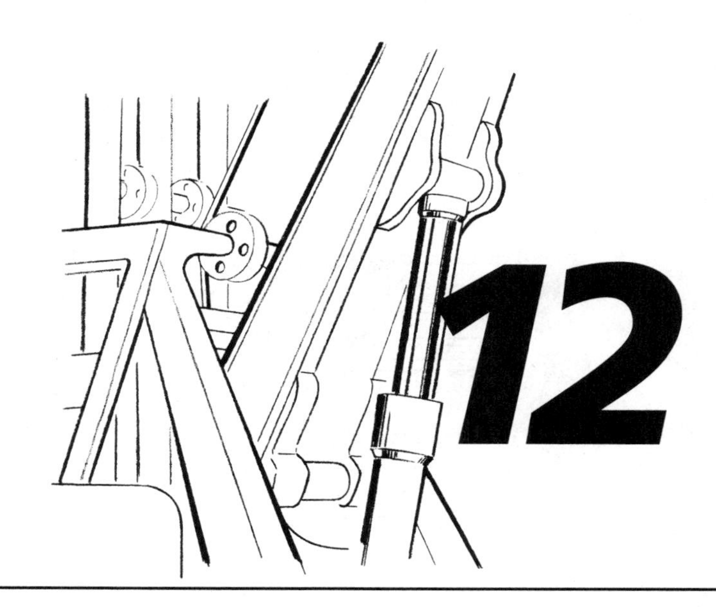

INTRODUCTION

No hydraulic circuit can operate without the proper seals to hold the fluid in the system. Seals also keep dirt and grime out of the system.

Hydraulic seals appear to be simple objects, but are complex, precision parts and must be treated carefully if they are to do their job.

THE USES OF SEALS

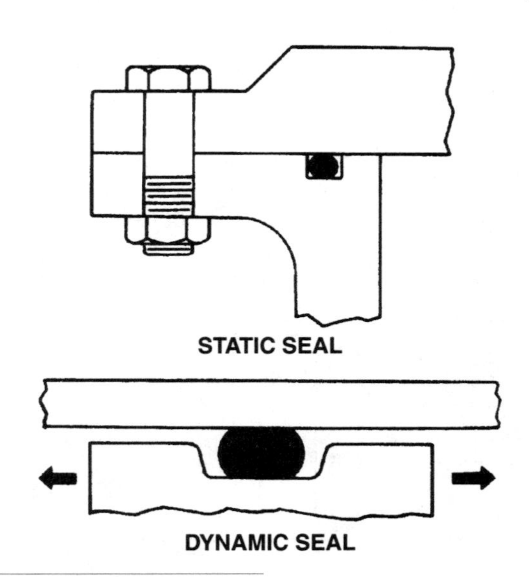

Fig. 1 — Use of Hydraulic Seals

Hydraulic seals are used in two main applications:

- **Static Seals - to seal fixed parts**
- **Dynamic Seals - to seal moving parts**

Static seals can be gaskets, O-rings or packings (Fig. 1).

Dynamic seals include shaft and rod seals and compression packings.

Later in this chapter we'll talk in more detail about the uses of seals and the problems with each one.

THE TYPES OF SEALS

Seals can be identified by their form or design (Fig. 2). Let's discuss each type of seal.

O-RINGS

The simple O-ring is the most popular seal in farm and industrial hydraulics. Usually made of synthetic rubber, it is used in both static and dynamic applications.

O-rings are designed for use in grooves where they are compressed (about ten to twenty percent) between two surfaces. In dynamic use, they must have a smooth surface to work against. O-rings are not used where they must cross openings or pass corners under pressure. They are not normally used on rotating shafts because of wear problems.

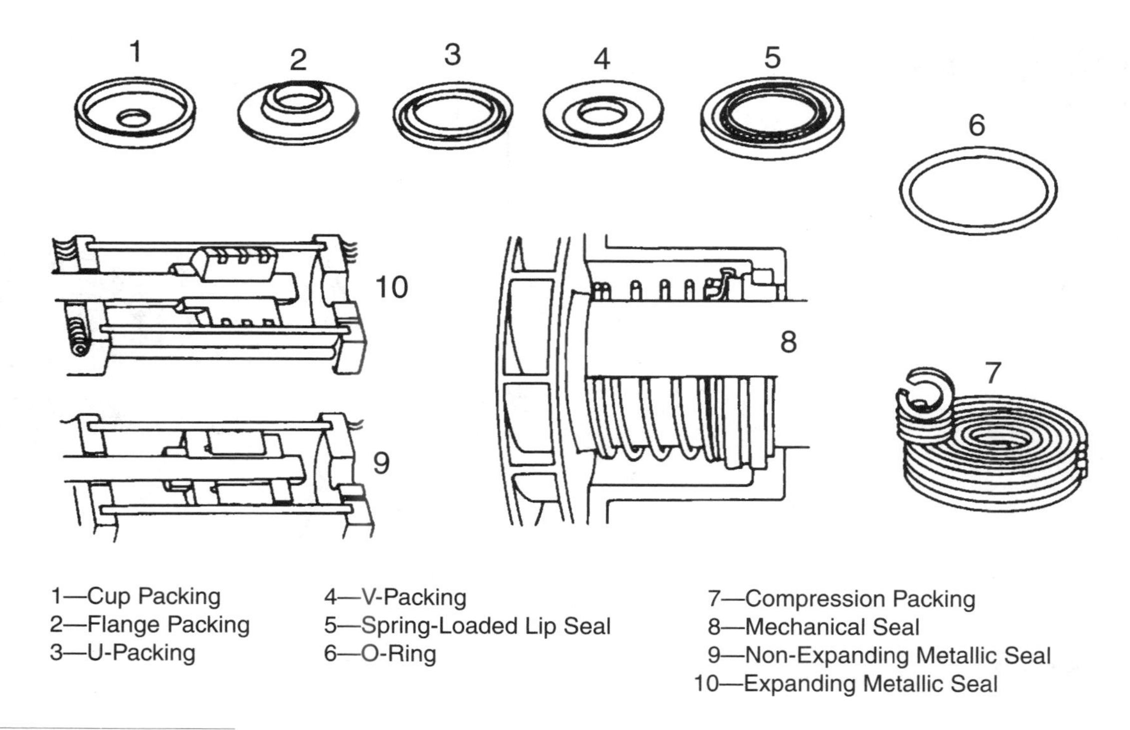

Fig. 2 — Types of Hydraulic Seals

Under high pressure, they are often strengthened by a back-up ring to prevent them from squeezing out of their grooves. The back-up ring is usually of fiber, leather, synthetic plastic, or rubber design. Leather or fiber should not be used in a cylinder.

In dynamic applications the O-ring groove must be wider than the O-ring. This allows the O-ring to roll and lubricate itself. Without this lubrication, the O-ring would erode. This can cause minimal seepage.

U- AND V-PACKINGS

U- and V-packings are dynamic seals for pistons and rod guides of cylinders, for pump shafts and gland nuts. They are made up of multiple rings of leather, synthetic and natural rubber, plastics, and other material.

It is most desirable to install these packings with the open side, or lip, toward the highest pressure. The pressure will then push the lip against the mating surface to form a tight seal.

For field installation, it may be necessary to have the V's or U's facing in the opposite direction. Without a special installation tool, the seals must face away from the direction of installation.

SPRING-LOADED LIP SEALS

These seals are a refinement of the simple U- or V-packing. The rubber lip is ringed by a spring that gives the sealing lips tension against the mating surface. Usually the seal has a metal case, which is pressed into a housing bore and remains fixed.

This seal is often used to seal rotary shafts making it one of the most used non-hydraulic seals. The lip normally faces toward the system oil. Double-lip seals are sometimes used to seal when fluids are on both sides of the seal.

CUP AND FLANGE PACKINGS

Cup and flange packings are dynamic seals and are made of leather, synthetic rubber, plastics, and other material. The surfaces are sealed by expansion of the lip or beveled edge of the packing. The higher the pressure, the tighter it seals (Fig. 3). They are used to seal cylinder pistons and piston rods.

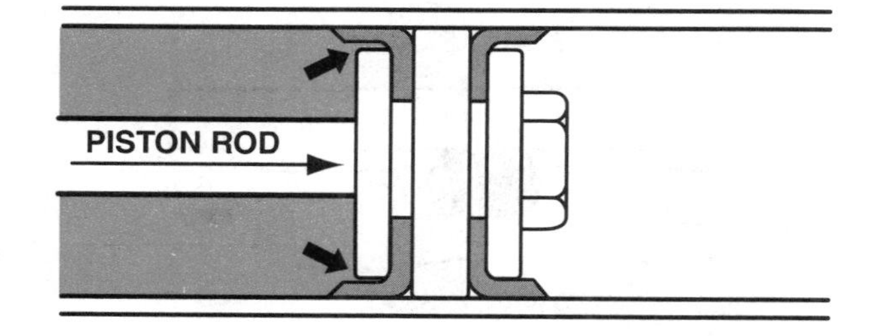

Fig. 3 — Cup Seal

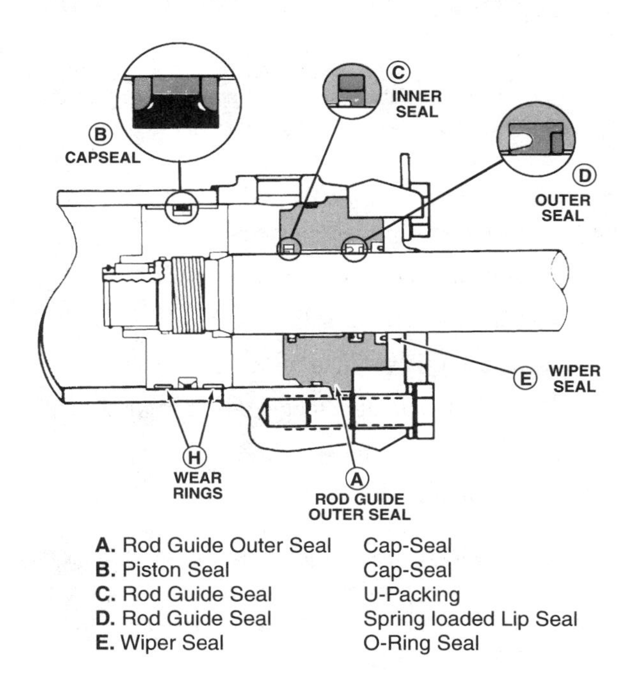

A. Rod Guide Outer Seal	Cap-Seal
B. Piston Seal	Cap-Seal
C. Rod Guide Seal	U-Packing
D. Rod Guide Seal	Spring loaded Lip Seal
E. Wiper Seal	O-Ring Seal

Fig. 4 — Typical Modern Cylinder

CAP-SEAL

The cap-seal is the most common seal used on larger high-pressure cylinders. Fig. 4 shows it used on the cylinder piston and on the rod guide (inner seal).

The cap-seal consists of a Teflon seal, a rubber expander ring and back-up washers. The rubber expander supplies a constant force, pushing the Teflon seal against the cylinder wall. The backup washers prevent the seal from being extruded into the clearance between the piston and cylinder.

The inner rod guide seal is also a cap-seal. It does not require a backup in this application.

MECHANICAL SEALS

These seals are designed to eliminate some of the problems in using chevron packings for rotating shafts. They are dynamic seals usually made of metal and rubber. Sometimes the rotating portion of the seal is made of carbon, backed up with steel.

The seal has a fixed outer part attached to the housing. An inner part is attached to the revolving shaft and a spring holds the two parts of the seal tightly together.

A rubber ring (flange-shaped) or a diaphragm is usually included to permit lateral flexibility, and to keep the rotating part of the seal in motion.

METALLIC SEALS

Metallic seals used on pistons and piston rods are very similar to the piston rings used in engines. Used as dynamic seals, they are usually made of steel. The ends overlap and are hooked together to form an endless seal.

They may be either expanding or contracting seals. Expanding seals (for use on pistons) and contracting seals (for use on piston rods) are subject to moderate friction and leakage losses.

Metallic seals are especially well adapted for use in extremely high temperatures.

Since metallic seals are subject to some leakage, wiper seals with drains are often used to capture the oil.

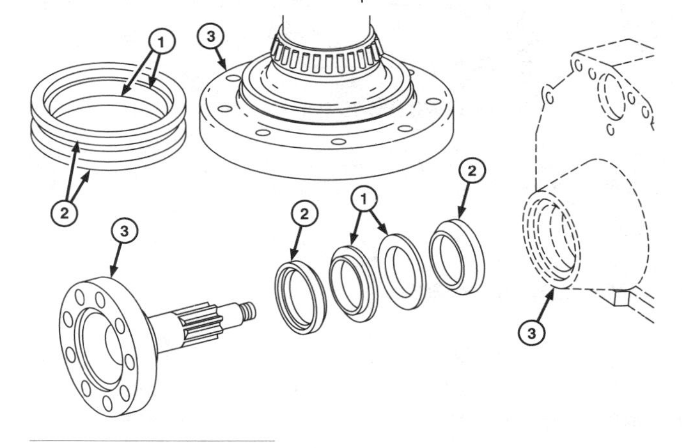

Fig. 5 — Metal Face Seal

METAL FACE SEAL

The metal face seal shown in Fig. 5 is used only as a dynamic seal. Sealing is done at the highly polished surfaces of two identical steel sealing rings (1).

The metal sealing rings are held in position by two identical rubber rings (2). They fit on a shoulder on each of the seals and into an internal bore on both the fixed and rotating member (3). The tight fit of the rubber seals prevent the seals from rotating relative to the supporting member, making the polished surface the only friction (wear) point.

COMPRESSION PACKINGS

Compression packings (jam packings) are used in dynamic applications. They are made of plastics, asbestos cloth, rubber-laminated cotton, or flexible metals.

Compression packings are often used in the same ways as U- and V-packings. They are designed as single coils or as endless rings from which sized pieces may be cut.

Compression packings are generally suitable only for low-pressure uses. Lubrication is very important, as they will score moving parts if allowed to run dry.

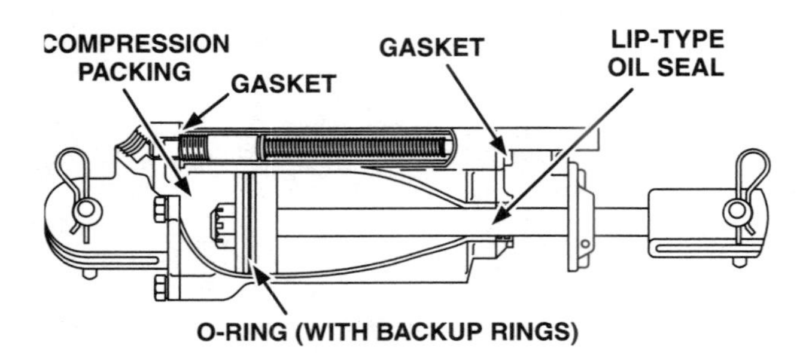

Fig. 6 — Use of Seals in a Hydraulic Cylinder

COMPRESSION GASKETS

Gaskets are suitable, of course, only for static uses. Gaskets seal by molding into the imperfections of the mating surfaces. This molding depends upon a very tight seal at all points.

Gaskets are made from many materials, both metallic and non-metallic, and come in any shape to conform to the surface being sealed.

HOW SEALS ARE SELECTED FOR EACH USE

The designer of a hydraulic system has many factors to consider when choosing a seal. Some of these are:

1. Will the seal resist expected pressures?
2. Can the seal withstand the heat of operation?
3. Will the seal wear too rapidly?
4. Will the fluid in use harm the seal?
5. Does the seal fit without dragging on the moving parts?
6. Will the seal score or scratch polished metal parts?

Each application for a seal presents a different set of problems. This is why there are so many seals on the market today. It is also why it is so crucial to follow the manufacturer's recommendation when replacing parts.

SEAL FAILURES AND REMEDIES

A top-notch hydraulic system, which is very complex, still depends upon the simple seal for good operation.

The perfect seal should prevent all leakage. This is not always practical. In dynamic applications, for example, slight leakage is necessary to provide an oil film to lubricate the moving parts. In practice, a seal is regarded as free of leaks if, after continued operation, any leakage is very hard to detect. In other words, there are no drips or pools of oil.

Internal leakage is always hard to detect and takes some testing to find out if and where the system is leaking.

To get the best use of seals, proper handling and replacement is vital. Most seals are fragile and can be easily damaged.

To prevent this, keep seals protected in their containers until ready for use. Store them in a cool, dry place free of dirt. Seals should be given the same care as precision bearings.

As a general rule, replace all seals that are disturbed during repair of the system. The price of a few seals is very cheap compared to an unsatisfactory repair job.

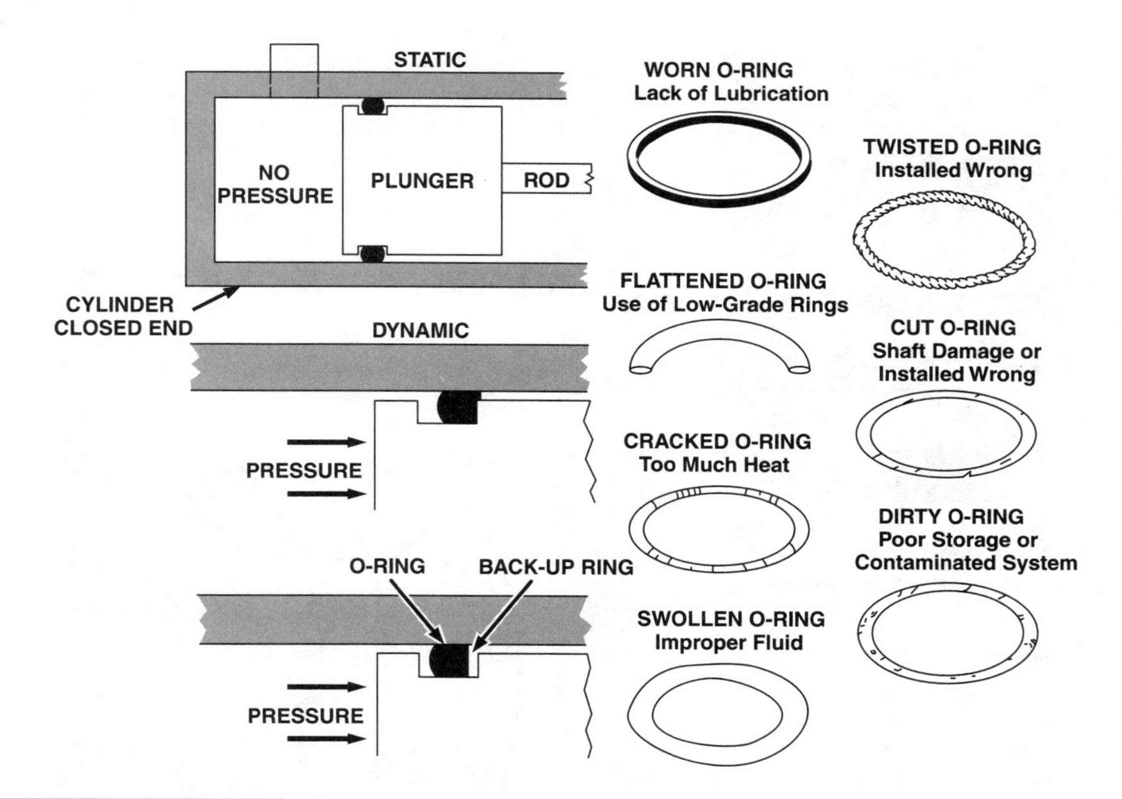

Fig. 7 — O-Rings and Backup Rings

O-RING MAINTENANCE

Cutting or nicks from sharp objects can easily damage o-Rings. Heat, improper fluids, inadequate lubrication, and improper installation (Fig. 7) can also damage them.

INSTALLATION OF O-RINGS

1. Be sure the new O-ring is compatible with the hydraulic fluid. Otherwise, the O-ring may corrode, crack, or swell in operation.
2. Clean the entire area of all dirt and grit before installing O-rings.

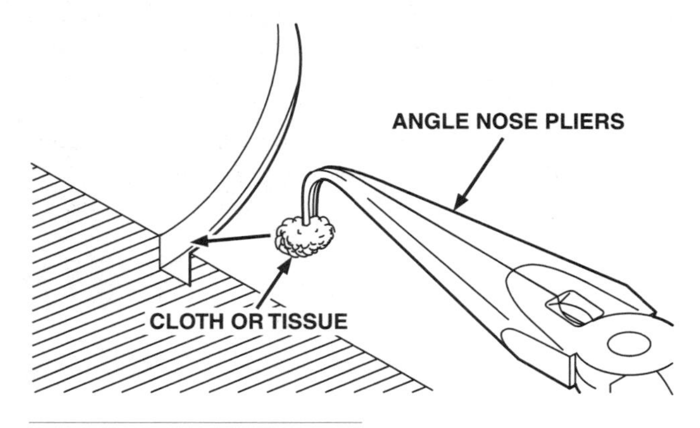

Fig. 8 — Cleaning O-Ring Groovs

3. Inspect O-ring grooves and remove any sharp edges, nicks, or burrs with a fine abrasive stone. Then clean the area again to remove all metal particles. Before installing rings, wipe grooves with a clean cloth (Fig. 8).
4. Inspect the shaft or spool (if used). Sharp edges or splines can cut O-rings. Remove any nicks or burrs with a fine abrasive stone. Then polish with a fine abrasive cloth. Clean the area again to remove all metal particles (Fig. 8).
5. Lubricate the O-ring before installing it. Use the same fluid as used in the system. Also, moisten the groove and shaft using the hydraulic fluid.
6. Install the O-ring, protecting it from sharp edges or openings. Be careful not to stretch it more than necessary.
7. Align the parts accurately before assembling to avoid twisting or damage to the ring.

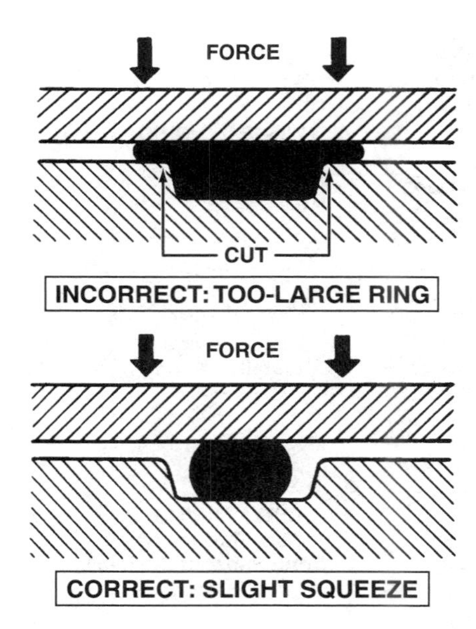

Fig. 9 — O-Ring in Static Use

8. Check to see that the O-ring is of the correct size to give only a slight "squeeze" in the installed position (Figs. 7 and 9). In dynamic use (Fig. 7), the O-ring should be able to roll slightly in its groove.

IMPORTANT: When installing spool valves, be especially careful of any O-rings. The sharp edges of the spool lands can cut the O-rings unless you are very careful.

O-RING, BACKUP RING APPLICATION

O-rings, in combination with Teflon rings, often replace flat and other types of seals in cylinders. An O-ring can be both a static and a dynamic seal when pressure is applied (Fig. 7). A back-up ring is usually used to keep the O-ring from extruding into the space between the mating parts.

Back and forth movement of the O-ring can create failure if O-rings are installed improperly. O-Rings which are the wrong size or the wrong material for the application, damaged cylinder walls, excessive heat and pressure, and fluid contamination can also damage O-rings (Fig. 7).

CHECKING O-RINGS AFTER INSTALLATION

Static O-rings, which are used as gaskets, should be tightened or torqued again after the unit has been warmed up and cycled several times.

Dynamic O-rings should be cycled (moved back and forth through their normal pattern of travel) several times to allow the ring to roll and assume a natural position.

All dynamic rings should pass a very small amount of fluid when rolling, which permits a lubricating film to pass between the ring and the shaft. Without this film, scuffing of the ring would result in short life.

MAINTENANCE OF OTHER SEALS

Modern seals use rubber, leather, plastics, and other materials that require special handling. Some of the maintenance rules are given below.

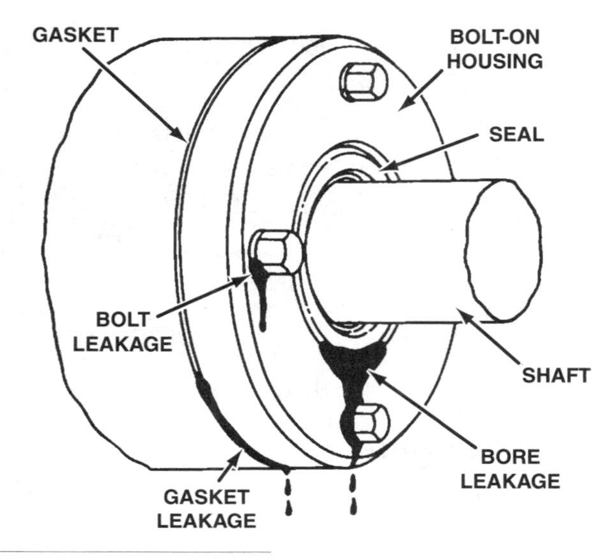

Fig. 10 — Types of Oil Leaks

CHECKING SEALS FOR LEAKAGE

Before disassembling a component, check out the causes of leakage. This may save a return job, by fixing the cause rather than the result of the problem.

Before cleaning the area around the seal, find the path of leakage (Fig. 10). Sometimes the leakage may be from sources other than the seal. Leakage could be from worn gaskets, loose bolts, cracked housings, or loose line connections.

Inspect the outside sealing area of the seal to see if it is wet or dry. If wet, see whether the oil is running out or is merely a lubricating film.

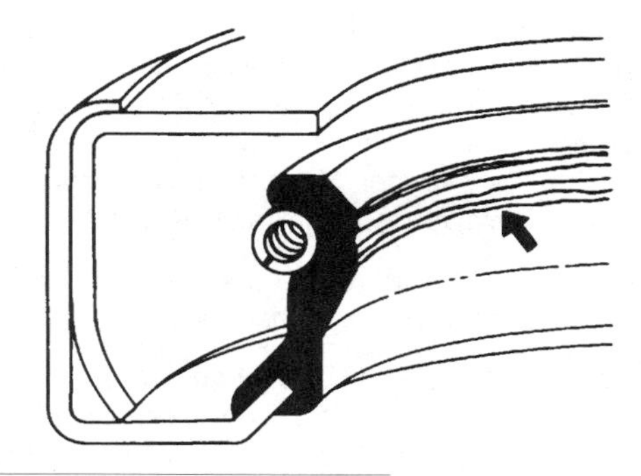

Fig. 11 — Seal Worn by Rough Shaft

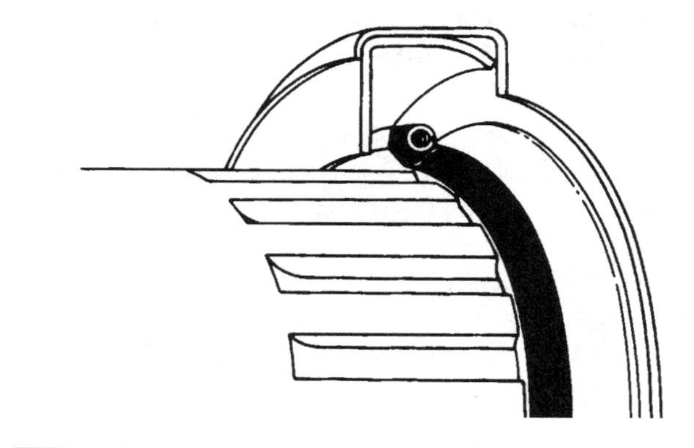

Fig. 13 — Shaft Splines or Keyways Can Damage Seals

REMOVING SEALS

During removal, continue to check for causes of leakage.

Check both the inner and outer parts of the seal for wet oil, which means leakage.

If O-ring picks or other sharp objects are used to remove seals, make certain that the metal surfaces are not scratched.

When removing the seal, inspect the sealing surface or lips (Fig. 11) before washing. Look for unusual wear, warping, cuts, gouges, and particles embedded in the seal.

On spring-loaded lip seals, be sure the spring is seated around the lip, and that the lip was not damaged when first installed.

Do not disassemble the unit any more than necessary to replace the faulty seals.

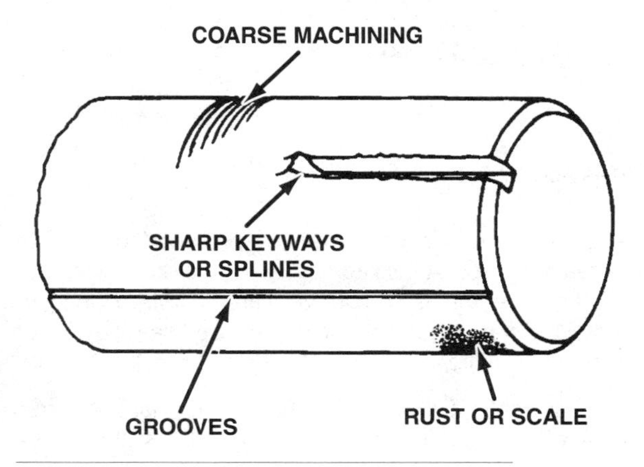

Fig. 12 — Shaft Conditions Which Damage Seals

CHECKING SHAFTS AND BORES

Check shafts for roughness at seal contact areas (Fig. 12). Look for deep scratches or nicks that could have damaged the seal.

Check shaft splines, keyway, or burred end to determine if they caused a nicked or cut the seal during installation (Fig. 13).

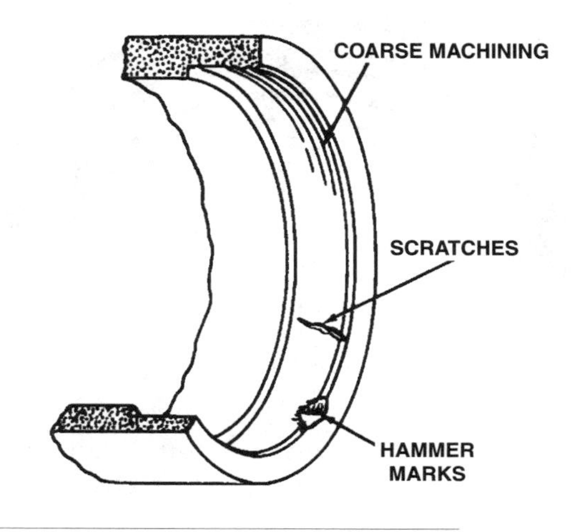

Fig. 14 — Bore Conditions Can Damage Seals

Inspect the bore into which the seal is pressed (Fig. 14). Look for nicks and gouges that could create a path of oil leakage. A coarsely machined bore can create a spiral path, allowing oil to seep out. Sharp corners at the bore edges can score the metal case of the seal when it is pressed in. These scores can make a path for oil leakage.

CHECKING SEALS FOR COMPATIBILITY WITH FLUIDS OR OPERATING TEMPERATURES

Some hydraulic oils are harmful to certain seals, especially rubber lips. Incorrect oil can either harden or soften the synthetic rubber in seals making them ineffective.

If the seal lip is "spongy," this probably means that the seal and the hydraulic fluid are not compatible. If the seal is factory-approved, then improper fluid has been used in the system (see Chapter 13).

Either heat or chemical reaction with an improper fluid can cause hardening of the seal lip.

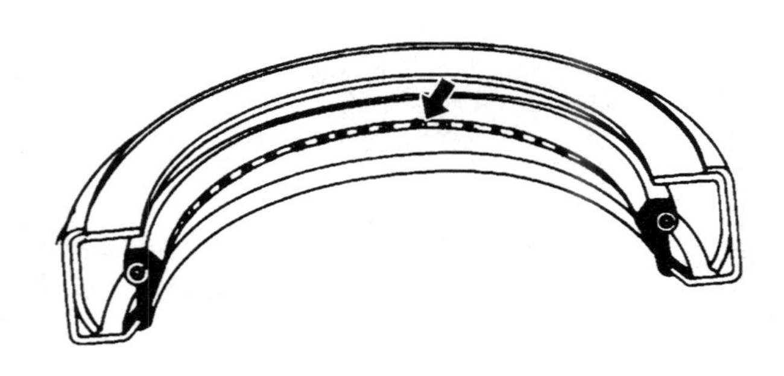

Fig. 15 — Seal Lip Damaged by Heat

Hardening of the seal lip on the area of shaft contact (Fig. 15) is generally the result of heat from either the shaft or the fluid.

INSTALLING SEALS

1. Install only seals recommended by the manufacturer of the machine.
2. Use only the proper fluids as stated in the machine operator's manual.
3. Keep the seals and fluids clean and free of dirt.
4. Before installing seals, clean the shaft or bore area. Inspect these areas for damage. File or stone away any burrs or bad nicks and polish with a fine emery cloth for a smooth finish, then the area **must be thoroughly cleaned** to remove metal particles.

   In dynamic applications, the sliding surface for the seal should have a mirror finish for best operation.
5. Lubricate the seal, especially any lips, to ease installation. Use the systems hydraulic fluid to lubricate the seal. Also, soak packings in the hydraulic fluid before installing them.
6. With metal-cased seals, coat the seal's outside diameter with a thin film of gasket cement to prevent bore leakage.

   NOTE: Pre-coated seals do not require sealant on the case.
7. Use a factory-recommended tool to install the seal properly. This is very important with pressed in seals. If using a universal driver, make certain the driver contacts the seal case at the outside diameter. Make certain the driver guide fits the lip diameter. ***Never use sharp tools such as punches, screwdrivers, etc. to install seals.***
8. Packings should fit snugly without the use of undue force. Be sure they are not too tight and that cased seals are not bent.
9. Use shim stock to protect seals when installing them over sharp edges such as shaft splines and keyways. Place rolled plastic shim stock (0.003-0.010 inch) over the sharp edge, then pull it out after the seal is in place.

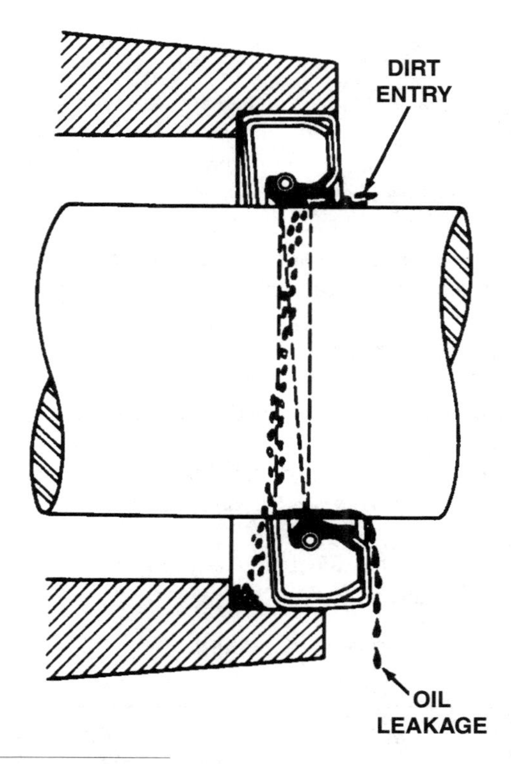

Fig. 16 — Cocked Seals

10. Be sure the seal is driven in evenly to prevent "cocking" of the seal (Fig. 16). A cocked seal can allow oil to leak out and dirt to enter as shown. Be careful not to bend or "dish" the flat metal area of metal-cased seals. This can distort the seal lips.
11. After assembly, always check the unit by hand for free operation, if possible, before starting up the system.
12. Try to prevent dirt and grit from falling on piston rods, etc. and being carried into the seal. This material can quickly damage the seal or score the metal surfaces.

RUN-IN CHECKING OF NEW LIP-TYPE SEALS

When a new lip-type seal is installed on a clean shaft, a break-in period of a few hours is required to seat the seal lip on the shaft surface. During this period, the seal polishes a pattern on the shaft. The shaft in turn seats the seal lip, wearing the knife-sharp lip to a narrow band.

During this period, slight seepage may occur. After seating, the seal should perform without any measurable leakage.

TEST YOURSELF

QUESTIONS

1. (Fill in the blanks with "dynamic" or "static".) "__________ seals are used to seal fixed parts, while ______________ seals are used to seal moving parts."

2. What is the most common seal used in farm and industrial hydraulic systems?

3. (True or false?) "Slight leakage is permissible in some dynamic seal applications."

4. (True or false?) "When repairing a component, replace only the seals that are damaged."

5. (True or False) "A wrench socket is an acceptable tool for the installation of a lip seal"

HYDRAULIC FLUIDS

Fig. 1 — Hydraulic Fluids are Highly Refined Petroleum Oils

INTRODUCTION

The hydraulic fluid is the medium by which power is transmitted from a pump to the mechanisms which produce work such as cylinders and hydraulic motors. The fluid is just as important as any other part of a hydraulic system. In fact, it has been estimated that 70 percent of hydraulic problems stem from the use of improper types of fluids, or fluids containing dirt and other contaminants.

When we speak of hydraulic fluid, in almost all cases we really mean a highly refined petroleum oil (Fig. 1) usually containing additives, some to suppress unwanted properties and others to give the oil desirable properties.

Here it might be well to stress a point of caution. **NEVER** use hydraulic brake fluid in a hydraulic system designed to use petroleum-base oils. Brake fluid is **not** a petroleum product and is completely incompatible with petroleum-base hydraulic fluids.

During the development of hydraulic equipment, engineers make careful studies of the available fluids to find one best suited to enable their product to give efficient, trouble-free operation. Sometimes it is even necessary to develop a new fluid which has just the right properties. This is why it is always essential to use the fluid recommended in the instructions that accompany a hydraulic machine or mechanism.

BASIC PROPERTIES OF HYDRAULIC FLUID

First of all, hydraulic fluid must be capable of transmitting the power applied to it. Of equal importance it must do several other things. It must provide lubrication for moving parts, be stable over a long period of time, protect machine parts from rust and corrosion, resist foaming and oxidation and be capable of separating itself readily from air, water, and other contaminants. The fluid must also maintain proper viscosity through a wide temperature range, and finally, be readily available and reasonably economical to use.

VISCOSITY

For proper power transmission, this is a most important property. Viscosity is a measurement of a fluid's resistance to flow. Said another way, it is a fluid's "thickness" at a given temperature. Viscosity is expressed by SAE (Society of Automotive Engineers) numbers; 5W, 10W, 20W, 30, 40, etc. All petroleum oils tend to become thin as temperature goes up, and to thicken as the temperature goes down. If viscosity is too low (fluid too thin), the possibility of leakage past seals and from joints is increased. This is particularly true in modern pumps, valves and motors which depend on close fitting parts for creating and maintaining oil pressure. If viscosity it too high (fluid too thick), sluggish operation results and extra horsepower is required to push the fluid through the system. Viscosity also has a definite influence on a fluid's ability to lubricate moving parts.

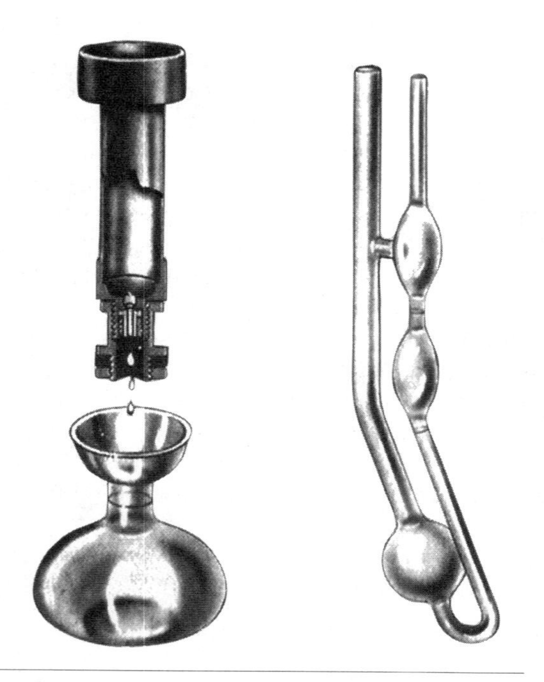

Fig. 2 — Saybolt Viscometer (Left) and Kinematic Viscometer (Right)

Viscosity is determined by measuring the time required for 3.66 cubic inches (60 cubic centimeters) of an oil at a temperature of 210°F (98.9°C) to flow through a small orifice in an instrument known as a Saybold Viscometer or another instrument called a Kinematic Viscometer (Fig. 2). The actual SAE number is determined by comparing the time required for the oil to pass through the instruments with a chart provided by the Society of Automotive Engineers.

Viscosity Index (VI)

This is simply a measure of a fluid's change in thickness with respect to changes in temperature. If a fluid becomes thick at low temperatures and very thin at high temperatures, it has a **low** VI. On the other hand if viscosity remains relatively the same at varying temperatures, the fluid has a **high** VI (Fig. 3). As pointed out earlier, in a fluid with good viscosity characteristics, there is a balance between a fluid thick enough to prevent leakage and provide good lubrication while, at the same time, being thin enough to flow readily through the system. Therefore, a fluid with a high VI is almost always desirable and must be an important consideration in hydraulic fluid recommendations.

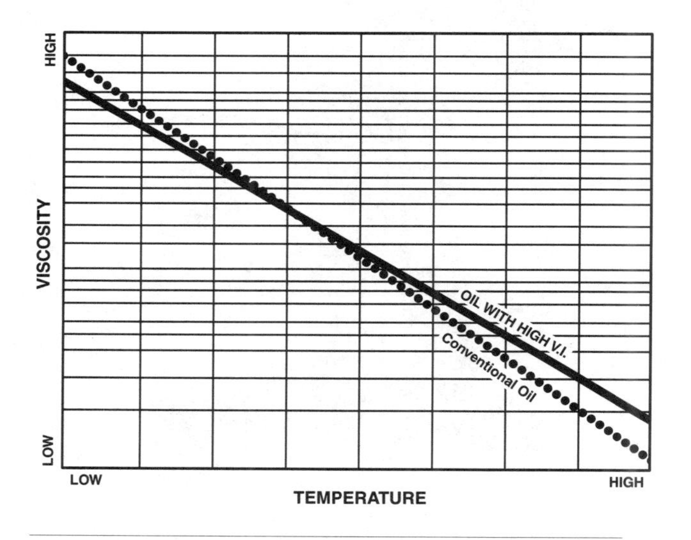

Fig. 3 — The Viscosity of an Oil with a High VI Contrasted with the Viscosity of a More Conventional Oil

Viscosity Index Improver

Even though carefully refined oils have a good viscosity index, a substance called a viscosity index improver is often added to hydraulic fluids. This substance increases the VI of the fluid so that its viscosity change over a wide range of temperatures is as little as practical.

WEAR PREVENTION

Most hydraulic components are fitted with great precision, yet they must be lubricated to prevent wear (Fig. 4). To provide this lubrication, a satisfactory fluid must have good "oiliness" to get in the tiny spaces available between moving parts and hold friction between the parts to a minimum.

A good fluid must also have the ability to "stick" to these closely fitted parts, even when they are quite warm.

Good lubricating qualities become even more important in many modern machines where the hydraulic fluid has a double purpose—to operate the hydraulic mechanisms, and also to lubricate the transmission, differential, and other parts of the machine.

The best hydraulic fluids contain an "extreme pressure" additive which assures good lubrication of very close fitting metal-to-metal parts operating at high pressures and temperatures. This additive reduces friction and helps to prevent galling, scoring, seizure, and wear.

Fig. 4 — Vane Pump Ring Worn Due to Lack of Lubrication

RESISTANCE TO OXIDATION

Everyone is familiar with the effects of air on a piece of shiny iron, especially in the presence of water; it combines with oxygen in the air to form rust and other foreign materials. Its chemical properties are also changed. Just like the iron, all oils combine to some extent with the oxygen in the air. This changes the oils chemical composition. Organic acids are formed which may be harmful to metal parts (Fig. 5 and 6) and many types of seals and packings in the system. This can also give oil a pungent odor. In addition to the acids, sludges are often formed through the reactions between fluids and air. Both reactions are speeded up by the presence of water and other contaminants such as dust, dirt and metallic particles in the fluid. This is one reason why an efficient filtering system is so important in any hydraulic system. Heat is also a very important factor in oxidation. It has been determined, for instance, that for every 18°F (10°C) rise in temperature, the rate of oxidation doubles. This is one of the reasons some hydraulic systems contain a cooler to hold temperatures to reasonable limits.

Fortunately, carefully refined fluids, plus the addition of a special chemical, successfully resist oxidation. With careful attention to prevent the entrance of dirt and other contaminants, most modern hydraulic fluids will operate for many hours without ill effects due to oxidation. However, oxidation can be a real problem unless high-quality hydraulic fluids, specifically recommended by the equipment manufacturer are used.

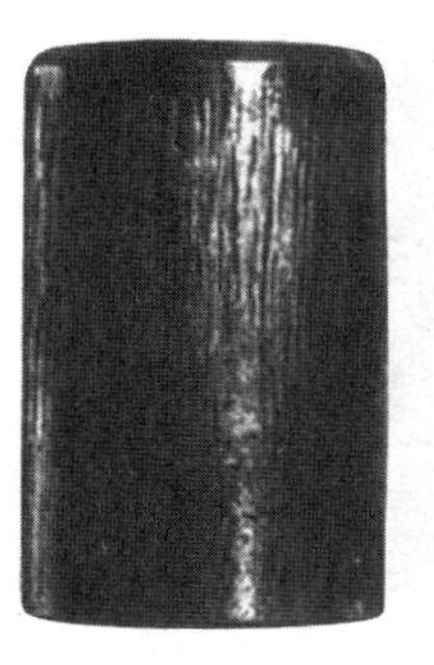
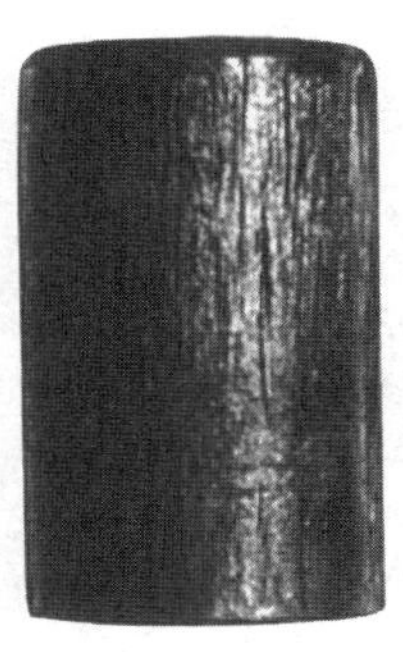

Fig. 5 — Radial Pump Pistons Scored by Contaminated Fluid

Fig. 6 — Vane Pump Rotor Ring Worn and Pitted by Contaminated Fluid

RUST AND CORROSION PREVENTION

Rust and corrosion are both related to oxidation, and a hydraulic fluid (providing it is kept clean) with good anti-oxidation qualities is likely to resist rust and corrosion. However, the possibilities of rust or corrosion developing are always present and they cannot be ignored. Rust and corrosion differ in that rusting adds to metal making the part larger while corrosion, caused by acids or local electrochemical cells, is an eating away of the metal. Either condition, of course, is highly detrimental to a hydraulic mechanism. Rust causes rough spots (Fig. 7) which damage seals and close fitting parts. It is most likely to form during storage periods, down-time, or even overnight. Corrosion affects the fit of closely machined parts and permits undesirable leakage. Both rust and corrosion cause erratic operation and untimely wear.

Good hydraulic fluids contain both rust and corrosion additives which neutralize corrosion-forming acids and cling to metal parts to protect them from rusting and corroding.

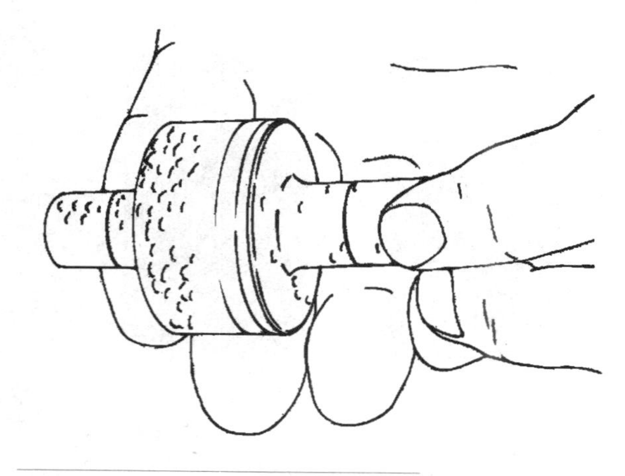

Fig. 7 — Pump Drive Cam Pitted by Rust

Fig. 8 — Balanced Vane Pump Rusted by Water in Fluid

RESISTANCE TO FOAMING

The proper operation of any hydraulic system is based on the fact that fluids cannot be compressed by pressures normally encountered in the system. In effect the fluid acts like a "liquid" steel rod. Any force applied to it at one end is transmitted to the other end without any "slack" due to compression. However, air which **is** compressible can be absorbed in the fluid. In many systems the fluid reservoir is directly exposed to the atmosphere which promotes entrance of air into the fluid. In addition, air can enter the system through defective packings, leaky lines, or if the fluid level in the reservoir is allowed to get too low. In many systems turbulence promotes the mixing of fluid and air.

Good fluids have the capacity to "dissolve" a small amount of air. The amount of air that can be dissolved increases as pressure and temperature increase. This dissolved air has no harmful effects on operation. But if the amount of air which enters the fluid is greater than the fluid's capacity to dissolve it, bubbles form which, since air is compressible, result in mushy, unsatisfactory operation. Furthermore, some air in solution under pressure comes out of solution when pressure is released. This air creates foam which seriously affects proper action, and especially, lubrication.

While most well refined oils are not subject to excessive foaming, most good hydraulic fluids contain a foam inhibitor additive which speeds up the rate at which bubbles break up. This improves the fluid's ability to do its work properly, and to increase its capacity for adequately lubricating moving parts.

ABILITY TO SEPARATE FROM WATER

Contrary to popular opinion oil and water **will** mix. This mixture is called an "emulsification." It is almost impossible to keep all water out of a hydraulic system. Water vapor enters the reservoir where it condenses into droplets. It also may enter through tiny leaks in the system. Because of the violent agitation, churning, and continual recirculation in a typical hydraulic system, the water and fluid quickly mix to form an emulsion. Any appreciable amount of water in the fluid is highly detrimental. The emulsion promotes rust (Fig. 8), increases oxidation which forms acids and sludges, and reduces the fluid's ability to lubricate moving parts properly. Also, emulsions often have a slimy, sticky, or pasty consistency which interferes with normal operation of valves and other parts.

LACK OF OTHER CONTAMINANTS

It should go without saying that a good hydraulic fluid should be as free as possible of contaminants such as metallic parti-

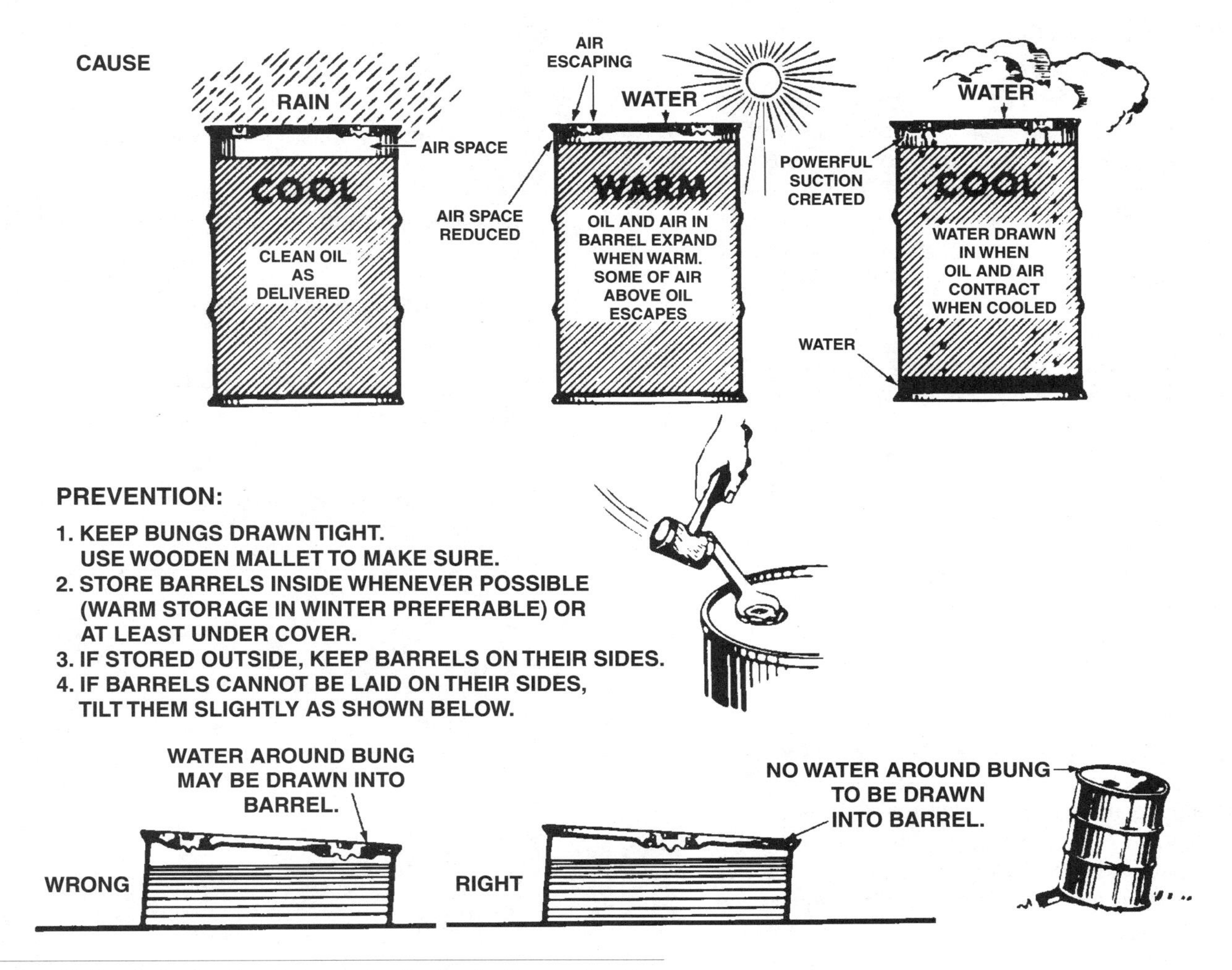

Fig. 9 — Oil Contamination Can Be Prevented By Careful Storage Practices

cles, dust, dirt and the like. Such materials are not only likely to damage closely fitted parts, but they also seem to help the undesirable oxidation process to take place.

Use of reliable fluids, careful storage, good filters, proper handling of fluids, and periodic cleaning of hydraulic systems all reduce the danger of contamination.

MAINTENANCE OF FLUID

As has been said many times, dirt and contamination are the worst enemies of any hydraulic system. Continued, long operation at high efficiency is very dependent upon proper fluid maintenance. First of all, **only** a fluid recommended by the manufacturer of the system should be used; it should be checked at the suggested intervals maintained at the correct level, properly filtered, and changed at the recommended intervals.

OIL CHANGE SCHEDULE

Periodic drainage of the entire system is very important. This is the only way to remove contaminants, products of oxidation such as sludge and acids, and other particles that may be injurious to the system. Actually, in modern hydraulic systems using approved fluids, the drain period is not frequent and it should work no hardship on the part of the owner to follow the manufacturer's directions. Many modern fluids are so highly refined, filtered, and fortified by additives that system flushing is not necessary. However if oil flushing is recommended by the manufacturer, it is always advisable to follow directions so as not to contaminate the new oil with flushing oil that cannot be drained from the system.

See Chapter 14 for details on draining, cleaning, and flushing systems.

KEEPING FLUIDS CLEAN

All good hydraulic fluids, which come in cans or barrels, are delivered perfectly clean and free from contaminants. It is when the containers are opened or stored that troubles develop.

When opening a can or barrel, be absolutely certain that the area around the opening is completely free of dust, dirt, lint, or water. If a container, funnel, or hose is required to fill the system, be sure that it is spotless.

When possible, always store barrels of hydraulic fluid indoors, or at least under cover, and be sure the bung is tight. If barrels are stored in the sun without a tight bung, the fluid will expand, forcing some air from the bung. Then as the fluid cools off, the fluid contracts, drawing any rain, dew, or moisture into the barrel and fluid. The detrimental effects of water in hydraulic fluids have already been discussed. For this reason, keeps the bungs in barrels as tight as possible, and tip the barrels in such a manner that water cannot collect around the bung (Fig. 9).

CHOOSE THE RIGHT VISCOSITY

Choosing the quality oil with the right viscosity will extend component life and help to ensure troublefree operation. Manufacturers conduct extensive performance tests on how specific oils function under different operating and climactic conditions in their equipment. So, always follow the manufacturer's recommendations. Be particularly attentive to the oil recommendations for different temperature ranges. See Fig. 10.

TEST YOURSELF

QUESTIONS

1. What are good hydraulic fluids composed of?

2. At higher temperature do oils become thicker of thinner?

3. Is a high-viscosity oil thicker of thinner than a low-viscosity oil?

4. What happens when a hydraulic fluid oxidizes?

5. Why is it important to follow manufacturer's recommendations?

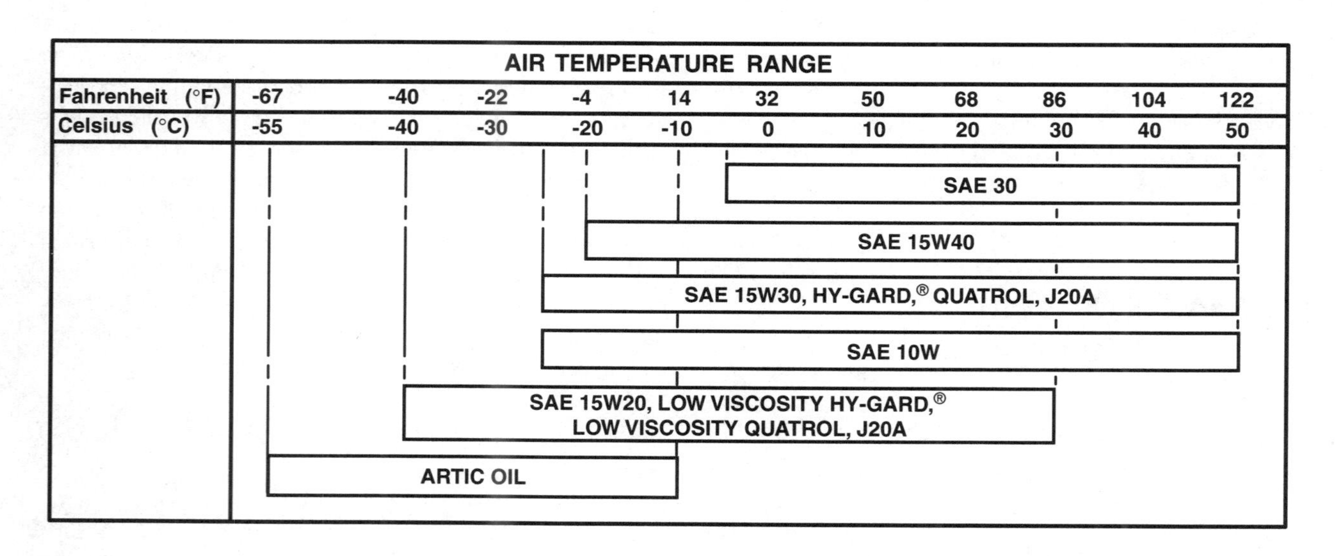

Fig. 10 — Example of Oil Recommendation

GENERAL MAINTENANCE

Fig. 1 — Proper Maintenance = Dependability

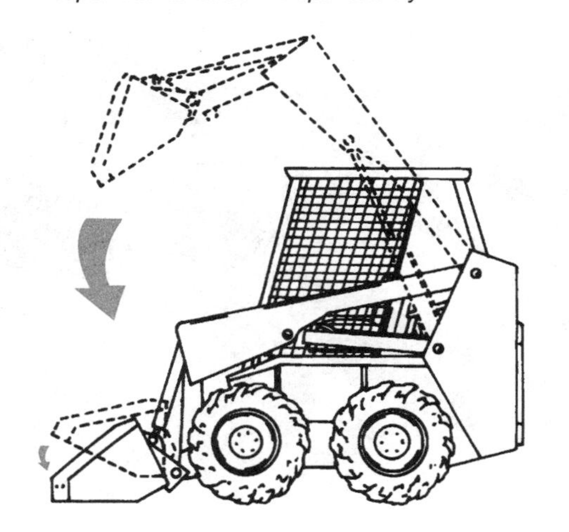

Fig. 2 — Lower Bucket

INTRODUCTION

Proper maintenance reduces the hydraulic system trouble (Fig. 1).

The number one ingredient in **PROPER MAINTENANCE** is an **OBSERVANT AND CONSCIENTIOUS OPERATOR**.

The number one tool in **PROPER MAINTENANCE** is the manufacturer **OPERATOR'S MANUAL**.

The operator's manual provides:

- **Safety precautions**
- **Pre operating inspection**
- **Periodic service intervals and procedures**
- **Tips for proper operation**
- **Simple problem diagnosis**

These maintenance procedures coupled with accurate and complete records will eliminate many common problems and make others easier to diagnose.

This chapter will explain general maintenance that will help keep hydraulic systems operating with minimum breakdowns.

SAFETY RULES FOR HYDRAULICS

First, think about safety:

- Always lower implements before working on them (Fig. 2).
- Park equipment where children cannot reach (Fig. 3).
- Block up hydraulic implements if you must work on them in the raised position (Fig. 4).
- Never service a hydraulic system while the engine is running, unless absolutely necessary, such as while bleeding the system or looking for leaks.
- Do not remove cylinders until the implements are resting safely on the ground or blocked up securely. Make sure the engine is off and the key is removed.
- Lock the cylinder stops to hold the working units in place solidly while transporting the equipment.
- Relieve all hydraulic pressure before disconnecting lines or hoses.
- Make sure that all line connections are tight and lines are good before starting equipment. Hydraulic fluid escaping under pressure from a loose connection or broken line can cause personal injury in several ways, including fire, slipping (falls), burns and very dangerous injection under the skin (Fig. 5).

Fig. 3 — Park In a Safe Place

Fig. 4 — Block Up Before Maintenance

AVOID HIGH-PRESSURE FLUIDS

CAUTION: Escaping fluid under pressure can penetrate the skin causing serious injury. Avoid the hazard by relieving pressure before disconnecting hydraulic or other lines. Tighten all connections before applying pressures. Search for leaks with a piece of cardboard. Protect hands and body from high pressure fluids.

If an accident occurs, see a doctor immediately. Any fluid injected into the skin must be surgically removed within a few hours or gangrene may result. Doctors unfamiliar with this type of injury should reference a knowledge able medical source.

Fig. 5 — Be Careful

- *Many hydraulic components are heavy. Get help to lift them. You may need a chain hoist (Fig. 6), jack, or blocks for ease and safety.*
- *Wash parts with safe cleaning solvents to reduce contamination.*
- *Always keep hydraulic equipment adjusted properly for good control.*

Fig. 6 — Use a Chain Hoist

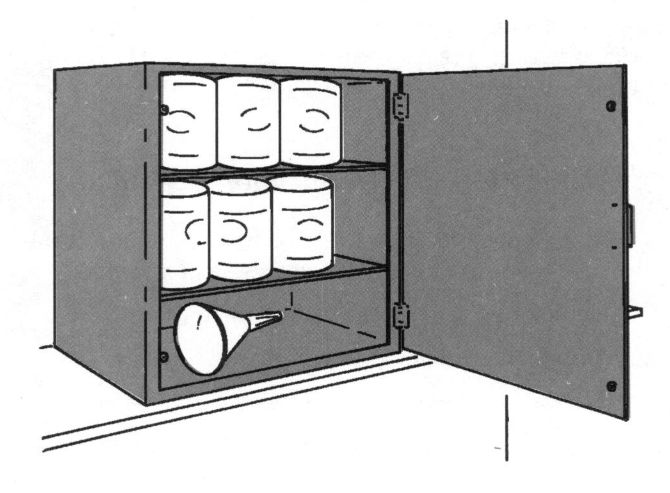

Fig. 7 — Keep Oil Clean

Fig. 8 — Clean Parts

GOOD MAINTENANCE PRACTICES

- *Keep all parts clean.*
- *Stop, look, listen, and touch before working on the equipment.*
- *Change fluid and service filters at intervals recommended by the equipment manufacturer or more often.*
- *Maintain complete, accurate records of all service and repair to the system.*

CLEANLINESS

The most important rule of hydraulic maintenance is: keep everything absolutely clean. **Keep everything except hydraulic fluid out of the system!** Even small, soft items can score valves, clog orifices, stick controls, and seize pumps. These contaminants may keep equipment from operating and cause expensive damage. Keep hydraulic systems clean:

- *Use clean hydraulic fluid.*
- *Wash dirt off the outside of hydraulic components.*
- *Keep work area clean.*
- *Do not contaminate system when adding or changing fluid.*

KEEP STORED HYDRAULIC FLUID CLEAN

Make every effort to keep the hydraulic fluid clean. Store fluid in a clean, dry place (Fig. 7). Wipe off and inspect containers before you pour fluid into a machine. Contamination from dirty oil cans is transferred to fluid when the can is opened and poured.

If a funnel is used, wipe it clean inside and out. Oily residue left on a funnel holds dirt until it is flushed into the system with the new fluid.

Do not use hydraulic fluid from a can that has been opened and left uncovered. Contaminants, including moisture, can enter an open can.

Use a lint-free cloth to wipe the dipstick when checking fluid level so lint isn't transferred from dipstick to fluid.

Wipe dirt from the area before you remove any cover, plug, or part. Use safe solvent or steam clean if necessary. This is just as important when checking the fluid level, removing the filler cap, or installing new filters as it is when removing a line or component.

NOTE: Cover breather caps, vents, and dipstick holes before cleaning with steam or solvent so moisture, solvent, and dirt cannot enter the system.

Cover the ends of disconnected lines with plugs and cover openings exposed when parts are removed.

Clean all removed parts with solvent and store them in plastic bags or clean containers until they are needed (Fig. 8).

Use extreme caution when cleaning precision parts. Prevent one part from nicking the finely machined surface of another. Use safe chemical solvents to clean metal parts. Do not allow these chemicals to contact seals, gaskets, or other nonmetal parts. Rinse off the solvent before assembling. Dry all parts with compressed air. Then coat them with oil to prevent rust.

KEEP THE WORK AREA CLEAN

The work area, including the tools that are used, must be clean (Fig. 9). A workbench with an aluminum sheet metal top is good. Use a vacuum cleaner to remove tiny metal and dirt particles. It does little good to repair a component, then have it fail because it was contaminated on a workbench.

Dirty tools should be cleaned, and worn tools should be repaired or replaced. Hammers used on hydraulic systems should be plastic, leather, or brass so there is less danger of striking chips.

Fig. 9 — Keep a Clean Shop

CHANGE OR ADD FLUID CORRECTLY

Immediately after filling a hydraulic system with fluid, normal wear, dust in the air, moisture, and heat generated by operation, begin contaminating the fluid. It is important to follow the equipment manufacturer's recommendation for changing fluid and filters. Do not stretch the intervals between service longer than recommended. Long stretches without use do not increase the service interval. Dust and moisture in the air may contaminate the fluid faster than when the system is being operated. Moisture in the air can be just as damaging as dust to the close fitting parts in a hydraulic system.

Clean around the dipstick and the filler cap first. Pull the dipstick and check the condition of fluid. Adding oil will not remove contamination. If the records or contamination in fluid on the dipstick indicate the fluid should be changed: drain the system, service the filters, and fill to the correct level with clean, new hydraulic fluid. Do not overfill! Overfilling causes foaming in some systems. Do not mix different kinds of hydraulic fluids in a system. Use only recommended hydraulic fluids.

CHECK THE SYSTEM BEFORE REPAIR

Check the system for proper fluid level, leaks, proper operation, and overheating (Fig. 10) before you put it to work.

1. Check for twisted, kinked, dented, or cut lines.
2. Operate the system until it is warm.
3. Operate all the functions several times.
4. Then check fluid level again, look for leaks, and check for overheating.

CHECK THE SYSTEM DURING OPERATION

Stop and check the fluid condition more often as the time for changing approaches. Do not extend the change interval.

Check for foaming or milky fluid. It indicates air or water in the fluid.

If there is a noticeable change in fluid level from one day to the next, check for external leaks.

Check lines and connections often. Repair or tighten leaking lines before continuing operation. Follow safe procedures (Fig. 11). Check for pinhole leaks safely (Fig. 12). Tighten fittings only until leak stops. Use two wrenches (Fig. 13) to loosen or tighten connections so lines and hoses don't twist.

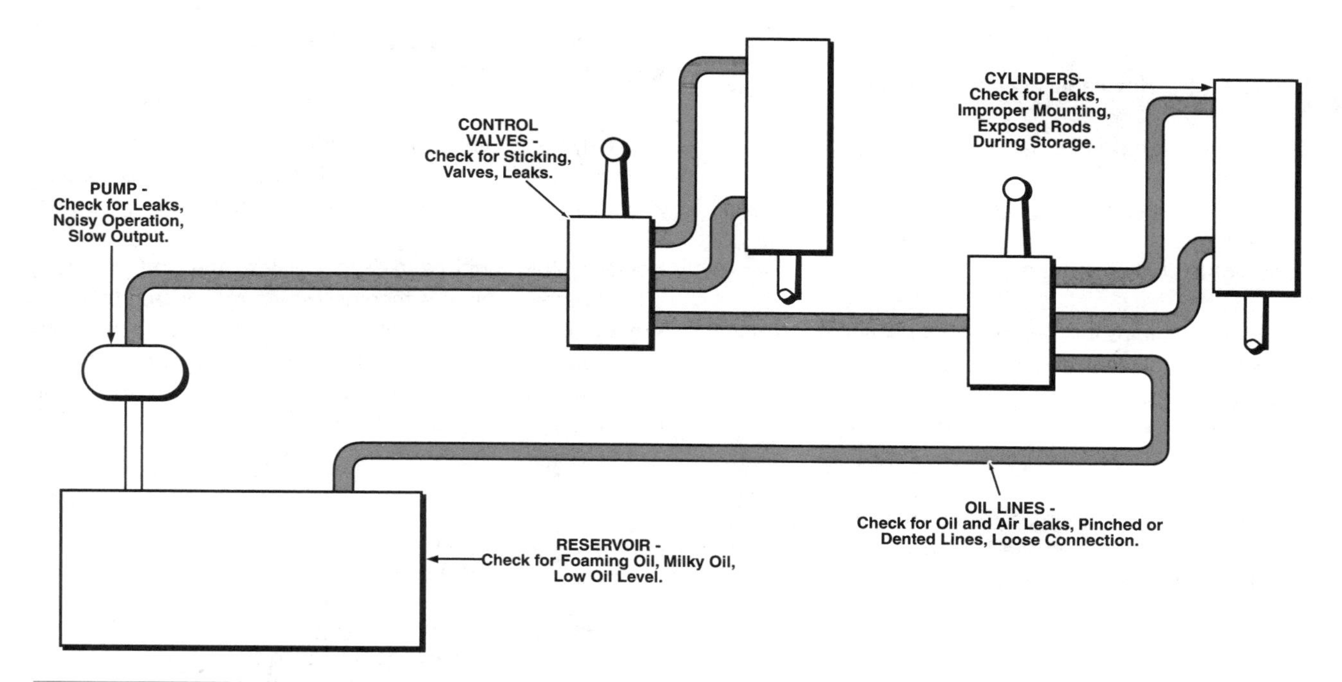

Fig. 10 — Check the System

Fig. 11 — Be Careful — Avoid High-Pressure Fluids

Fig. 12 — Check for Leaks

External leaks leave visible puddles and drips. Internal leaks are harder to detect. A continuing low oil level, sluggish operation, and excessive heat are indications of internal leaks. Feel the valves for heat caused by fluid leaking (Fig. 14).

CAUTION: Escaping fluid under pressure can penetrate the skin causing serious injury. Avoid the hazard by relieving pressure before disconnecting hydraulic or other lines. Tighten all connections before applying pressure. Search for leaks with a piece of cardboard. Protect hands and body from high pressure fluids.

If an accident occurs, see a doctor immediately. Any fluid injected into the skin must be surgically removed within a few hours or gangrene may result. Doctors unfamiliar with this type of injury should reference a knowledgeable medical source.

 **CAUTION: Hot valves can burn hands.**

Replace valves that have internal leaks.

Cylinders should always be mounted so they can't twist and bend a rod. Watch them during operation for movement. Stop and tighten loose cylinders.

Watch hydraulic pumps for leaks during operation. If you suspect a leak, test to see if enough fluid is being delivered to operate the equipment correctly. Refer to the technical manual for test equipment and flow and pressure specifications.

Never allow a hydraulic motor to overheat. If allowed to run hot, damage will result. If a motor is overheating, make sure the fluid level is correct. Then check for restrictions and contaminated fluid. Check for leaks from connections and seals. Correct the condition before continuing operation.

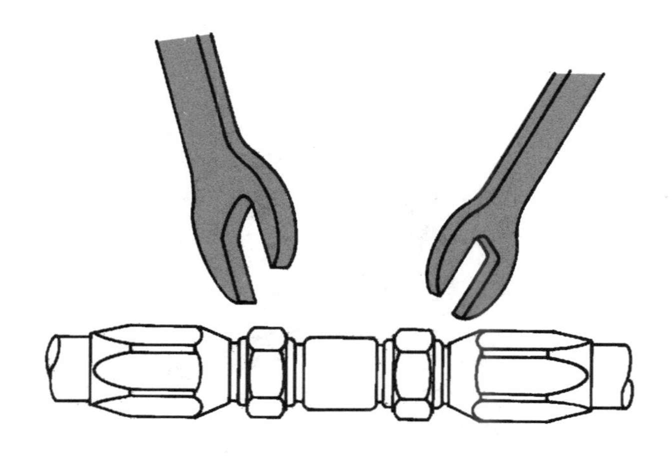

Fig. 13 — Don't Twist Connections

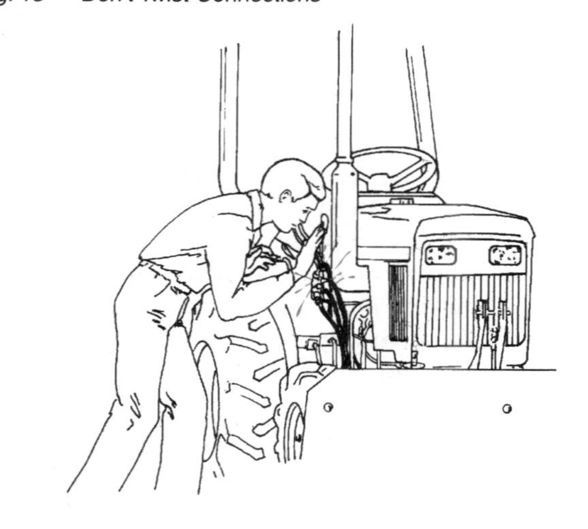

Fig. 14 — Hot Valve Indicates Internal Leak

GENERAL FLUID AND FILTER MAINTENANCE

Good hydraulic fluids hold contaminants in suspension so they are trapped in filters. Most hydraulic fluids also contain additives which help slow the formation of sludge and acids.

However, the additives in fluid will lose their effectiveness after a certain time. It is then necessary to drain and refill the system. Hydraulic fluid filters also need maintenance. As described in the chapter on filters, filter elements collect contaminants until they are saturated. When saturated, they must be replaced or cleaned.

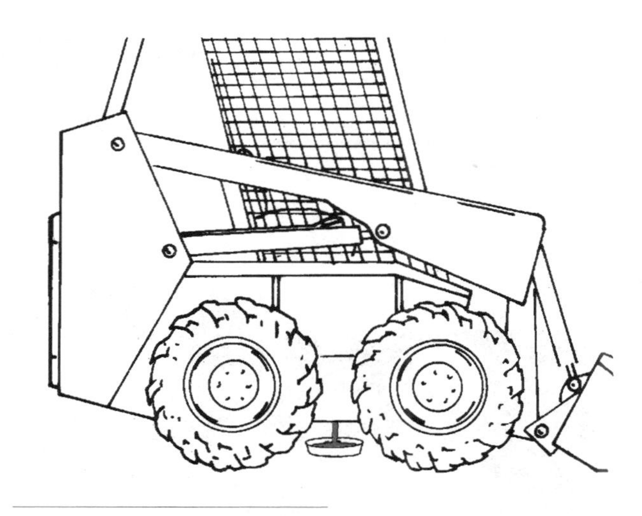

Fig. 15 — Drain the Hydraulic Oil

Fig. 16 — Wipe Off Cap and Container

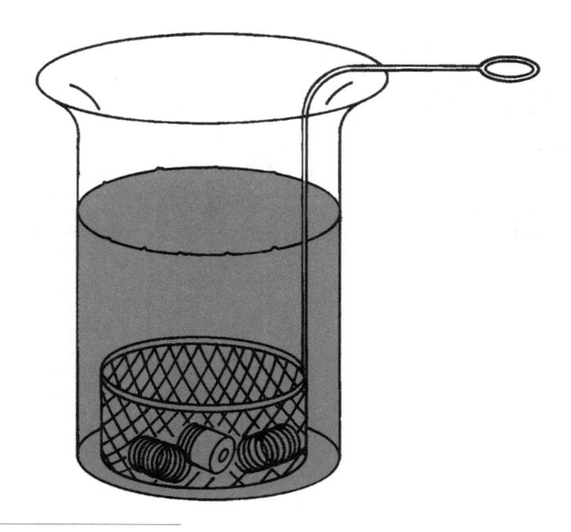

Fig. 17 — Clean Parts

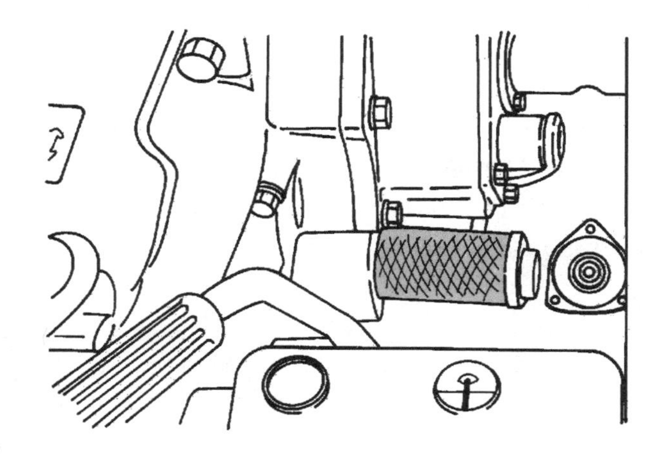

Fig. 18 — Remove and Clean Filter Screen

FLUID MAINTENANCE

It is especially necessary to drain and change the hydraulic fluid if a pump fails and dumps contamination in the system or the system is overheating. The only positive way to assure fluid is free of contaminants is to drain the system and fill it with clean fluid (Fig. 15). Follow the procedure in the operator's manual provided by the equipment manufacturer. If component parts are sticking, draining and flushing will often correct the sticking problem.

CAUTION: Before draining the hydraulic system, shut off the machine's engine, put the machine in park, and remove the ignition key.

Flushing may wash out residual contamination that a simple fluid change would leave in the system. Flush a hydraulic system by changing the fluid twice:

1. Run the engine and operate the hydraulic components to warm up fluid and get air out of the system.
2. Open the drain plug and drain the fluid.
3. Clean sediment off the drain plug.
4. Clean or replace all the filters.
5. Clean around the filler cap and wipe off the hydraulic fluid container (Fig. 16).
6. Fill the reservoir with the recommended hydraulic fluid. Do not overfill. Overfilling causes foaming.

IMPORTANT: Chemical solvents and cleaners are not recommended for flushing systems. Even a trace of these cleaners left in the system can break down the lubricating ability of hydraulic oil.

7. Operate the machine to circulate the new oil through the system. It is important to operate all of the system valves to pump new oil through the entire system. Bleed cylinders to get air out of the system.
8. To flush, repeat the drain and fill procedures, steps 2 - 6. Drain the flush oil, replace or clean the filter(s), and fill the reservoir with the recommended hydraulic oil.
9. Operate the system again until it is operating properly or until it is obvious that the system will have to be disassembled and cleaned.

STICKING HYDRAULIC PARTS

If a hydraulic component is sticking, and flushing does not correct the sticking, remove, disassemble and clean the component.

CAUTION: Relieve all pressure in the hydraulic system by shutting off the machine and moving control levers in both directions several times before removing components or lines.

1. Remove and disassemble the sticking component.
2. Clean the parts in gum solvent or noncorrosive chemical cleaner (Fig. 17).
3. Do not damage the close fitting parts while cleaning.
4. Do not permit solvent to contact seals or packings.
5. Rinse each cleaned part thoroughly in clean solvent and dry it with compressed air.
6. Inspect for wear, nicks, and corrosion while cleaning.
7. Replace bad parts.

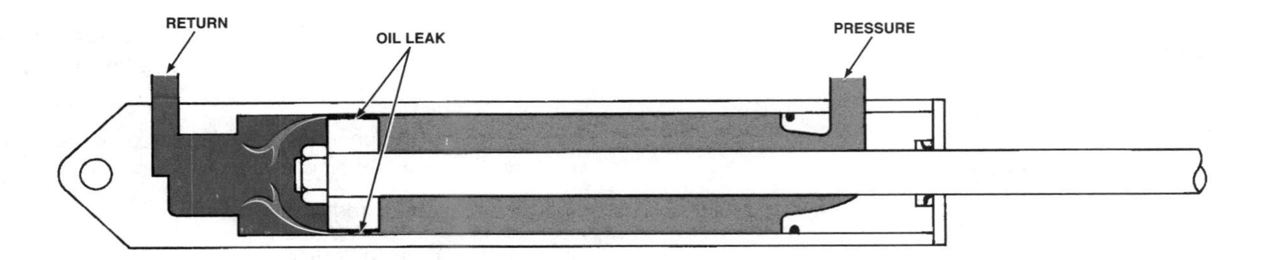

Fig. 19 — Internal Leaks

8. Apply hydraulic oil to parts to prevent rust.
9. Assemble the component parts using new seals and packing. Make sure the parts are assembled without dirt, lint, sealer, or other contaminants.
10. Install the cleaned component and test.

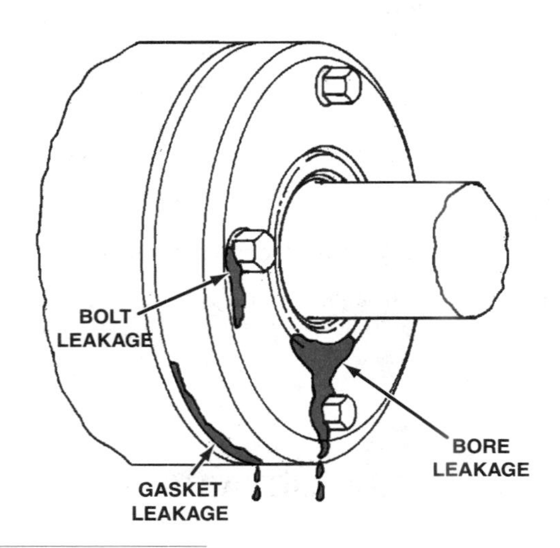

Fig. 20 — External Leaks

FILTER MAINTENANCE

Hydraulic oil filters may be a screen (Fig. 18) that can be cleaned. Or it may be a spin on filter that can be replaced.

LEAKS

Internal leaks are not a direct loss of fluid from the system, but they reduce power and create heat. External leaks drain fluid from the system, reduce power, create a fire hazard, pollute the environment and make footing slippery.

INTERNAL LEAKS

Internal leaking oil is not lost, but returned to the reservoir, making detection difficult.

Leaking fluid will return to the reservoir and not show up as a drip or spot like external leaks. But, internal leaks (Fig. 19) will erode tight fitting parts and produce heat that will breakdown hydraulic oil, hoses, seals, and packing.

Internal leaks are difficult to detect, however some symptoms are:

- **Sluggish action**
- **Creeping or drifting**
- **Localized hot spots**

If these symptoms occur, test the system to find out why before they get worse. Components must be removed and disassembled, and parts replaced to correct internal leaks. Follow disassembly steps in the technical manual.

EXTERNAL LEAKS

Dirty, oily spots are a sign of an external hydraulic fluid leak (Fig. 20) and an environmental hazard. A small leak may be the first indication of a hydraulic line about to rupture. When a hydraulic line ruptures, it not only puts the equipment out of order, it can cause physical injury (Fig. 21).

AVOID HIGH-PRESSURE FLUIDS

CAUTION: Escaping fluid under pressure can penetrate the skin causing serious injury. Avoid the hazard by relieving pressure before disconnecting hydraulic or other lines. Tighten all connections before applying pressure. Search for leaks with a piece of cardboard. Protect hands and body from high pressure fluids.

If an accident occurs, see a doctor immediately. Any fluid injected into the skin must be surgically removed within a few hours or gangrene may result. Doctors unfamiliar with this type of injury should reference a knowledgeable medical source.

Every joint and seal in a system is a potential leak point. Only constant attention and routine inspection can detect leaks before they cause major damage.

Fig. 21 — Be Careful

FIND EXTERNAL LEAKS

If you have to keep adding fluid, look for leaks around seals, gaskets, bolts, and piston bores.

The rubber surface covering of flexible hoses may crack and split without leaking. It is the depth of a crack that determines whether the hose leaks. Replace hoses with deep cracks if they show any sign of oil dampness.

Pin-hole leaks are hard to detect, yet they can be dangerous. A "vapor" of oil from a small leak can create a fire hazard, or a fine spray of oil against a hot engine can ignite.

Leaks in suction lines are very difficult to locate. They can quickly damage a system by sucking in air and dirt. Clean the suction line carefully with the equipment stopped. Dry the cleaned hose completely. Then start the machine. Squirt a little oil over the hose (Fig. 22). Watch! If the oil is drawn into the hose through a hole, replace the hose. There may also be a change in sound as the oil is drawn in.

If connections in lines are leaking, tighten them only until the leak stops. If the connection will not stay tight or if the leak persists, disconnect and check carefully for stripped threads, a cracked flare or a damaged seal. But remember: *More damage has been done to line connectors by overtightening than from any other cause.*

Check the system again after correcting the leak. Fill the system to the correct level with the recommended fluid. Start the equipment and warm up the system fluid. Check all the connections to be sure you didn't loosen one connection while tightening another.

OVERHEATING

Heat causes hydraulic fluid to break down and lose its lubrication properties. Heat also hardens the seals, coats parts with varnish, causes leaks, and reduces power.

All systems have some form of cooling designed in the system. Correct cooling may depend on the volume of fluid in the reservoir, heat dissipated by oil lines or oil coolers (if equipped) and heat dissipated by system components.

To Prevent Overheating:

- Keep fluid at proper level in reservoir.
- Clean dirt and mud from lines, cooler, reservoir, and components.
- Check and correct kinked or dented lines.
- Do not overspeed or overload the system.
- Check relief valves for proper adjustment.
- Do not hold control valves in power position longer than necessary.

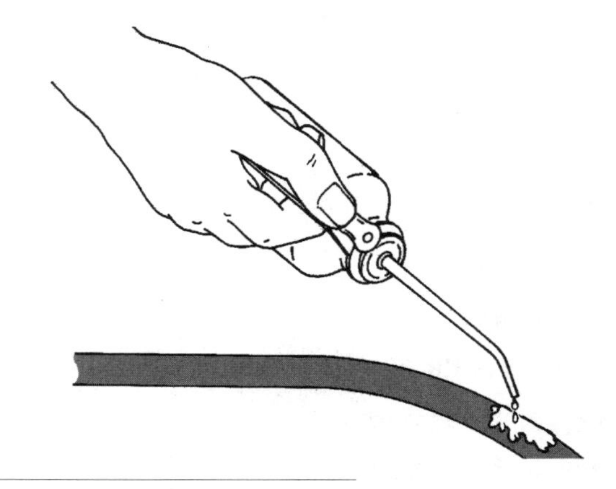

Fig. 22 — Find Suction Line Leaks

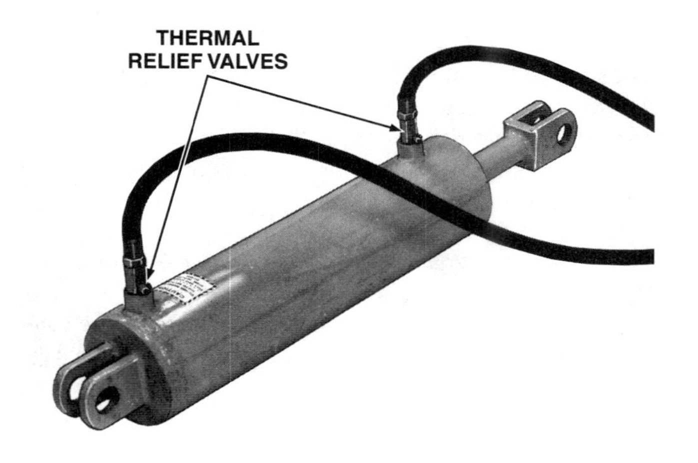

Fig. 23 — Thermal Relief Valves

THERMAL (HEAT) EXPANSION

Expansion of the fluid trapped in the system increases pressure. A one degree increase in temperature can cause a pressure increase of 50-60 psi (345-414 kPa) in a tightly sealed system.

When a hydraulic system is operating, pressure surges are relieved by relief valves. But if a machine is being stored in a hot environment or in full sun, pressure can build up in portions of the system that are blocked.

For example, the pressure in hydraulic cylinders can build up and cause damage. Thermal relief or circuit valves (Fig. 23) are sometimes installed on cylinder circuits to protect them from thermal expansion. Do not replace a thermal relief valve with a regular valve.

Another solution to thermal (heat) expansion is for the operator to retract the cylinders before storage. This allows the cylinders to extend as the fluid expands.

AIR-IN-OIL

Air in the system will result in:

- **"Spongy" action**
- **Chattering in the system**
- **Noisy pump operation**
- **Pumping action will stop**

Air collects in the reservoir if the fluid level is too low. Air can also enter the system through a leak in a suction line (Fig. 24), when lines are disconnected for making repairs or when the system is drained and filled.

To keep air out of the system:

- Be sure fluid in the reservoir is maintained at the correct level.
- Repair or replace leaking suction lines.
- Tighten line connections that leak. Tighten only until leakage stops. Do not overtighten.
- Cycle all systems at least four times to remove air from the system after draining and refilling the reservoir. Be sure to check fluid level in reservoir after cycling the system.
- Bleed air from remote cylinders after attaching them to the system (Fig. 25).

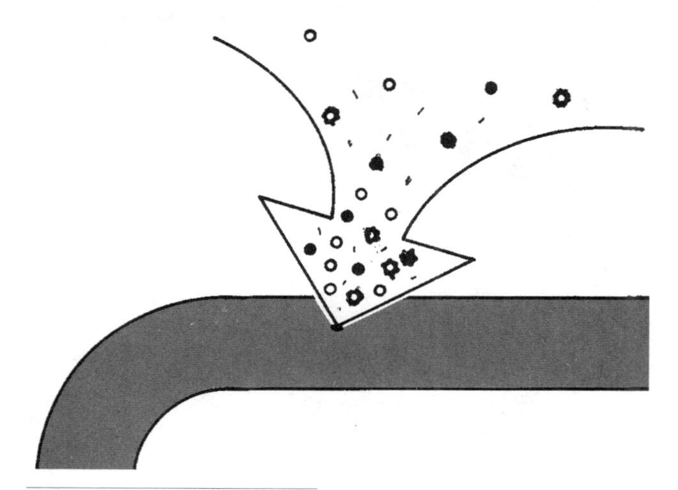

Fig. 24 — Hole in Suction Line

Fig. 25 — Bleed Air From Cylinders

TEST YOURSELF

QUESTIONS

1. (True or False) "Always lower hydraulic equipment to the ground or blocks before removing components."
2. (True or False) "Hydraulic pressure in a system is removed by loosening fittings."
3. (True or False) "Small hydraulic leaks can best be detected by feeling for them with your hand."
4. (True or False) "When flushing a hydraulic system, it is best to use a nonpetroleum solvent or a chemical cleaner."
5. (Fill in the blank) "A component with an internal leak will run ____________ than normal."
   (hotter or cooler)
6. (Fill in the blank) "Thermal expansion is the expanding of oil due to ____________ ."
7. (Fill in the blank) "If a cylinder has sluggish operation, is running hot, and there is no leakage at the rod, you would diagnose the problem as ____________."
8. Match the items in the left-hand column with the likely cause in the right-hand column:

a. Foaming Oil	1. Water
b. Milky Oil	2. Heat
c. Scorched Oil	3. Air

DIAGNOSIS AND TESTING

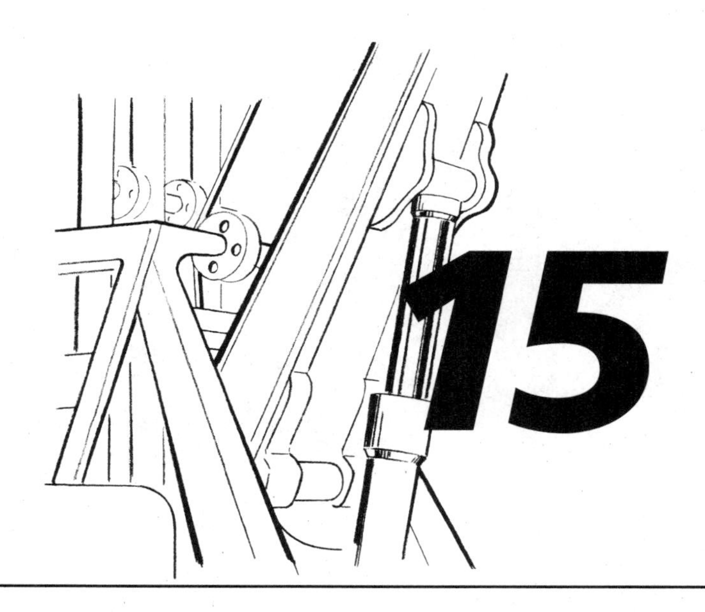

INTRODUCTION

Successful operation of today's high-performance vehicles depends totally on the proper operation of their on-board hydraulic systems. These systems, which first appeared more than half a century ago, were then used only for the control of a few integral implements. Through the years, however, the demonstrated convenience of hydraulic control has encouraged the expanded use of hydraulics. Today, the engines on many machines transmit 100% of their power through hydraulic systems to control just about every aspect of machine performance (Fig. 1).

As a result, if any portion of these systems malfunctions, the vehicle typically "just sits there," completely useless.

Machine manufacturers make a determined effort to provide the best possible long-term reliability of their hydraulic systems. However, in the real world of machine operation, no system will operate forever without some sort of malfunction.

When this day arrives, understandably the emphasis is "Get this problem fixed now," so that the vehicle can be returned just as soon as possible to its intended operation. Unfortunately, most modern machines are just one of many in an operation. The loss of this machine shuts down an operation that consists of many other pieces of equipment and the personnel to operate them.

This puts tremendous pressure on the service technician to get the machine back in operation.

As a service technician, confronted with such a challenge, you realize that before the problem can be fixed, it has to be identified.

It's easy to fall into the trap of using the hit-and-miss guessing-game approach of replacing components until the problem is fixed. But in these times, the time and expense required for this out-dated approach cannot be tolerated.

Instead, an informed, knowledgeable approach to identifying the problem is essential. This chapter won't tell you how to locate every malfunction in every hydraulic system that comes into the shop. It will give you the background information to use an informed approach to troubleshooting. It will enable you to get more done by working smarter—not just working faster.

Before getting into the troubleshooting, let's take another look at safety.

TROUBLESHOOTING SAFELY

The same safety precautions apply to troubleshooting as in the general maintenance and repair chapter (Chapter 14).

Block up loaders, booms, blades, and other implements when you work on them (Fig. 2).

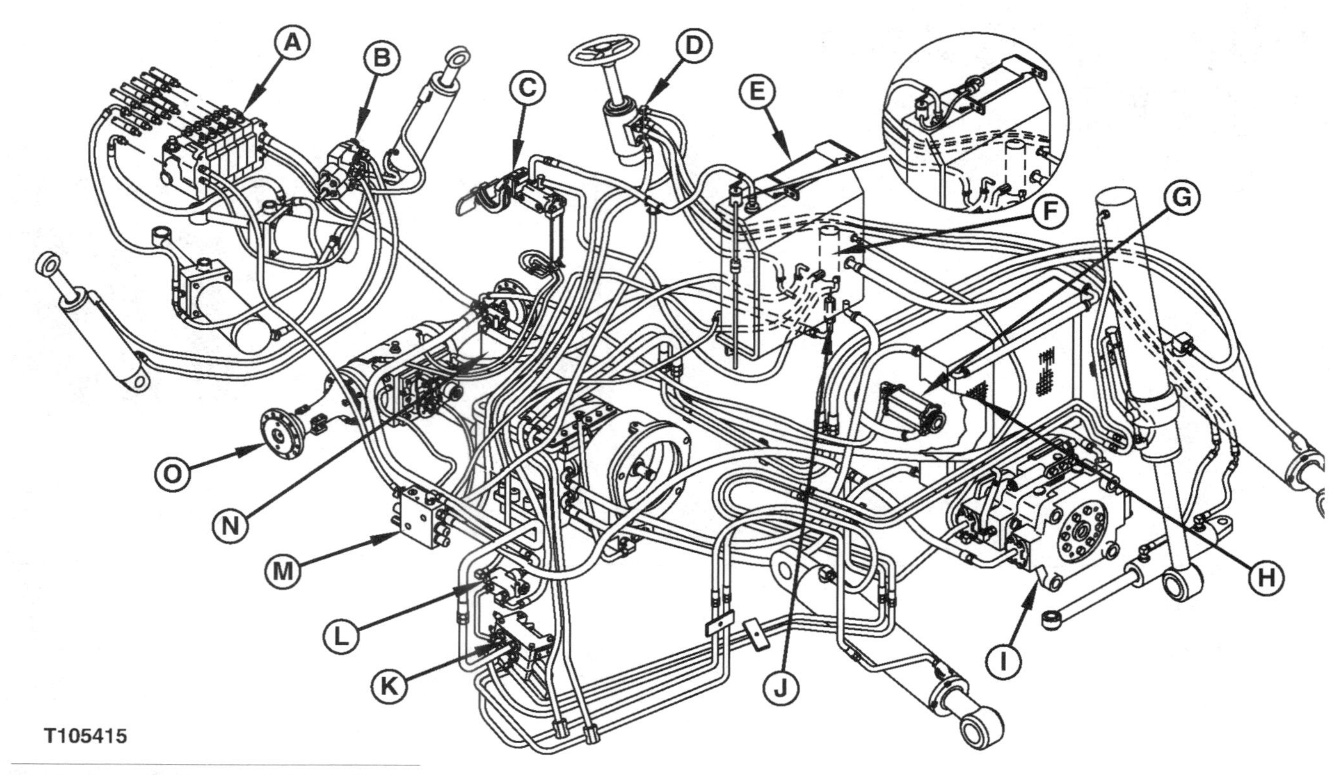

Fig. 1 - Complex Modern Hydraulic System

Fig. 2 - Block Up Hydraulic Equipment

Fig. 3 - Avoid Pressurized Fluids

Do not allow pressurized fluid to spray on your skin (Fig. 3).

CAUTION: Escaping fluid under pressure can penetrate the skin causing serious injury. Avoid the hazard by relieving pressure before disconnecting hydraulic or other lines. Tighten all connections before applying pressure. Search for leaks with a piece of cardboard. Protect hands and body from high-pressure fluids.

If an accident occurs, see a doctor immediately. Any fluid injected into the skin must be surgically removed within a few hours or gangrene may result. Doctors unfamiliar with this type of injury should reference a knowledgeable medical source.

Clean up leaks before they accumulate and cause slippery floors, fire hazards or environmental damage.

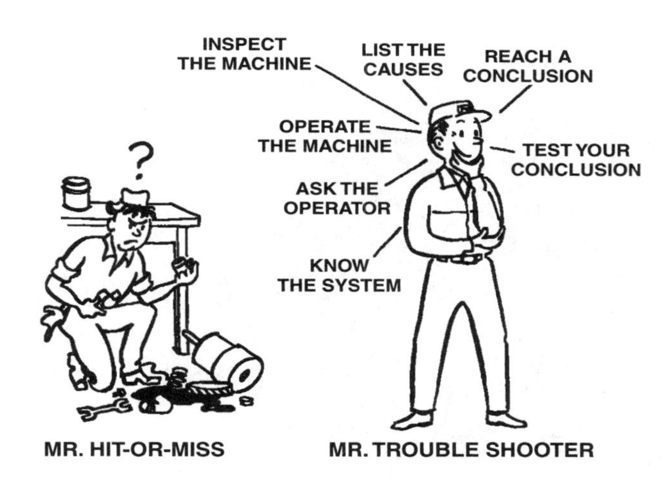

Fig. 4 - Which Would You Rather Be

SEVEN BASIC STEPS

Hit-or-Miss or a TroubleShooter — which would you rather be? Both have the title of technician but don't be fooled by that.

Hit-or-Miss is a parts exchanger who dives into a machine and starts replacing parts helter-skelter until the trouble is found — maybe — after wasting a lot of the customer's time and money.

TroubleShooter begins using the brain before the wrench. The facts are collected and examined to pinpoint the trouble. The diagnosis is then checked out by testing and only then does the technician start replacing parts.

Hit-or-Miss is fast becoming a person of the past. A dealer cannot afford to keep this person around at today's prices.

With the complex systems of today, diagnosis and testing by a TroubleShooter is the only way.

The difference between troubleshooting and hit or miss repair is the step by step process used in troubleshooting.

Many of todays machines have on board computers which are able to provide fault codes to assist in troubleshooting. Other machines have operational checkouts that have been expanded to provide considerable diagnostic information. On many machines, it will be your ability to use the information and tools available to diagnose the problem.

Regardless of the equipment, good troubleshooting requires the following seven basic steps:

1. **Know the system**
2. **Ask the operator**
3. **Operate the machine**
4. **Inspect the machine**
5. **List the possible causes**
6. **Reach a conclusion**
7. **Test your conclusion**

Let's see what these steps mean.

KNOW THE SYSTEM

Fig. 5 - Know the System

In other words, "Do your homework." Study the machine technical manuals. Know how the system works, whether it's open- or closed-center, what the valve settings and pump output should be.

It is also important to understand the limitations of the machine. There is nothing more frustrating to all concerned when time and money are wasted trying to "fix" a machine that is operating normally.

Keep up with the latest factory service bulletins. Read them and then file in a handy accessible place. The cause and remedy of the problem on this machine may be in a recent publication.

You can be prepared for any problem by knowing the system.

Fig. 6 - Ask the Operator

ASK THE OPERATOR

The operator knows more about how the machine behaves than anyone. A good reporter gets the full story from a witness - the operator.

He/she can tell you how the machine acted, when it started to fail and what was unusual about it.

Find out just what was being done with the machine at the time it failed. - Was the failure sudden or did it happen gradually?

Try to find out too if any adjusting or servicing had been performed recently. - When was it last serviced? - What did the filter look like the last time it was changed?

Ask about how the machine is used and when was it last serviced. Many problems can be traced to poor maintenance or abuse of the machine.

Know the operator. If it is an operational type complaint get all the details. Remember that you do not make your living operating that machine and you may not be able to duplicate the complaint when you operate it.

Fig. 7 - Operate the Machine

OPERATE THE MACHINE

Get on the machine and operate it. Warm it up and put it through its paces. Always verify the operator's story by checking it out yourself.

At this point, it would be wise to go through an operational checkout if it is available for the machine. These will be covered later in this chapter.

If there is no procedure available, try to answer the following questions:

How's the performance? - Is the action slow, erratic, or nil?

Do the controls feel solid or "spongy"? - Do they seem to be "sticking"? - Are the gauges reading normal?

Any signs of smoke? - Smell anything?

Hear any unusual sounds? - Where? - At what speeds or during what cycles?

Fig. 8 - Inspect the Machine

Fig. 9 - List the Possible Causes

INSPECT THE MACHINE

Now get off the machine and make a visual check. Use your eyes, ears, and nose to spot any signs of trouble.

First inspect the oil in the reservoir. How is the oil level? Is the oil foamy? Milky? Does it smell scorched? Does it appear to be too thin or too thick? How dirty is it?

Inspect the filter (even if they have just changed it). Cut it open and inspect it. What type of contamination does it contain? - Knowing this often leads you directly tō the problem.

Feel the reservoir. Is it hotter than normal? Is it caked with dirt and mud? Does the paint show signs of excessive heat? Check for a collapsed pump inlet line.

Follow the circuit looking for oil leaks.

CAUTION: Do not feel for leaks with your hands. Use a piece of paper or cardboard. Escaping fluid under pressure can penetrate the skin causing injury.

Check the oil cooler. Is it free of trash and mud?

Look closely at the components. Inspect for cracked welds, hairline cracks, loose tie bolts, or damaged linkage.

While you inspect, make a note of all the trouble signs and think how they might affect the problem.

LIST THE POSSIBLE CAUSES

Now you are ready to make a list of the possible causes.

What were the signs you found while inspecting the machine? And what is the most likely cause?

Are there other possibilities? Remember that the cause of a problem can be much different than the result of that problem.

Fig. 10 - Reach a Conclusion

REACH A CONCLUSION

Look over your list of possible causes and decide which are most likely and which are easiest to verify.

Use the Troubleshooting Charts in the machine technical manual as a guide to help you. General troubleshooting charts are found at the end of this chapter.

Reach your decision on the leading causes and plan to check them first.

Fig. 11 - Test Your Conclusion

TEST YOUR CONCLUSION

Now for the final step: Before you start repairing the system, test your conclusions to see if they are correct.

Some of the items on your list can be verified without further testing. Analyze the information you already have:

> Were all the hydraulic functions bad? Then probably the failure is in a component that is common to all parts of the system. Examples: pump, filters, system relief valves.
>
> Was only one circuit bad? Then you can eliminate the system components and concentrate on the parts of that one circuit.

Now your list is beginning to narrow so that you can point your tests at one or two components.

The next part of this chapter will tell you how to test the system and pinpoint these final troubles. But first let's repeat the seven rules for good troubleshooting:

1. Know the system

2. Ask the operator

3. Operate the machine

4. Inspect the machine

5. List the possible causes

6. Reach a conclusion

7. Test your conclusion

TYPES OF SYSTEMS

Know your system - There is no substitute for understanding the general principles of hydraulics when troubleshooting.

It is also necessary to know and understand the hydraulic system on the machine.

Since hydraulic power is composed of flow and pressure, one or both of these must be varied to control it. Each of the basic arrangements for doing this has its own advantages and limitations.

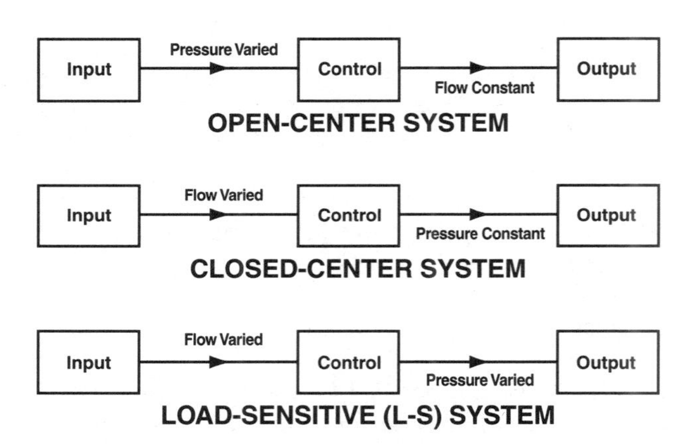

Fig. 12 - Three Types of Hydraulic Systems

In Chapter 1, many types of hydraulic systems were discussed. They all fall into one of the following three general systems.

- **Open-Center Systems: pressure varies, but flow remains constant.**
- **Closed-Center Pressure-Sensing Systems: flow varies, but pressure remains constant.**
- **Closed-Center Load-Sensing (L-S) systems: both pressure and flow vary.**

Early hydraulic systems were open-center. As the number of hydraulic functions increased, closed-center and, most recently, load-sensitive circuits have become more attractive.

OPEN-CENTER SYSTEMS

In the simpler open-center system, the pump produces a continuous flow that must be returned to the reservoir when the cylinder or other actuator is not in operation. During this standby condition, flow is high, but pressure is low. When flow is diverted by the control valve to the cylinder or other actuator, flow remains constant and the pressure increases to the level necessary to move the load or to the relief-valve setting, whichever is lower.

When the control valve is returned to neutral, fluid trapped in the cylinder supports the load, and pump output returns to the reservoir at low pressure.

For the operation of one function, this arrangement is quite satisfactory. It is also very efficient when used in a system where all the functions in that system can use the full output of the pump. A series-parallel valve (See Chapter 5) usually controls them. It does require that the operator meter the oil flow to the functions when more than one is used at the same time. If not, the function that requires the lowest pressure moves first, and must complete its stroke before the pressure will increase to start the next function.

All components in the system must be large enough to handle the full flow of the pump.

Where it is necessary to decrease the pump flow to any of the functions or to divide the flow to two or more functions, the system becomes very inefficient. The pump has to pressurize all the oil, whether it is used or not. That portion of the oil not used will have to drop in pressure without doing work and a *pressure drop without doing work turns into heat.*

With the systems having many functions, all requiring different flow, the closed-center systems were the answer to overcome these problems.

CLOSED-CENTER PRESSURE-SENSING SYSTEMS

In the typical closed-center system, a central source of hydraulic power, supplied by one pump, is used to power multiple functions.

Pump displacement, and thus its flow, changes to meet the demands of the system. When no functions are in use, oil flow is blocked at the valve (closed-center). The pump is at minimum flow but pressure is maintained to the valves at a preselected maximum pressure.

This pressure is available when needed simply by opening the valve and connecting the actuator to the circuit. When one or more control valves are opened, the pump automatically adjusts its delivery rate to satisfy the fluid-volume demands of the circuits. Pressure to the valves will be maintained as long as the pump flow is adequate to meet the combined demand created by the simultaneous operation of the functions.

In a closed-center system, each actuator circuit can be designed individually. The required displacement of each rotary actuator or the diameter of each cylinder is determined by its workload (torque or force) in relation to the system pressure. The size of the control valves and their connecting lines are selected to provide the flow rate required in each circuit to produce the required actuator speed.

This eliminates the need for any mechanical means to limit the driving torque or speed. The components can be of minimum size because they only need to handle the flow needed in that circuit.

A controlling orifice can "fine tune" the flow to the desired rate. These orifices can be in the pressure lines, the control valve or the actuator.

Power loss in standby is caused by pump and valve leakage. The power loss while working occurs when the oil pressure drops from standby to the working pressure at the valve or restricting orifice.

This is one of the reasons that the load sensing system has replaced the pressure sensing system in many applications. Pressure sensing is still the system of choice where it is necessary to maintain a high constant standby pressure against an actuator, such as a parking brake, differential lock, motor grader saddle pin, log skidder tongs, etc.

LOAD-SENSITIVE OR L-S SYSTEMS

A more recent development in the design of closed-center systems permits the standby pressure to be relatively low when the valve(s) is in neutral. When the valve is operated, flow is controlled to maintain a pressure slightly higher than the highest pressure needed in the system. This arrangement, known as the load sensitive (L-S) closed-center system, varies both the volume and the pressure with the demands of the load.

This system regulates flow based on the pressure required to move the load (load-sensitive) rather than based on the pump outlet pressure as it does on the pressure sensing system.

This system has all the advantages of the pressure compensated system plus several others.

Pressure drop and thus heat buildup is reduced across the valve. Valves used on these systems can easily be equipped with individual compensated flow valves. This means that actuator speed can be precisely controlled without the operator having to meter the valve. With each of the previous systems, the operator must meter the oil when operating multiple functions on machines like a backhoe. With this system, oil flow is proportioned to all the opened functions regardless of the load.

This system is used today in many farm and construction equipment applications because of its versatility and convenience.

COMPARING THE SYSTEMS: ADVANTAGES AND LIMITATIONS

Each system has its own advantages and limitations

OPEN-CENTER ADVANTAGES

Inexpensive and efficient when all functions use the full output of the pump.

Good valve metering.

Smooth engagement of load.

Maximum system pressure will be equal to, and determined by, the pressure that is required to move the load.

Usually has cooler operation.

OPEN-CENTER LIMITATIONS

All valves and cylinders in the system must be large enough to handle the full output of the pump. Does not allow components to be designed for the job, adding expense to the system.

If smaller valves are used, flow valves or flow dividers are required. This can be very inefficient and cause heat.

It is very inefficient when operating at or near the system relief valve setting. When relief pressure is reached, all of the pump oil can go over relief, creating excessive heat.

Cannot maintain a constant pressure in a circuit without going over a relief valve (brakes, differential lock, etc.) which would cause excessive heat.

CLOSED-CENTER ADVANTAGES

Can supply fluid to many functions of different sizes and capacities without the use of flow devices. This allows the components to be designed just large enough to get the job done.

Can hold a pressure against a load without creating heat or using additional energy.

Operates efficiently at maximum system pressure.

CLOSED-CENTER LIMITATIONS

More costly than other systems.

When metering the pump flow to a function requiring a low pressure, there is considerable pressure drop and thus, heat is created at the valve. The load-sensing system eliminates this because of the lower standby pressure.

CHECKING FOR INTERNAL LEAKS

If there is no operational checkout available for the machine, there are still some tests that can be done without gauges.

OPERATIONAL CHECKOUT

Some companies have designed a system of diagnosing machine performance called "Operational Checkout." This diagnostic approach is helpful in pinpointing many machine malfunctions without the use of gauges, sensors or other diagnostic tools. Operational checkout is also referred to as the Look— Listen—Feel approach.

HYDRAULIC SYSTEM CHECKS

LOOK-LISTEN-FEEL

The operational checkout procedures must be followed step-by-step to be effective and efficient. The following operational checkout is for a wheel loader backhoe with pressure compensated closed-center system. It will give you an idea how simply some of the hydraulic system malfunctions can be detected and isolated. So, look, listen and feel your way to more rapid trouble shooting.

HYDRAULIC PUMP PERFORMANCE CHECK	*NOTE: If hydraulic oil is not at operating temperature, heat hydraulic oil until loader and backhoe feel warm to touch using following procedure:* Put backhoe in transport position and engage boom and swing lock. Activate boom down function and run engine at 2000 rpm. *NOTE: If activating boom down does not load engine, boom down relief valve is set too high or hydraulic pump stand-by pressure is too low.* Operate all functions periodically to distribute heated oil to all cylinders. *Continued on next page*

HYDRAULIC PUMP PERFORMANCE CHECK - CONTINUED		Put backhoe at maximum reach with bucket fully dumped at ground level. Run engine at 2000 rpm. Measure cycle time by simulating loading the bucket, retracting the dipperstick and raising the boom to the boom cylinder cushion. Do not time boom cylinder through cushion. *LOOK: The maximum cycle time is as follows:* 410D - 9 seconds 510D - 12 seconds *NOTE: Take the average cycle time for at least 3 complete cycles. This average cycle time will give a general indication of hydraulic pump performance.*	**OK:** Go to next check **NOT OK:** Replace hydraulic filter and check for types of contamination. Rerun this check. **NOT OK:** Cycle times still slow. Go to Backhoe Relief Valve Test.

BACKHOE CIRCUIT LEAKAGE CHECK		Put backhoe in transport position and engage boom lock. Retract extendible dipperstick (if equipped). Raise stabilizers to full up position. Run engine at slow idle. Fully activate functions, one at a time: Boom up Bucket load Dipperstick retract Extendible dipperstick (if equipped). *LISTEN: When these functions are activated, NO decrease in engine rpm must be noted.* Fully activate functions, one at a time: Boom down Swing left then right Raise both stabilizers. *LISTEN: Boom down must cause rpm to decrease since relief valve setting is below standby pressure.* *LISTEN: Swing left and right may cause rpm to decrease slightly because relief valve setting is close to standby pressure.* *LISTEN: Stabilizer circuit can cause rpm to decrease slightly because of normal valve leakage.*	**OK:** Go to next check. *NOTE: Pressure seal passages are used in boom valve and leakage return passages are used in swing valve. When boom down or swing functions are bottomed and control valves "metered" rpm will decrease and leakages within circuit will be apparent. This is normal.* **NOT OK:** Continue on.

BACKHOE CIRCUIT LEAKAGE CHECK — CONTINUED		Lower stabilizer to maximum down position, extend dipperstick to maximum reach, extend extendible dipperstick (if equipped) and put bucket in dump position 1 m (3 ft) off ground. Fully activate functions, one at a time: Extend dipperstick Extend extendible dipperstick (if equipped) Bucket dump. Fully activate functions, one at a time: Stabilizer down left Stabilizer down right. *LISTEN: When these functions are activated, NO decrease in engine rpm must be noted.*	**OK:** Go to next check. **NOT OK:** If rpm decreases with a function bottomed and control lever fully open, a leak is indicated in the circuit. **NOT OK:** If rpm increases when a function is bottomed and control valve is fully open, a neutral leak indicated. **NOT OK:** A rpm decrease with a function bottomed in both directions is normally cylinder leakage. A rpm decrease in one direction is normally circuit relief valve leakage. Go to Hydraulic Component Leakage Test.
LOADER CIRCUIT LEAKAGE CHECK		Raise loader to full height and put bucket in dump position. Run engine at slow idle. Fully activate functions, one at a time: Loader boom up Bucket dump. Put bucket in rollback position and lower loader to full down position. Fully activate functions, one at a time: Loader boom down Bucket rollback. LISTEN: When functions are activated, engine rpm must NOT decrease. NOTE: Leakage return passages are used in boom valve. When boom cylinders are retracted and valve is "metered" rpm will decrease and leakage within the circuit will be apparent. This is normal.	**OK:** Go to next check. **NOT OK:** If rpm decreases with a function bottomed and control lever is fully open, a neutral leak indicated. **NOT OK:** A rpm decrease with a function bottomed in both directions is normally cylinder leakage. A rpm decrease in one direction is normally circuit relief valve leakage. Go to Hydraulic Component Leakage Test.

CYLINDER CUSHION CHECK		Run engine at approximately 1000 rpm. Activate backhoe swing left and right and boom raise. Note sound and speed as cylinders near end of their stroke. *LOOK: Speed of cylinder rod must decrease near the end of its stroke.* *LISTEN. Must hear oil flowing through orifice as cylinder rod nears the end of its stroke.*	**OK:** Go to next check. **NOT OK:** Remove and repair cylinder cushion Go to repair manual.
BACKHOE AND LOADER FUNCTION DRIFT CHECK		*FEEL Backhoe cylinders. Cylinders must be warm to touch 38-52°C (100-125°F). If cylinders are not warm, heat hydraulic oil to specification, (see Group 9025-25).* Raise unit off ground with stabilizers. Put backhoe bucket at a 45° angle to ground. Lower boom until bucket cutting edge is 50 mm (2 in.) off ground. Position loader bucket same angle and distance off ground as backhoe bucket. Run engine at slow idle and observe bucket's cutting edges. *LOOK: If bucket cutting edges touch the ground within 1 minute, leakage is indicated in the bucket or boom cylinders or control valves.*	**OK:** Go to next check. **NOT OK:** If rpm decreases with a function bottomed and control lever is fully open, a neutral leak indicated. **NOT OK:** A rpm decrease with a function bottomed in both directions is normally cylinder leakage. A rpm decrease in one direction is normally circuit relief valve leakage. Go to Hydraulic Component Leakage Test.
LOADER BOOM FLOAT AND R RETURN-TO-DIG CHECK		Put loader at maximum height position with bucket dumped. Run engine at approximately 2000 rpm. Move the loader control lever forward into boom float detent position, and at the same time into bucket rollback detent position. Remove hand from control lever. *LOOK: Loader control lever must remain in the boom float detent position.*	**OK**: Continue On. **NOT OK**: If level jumps out of detent, inspect detent spring and balls. Go to repair manual.

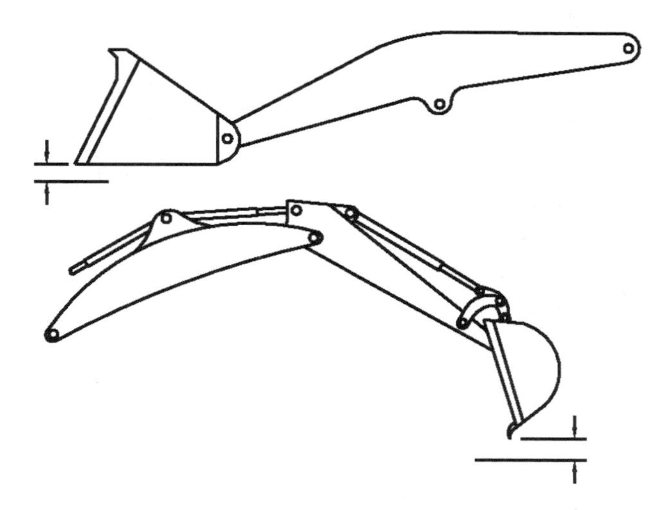

Fig. 13 - Checking Cylinder Drift

The normal problem with leakage is that the cylinder will collapse or drift when not in use.

Before proceeding, however, check the technical manual for acceptable cylinder drift or leakdown. Most specifications will be given for movement during a specific time, with a specific oil temperature, with a specific load, and with the machine in a certain position. This specification (Fig. 13) can call for a certain amount of settling measured at a point on the machine or a certain movement of the cylinder rod.

If excessive drift exists, it can be caused by a leaking cylinder packing, a leaking valve or a leaking circuit relief valve, if so equipped.

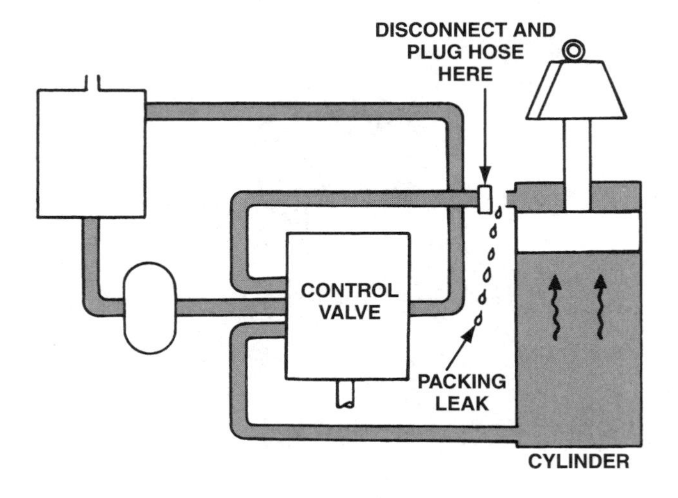

Fig. 14 - Leakage Check on Double-Acting Cylinder

To check a double-acting cylinder, run the cylinder to one end of its stroke. Support the equipment if it is raised, then shut off the engine. Remove the hose from the end of the cylinder that was not pressurized (Fig. 14). Start the engine again, pressurize the cylinder, and see if any oil comes out of the open port. If no oil comes out, repeat the test in the opposite direction since it may be possible for the cylinder to leak in only one direction.

Again it is important to check the technical manual for acceptable leakage amounts. This will normally be given in drops per minute.

If the leakage specifications are exceeded, the packings in the cylinder must be replaced.

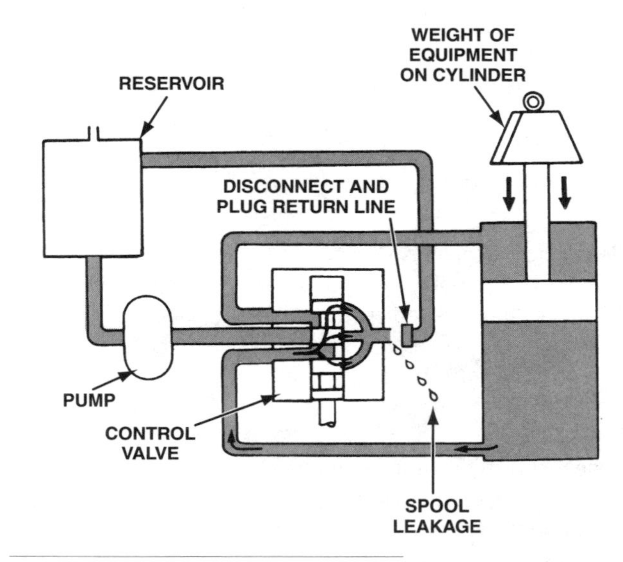

Fig. 15 - Checking Control Valve Leakage

Now that the cylinder has been checked, the valve can be checked for leakage.

Raise the hydraulic equipment a few feet off the ground, return the control lever to neutral, and shut off the engine.

Support the equipment and disconnect the return line between the control valve and the reservoir, then plug the line to the reservoir (Fig. 15).

Remove the support and examine the open port in the control valve as the equipment settles. If oil leaks from the port, the control valve spool or the circuit relief valve is leaking.

Again support the load and exchange the circuit relief valve with one from a good circuit. If the leakage is still present, it is the valve that is leaking.

Before replacing the valve, make certain that the leakage is not within specifications. Many valves, especially spool valves are not intended to be leakproof unless equipped with lock-out valves (See Chapter 5).

TEST EQUIPMENT

Good testing helps pinpoint the problem and prevent unnecessary repair work.

Hydraulic systems are designed to function at specific fluid flow rates, fluid pressures and fluid temperatures. It is also necessary to measure speed because of its affect on flow.

The four things requiring testing in the hydraulic system are:

- **Pressure**
- **Flow**
- **Speed**
- **Temperature**

Measuring these items at certain points in the system is a very effective method of locating and identifying problems but not necessarily the most efficient.

Be sure to perform the first six basic steps of troubleshooting before installing test equipment.

Some malfunctions, such as broken hoses, ruptured seals, and seized pumps are so obvious that further testing is needless. But other malfunctions require some testing to pin down the exact problem.

USING TEST EQUIPMENT

Before connecting test equipment to a machine, clean hoses, lines, and ports so the hydraulic fluid will not be contaminated.

 **CAUTION: Relieve pressure before loosening lines.**

Consult the technical manual and system diagram to be sure the test equipment is connected to the correct points in the system.

NOTE: Both English and metric fittings are used. Be sure threads are compatible or damaged threads will result.

Read the test equipment's instruction manual. Know the test equipment's function. And, know how to operate test equipment before testing.

Use a test form (Fig. 22) to record test data.

To get accurate test readings, hydraulic fluid must be at normal working temperature. Operate the machine for several minutes or partially close the flow meter or hydraulic tester load valve to load the system until normal operating temperature is reached.

PRESSURE TESTING

Pressure testing is used to check and adjust relief, priority, and pressure regulating valves. They are also used to determine pressure drops and the pressure needs of the system.

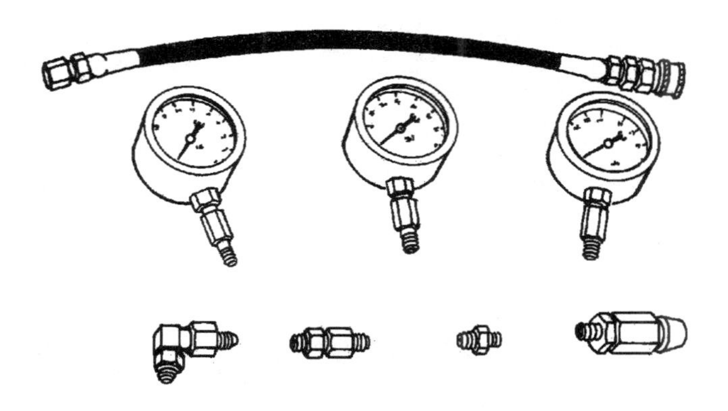

Fig. 16 - Pressure Gauges

Pressure gauges are available individually or in sets as shown in Fig. 16. To test most hydraulic systems, it requires gauges in several pressure ranges. For the most accurate readings, select a gauge that will register in the upper portion of the gauge but will have a safe margin in case the expected test pressure is exceeded.

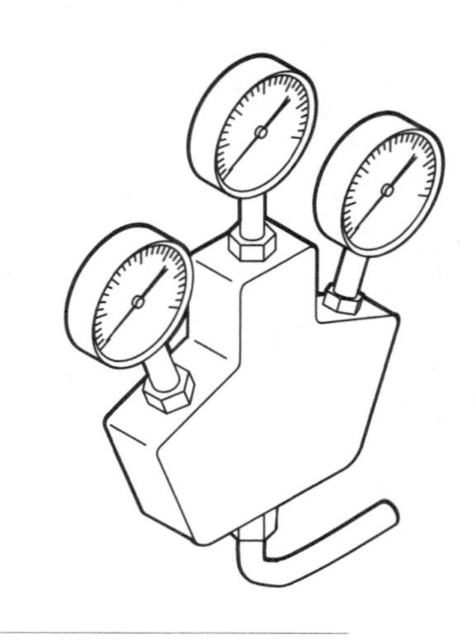

Fig. 17 - Compound Pressure Gauge

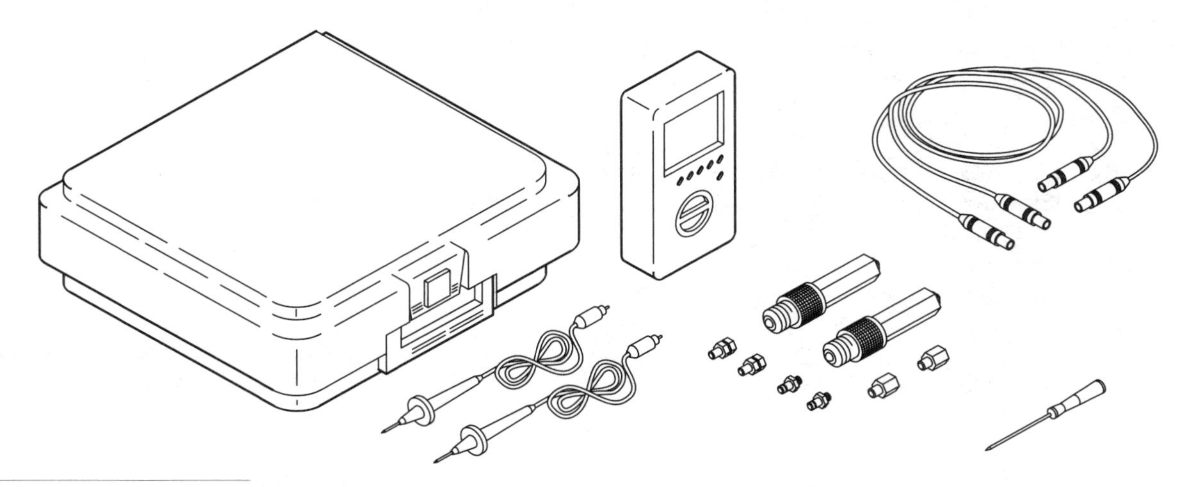

Fig. 18 - Digital Pressure Gauge

The compound pressure gauge in Fig. 17 has gauges with three different pressure ranges, all connected to the same source. A mechanism in the assembly blocks pressure from the lowest pressure gauge when its top limit is reached. Pressure is then read on the intermediate gauge until its limit is reached. Pressure is then blocked to the intermediate gauge and the higher pressures read on the high pressure gauge.

These gauges eliminate the need for changing gauges when testing circuits with both high and low pressures. They are also safe for testing ones with unknown pressures.

The digital pressure gauge is shown in Fig. 18. A transducer is installed into a tee fitting or test port of the circuit to be tested. The meter is then connected with wire leads to the transducer. Pressurized oil, acting on the transducer varies the current flow to cause a reading in the meter.

Although there are transducers to cover different pressure ranges, these meters are very accurate through most of their range. This eliminates the need to change gauges when testing a wide range of pressures in the same circuit.

The meter can be connected to several transducers so that several pressures or differential pressures can be read.

FLOW TESTING

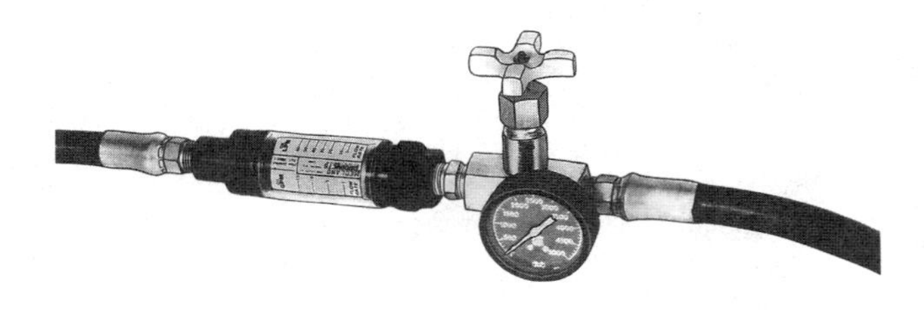

Fig. 19 - Flow Meter

An inline flow meter is shown in Fig. 19. This consists of a tapered plastic tube with a spring-loaded ball or poppet. The more oil that is forced through the tube, the higher into the larger portion of the tube the poppet will be pushed. Calibrated graduations on the tube indicate the flow through the tube in gallons or liters-per-minute.

A pressure gauge and load valve is coupled with the flow meter. The valve is closed to restrict flow to build a certain pressure while the flow is measured.

IMPORTANT: The load valve on flow meters must always be open before the engine is started. A closed valve connected to the outlet of a fixed displacement pump does not provide a flow path for the oil. Pressure will build until something fails.

Fig. 20 - Hydraulic Tester (Analyzer)

The hydraulic tester or analyzer shown in Fig. 20 is much like the flow meter, pressure gauge and load valve combination shown in Fig. 19.

The tester has a high and low pressure gauge with automatic switching between the two. It also has a temperature gauge.

Portable testers, designed for shop or field use, apply a controlled load to the vehicle hydraulic system and measure the effect of that load. A manually controlled valve provides the adjustable load. Gauges show flow, pressure, temperature, and, in some cases, shaft speed. A complete test kit consisting of a selection of fittings and hoses is often available with the tester.

Most hydraulic testers measure volume although some measure weight. Those measuring volume are calibrated to read in gpm (gallons per minute) or L/min (liters per minute).

The flow meter in these testers operates on one of three principles:

- ***Fixed-area flow meters*** measure the pressure drop across an orifice, or opening, and convert this pressure measurement to flow.
- ***Turbine flow meters*** use high-speed axial-flow rotors that are spun by the passing fluid. An external sensor counts the revolutions of the turbine, and from those, computes the flow rate.
- ***Positive-displacement flow meters*** are essentially small hydraulic motors. As the fluid passes through, it turns the motor. The meter then converts motor rpm's to the flow rate. These meters produce somewhat higher accuracy than the other types, but are more sensitive to fluid contamination.

Pressure that the flow meter housing can withstand varies among units offered. Some units will withstand only a relatively low pressure. On these units, flow must be measured down stream from the loading valve and returned directly to the reservoir. *There can be no backpressure at the tester outlet.*

Most modern units, however, can withstand full system pressure. They can be connected in series, directing output flow to the next component in the system.

Flow capacity of testers, range from 20 gpm in the smaller models to 150 gpm in the larger ones. Accuracy of flow measurements may be within three percent.

The flow meter is the most thorough method of testing and analyzing a system. Most problems, however, can be found with other methods. To avoid the time consuming job of hooking up test equipment, always perform the other six troubleshooting steps first.

The tester will provide readings for temperature, flow and pressure. Because flow is directly proportional to speed, it is always necessary to use an engine tachometer when using the tester.

Read the hydraulic tester's operating manual and the test machine's technical manual for the proper test procedure. All fittings must be connected correctly and tightly. Tie down loose hoses.

Using the tester requires the disconnecting of oil lines. So remember, *DIRT IS THE WORST ENEMY OF A HYDRAULIC SYSTEM.* Before disconnecting oil lines, clean the machine and be sure to cap all openings.

PUMP TESTING

Because the pump is the generating force for the whole hydraulic system, it is necessary to test the pump first.

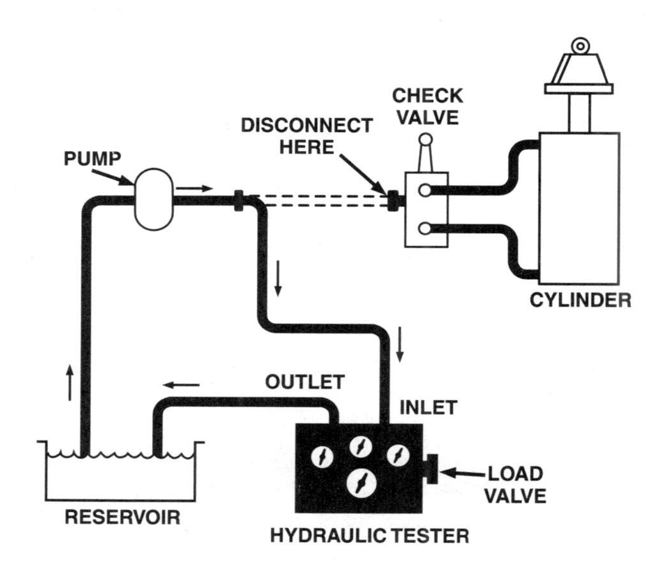

Fig. 21 - Testing Hydraulic Pump

Installing the Hydraulic Tester

1. Relieve any pressure in the system and disconnect the pressure line between the pump and the control valve. Connect the pump outlet to the tester INLET port (Fig. 21).
2. Connect hydraulic tester OUTLET port to the reservoir. Whenever possible, connect it to the reservoir return line because it usually has a return filter. On systems that have a charge pump to supply oil to the main pump, always return hydraulic tester oil to a point between the charging pump and the main hydraulic pump.

Owner ______________ Serial No. ______________

Tractor Model ______________ Equipment ______________

Comments ______________

		Gallons Per Minute @ Psi (kPa)									
		0	*250*	*500*	*750*	*1000*	*1250*	*1500*	*1750*	*2000*	*2250*
Pump Test		*31.0*	*28.0*	*25.0*	*22.0*	*19.0*	*16.0*	*13.0*	*10.0*	*7.0*	
Circuit Test	Direction Cylinder Travel										

Fig. 22 - Pump Test Results

3. Install tachometer on engine. Pump flow is dependent on speed so it is necessary that tests be run at a constant engine (pump) speed.
4. Check the oil level and *make certain the tester valve is open* before starting the engine. Start engine and slowly close the tester load valve to load the system. (Do not exceed the system's maximum rated pressure.) Continue to run under load until the normal operating temperature of the system is reached (see machine specifications).

Operating the Hydraulic Tester

1. Adjust engine to normal operating speed.
2. With the tester load valve open, record maximum pump flow at zero pressure.
3. Slowly close the load valve to increase pressure and record the flow at 250 psi increments from zero pressure to maximum system pressure. Maintain the same engine speed for each reading.

Record your test results on a form such as the one shown in Fig. 22, so you can refer to them later.

PUMP TEST DIAGNOSIS

Pump flow efficiency, at maximum pressure, should be at least 75 percent of pump flow at zero pressure. (On modern variable displacement pumps, 90 percent can be expected.)

To calculate the efficiency, divide the maximum pressure flow by the zero pressure flow. (In the example used in Fig. 22, 7.0 divided by 31.0 equals 0.22 or 22%) The 22% indicates that the pump is damaged or very badly worn and would require repair or replacement.

It is important that the test be run at normal operating speed. In the test above, the pump was pressurizing 31 gallons of oil per minute through the entire test. The oil not going through the flow meter was leaking around the gears, vanes or pistons in the pump.

The internal leakage, in this case, is 15 gpm at 1250 psi, and 24 gpm at 2000 psi.

The leakage would be the same at each pressure, even if the pump output were less. If the test were run at 1/2 throttle with a delivery of 15 gpm at 0 psi, the percentage would be a lot lower. This would not give a true indication of the pump's efficiency.

If pump flow is poor during the free flow test as well as the pressure tests, the pump probably is not getting enough oil. This problem could be caused by low oil supply, air leaks, a restricted pump inlet line, or a dirty reservoir filter.

If flow is unsatisfactory on a variable displacement pump:

Poor flow at all pressures - check reservoir level, charge pump operation, and charge and return filters if so equipped.

Even drop in flow at each pressure interval - worn pump.

Lower pressure readings OK and destroking range normal (approximately 200 psi) but too low - adjust pump standby.

Lower pressure readings OK but destroking range too great or if standby cannot be reached - high pressure leak in destroking valving (probably seals or stroke control valve seat).

If the pump tests okay, then start checking the system components for trouble.

SYSTEM TESTING

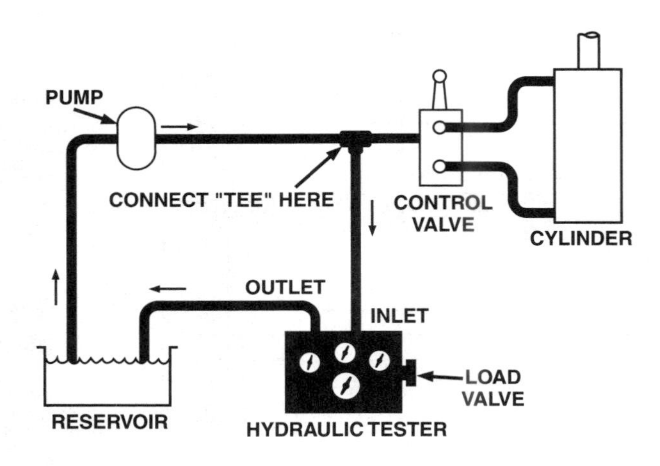

Fig. 23 - Testing the System Components

INSTALLING THE HYDRAULIC TESTER

Install a tee fitting in the line between the pump and the control valve and attach the hydraulic tester INLET port to this tee (Fig. 23).

Leave the return line from the hydraulic tester OUTLET port connected in the same way as it was for the pump test.

Operating the Hydraulic Tester

1. Open the hydraulic tester valve.
2. Start the engine and adjust it to the manufacturer's recommended operating speed (must be same speed as used in the pump test).
3. Bring the oil to the same temperature as used in the pump test (normal operating temperature).
4. Operate the control lever to activate the valve in one of its power positions.
5. Open load valve to record the flow at zero pressure (Fig. 24).
6. Slowly close the hydraulic tester load valve and record flow in 250 psi increments from zero pressure to maximum system pressure.
7. Open the load valve until maximum flow is again at zero pressure and repeat the test on the rest of the control valve power positions.

Be sure to make all the tests at the same oil temperature to get readings that can be compared. If oil is too hot from the previous test, allow it to circulate through the system for cooling before continuing.

SYSTEM TEST DIAGNOSIS

The difference between the readings recorded in the pump and the system tests represent the leakage in each of the circuits.

If the flow at each pressure is the same as for pump the test, the circuit is okay.

1. If flow is good at low pressures, then drops the same amount at each pressure interval with the control valve in all positions at higher pressures: The system relief valve is probably at fault.
2. If the flow in all the circuits is low and has dropped the same amount at each pressure interval, there is a leak, probably in the system relief valve seal or possibly a cracked housing.

Owner ______________ Serial No. ______________

Tractor Model ______________ Equipment ______________

Comments ______________

		Gallons Per Minute @ Psi (kPa)									
		0	*250*	*500*	*750*	*1000*	*1250*	*1500*	*1750*	*2000*	*2250*
Pump Test		*32.0*	*31.7*	*31.4*	*31.0*	*30.6*	*30.2*	*29.8*	*29.3*	*28.8*	
Circuit Test	Direction Cylinder Travel										
Boom Circuit	***Lower***	*32.0*	*31.7*	*31.4*	*31.0*	*30.6*	*30.2*	*29.8*	*29.3*	*28.8*	
	Raise	*32.0*	*31.7*	*31.4*	*31.0*	*30.6*	*30.2*	*29.8*	*29.3*	*28.8*	
Bucket Circuit	***Dump***	*32.0*	*30.7*	*29.4*	*28.0*	*26.4*	*25.2*	*23.8*	*22.3*	*20.8*	
	Roll-Bk.	*32.0*	*31.7*	*31.4*	*31.0*	*30.6*	*30.2*	*10.0*	*0.0*		

Fig. 24 - Recording System Test Results

3. If a circuit has good flow at lower pressures and drops off at higher pressures, it is probably a circuit relief valve (See the bucket rollback circuit in Fig. 24).
4. If the flow of an individual circuit is low and has decreased the same amount at each pressure interval, it is an indication of leakage in that circuit, probably a circuit relief valve seal (See bucket dump circuit in Fig. 24).
5. If the flow on the two circuits of one function decrease the same amount at each pressure interval, it is an indication of either cylinder packing or valve leakage. Perform the cylinder leakage test described earlier in this chapter.

Diagnostic Summary

1. Leakage will cause the flow to decrease about the same amount at each pressure interval.
2. A leaking relief valve will give the normal leakage indications.

Relief valves improperly adjusted will have proper flow at lower pressures, then suddenly decrease as the pressure setting is reached. Some low flow circuit relief valves will only pass about 3 gpm at 200 psi above their cracking pressure. In these cases, the flow would suddenly drop about 3 gpm when they begin to open, then increase the leakage gradually as pressure increased.

Faulty SYSTEM relief valves affect readings in all tests while CIRCUIT relief valves affect only the circuit they serve.

TEMPERATURE TESTING

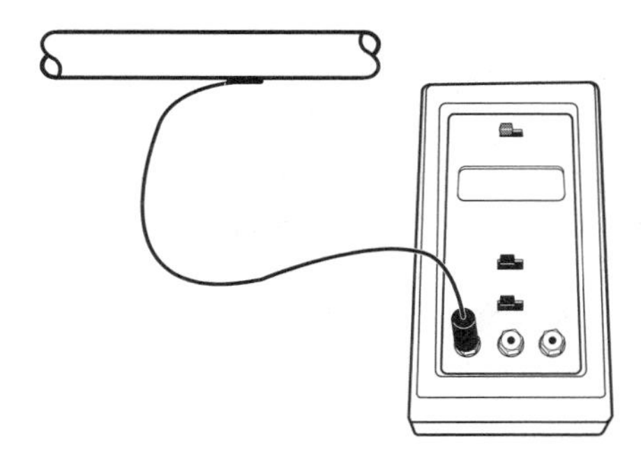

Fig. 25 - Temperature Meter

The mercury thermometer can still be used to check hydraulic oil temperature. Limited access to reservoirs, sealed reservoirs and the need to check line temperatures make the meter shown in Fig. 25 more practical. There are also attachments available to convert digital volt-ohm meters to read temperature.

Both consist of a thermocoupler sensor (probe), an electric meter and leads.

The probe can be immersed in the reservoir or attached to the reservoir or metal hydraulic lines. When attached externally, the probe must be held firmly against the metal and covered with insulation for accurate readings.

SPEED ADJUSTMENT

Because hydraulic flow is proportional to pump speed, it is important to have the proper speed adjustment while conducting hydraulic tests.

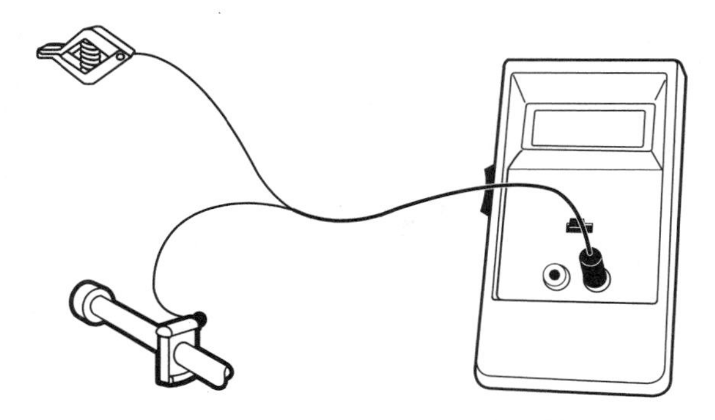

Fig. 26 - Electronic Tachometer

The tachometer shown in Fig. 26 consists of an impulse probe, a meter and leads. The probe is attached to one of the engines diesel injection lines. As that cylinder is fired, the probe senses the expansion of the line and sends a signal to the meter. The meter counts the pulses and converts the signal to indicate the engine rpm.

Another type of probe uses a magnetic pickup. This is placed in close proximity to the teeth of the engine's flywheel ring-gear. The pickup signals the meter each time a ring-gear tooth passes. The meter counts the teeth and displays the correct engine rpm. It is necessary to adjust this meter to correspond to the number of flywheel teeth on the machine.

The strobe light is seldom used because of the difficulty in accessing the engine on modern machinery.

SUMMARY: - TESTING THE MACHINE

The tests we have given you are only general and can serve only as basic guidelines. When you start testing actual machines, use the machine technical manual as your guide for detailed tests and test results.

And remember that the best testing equipment has no value unless the person at the controls knows how to install it, use it and interpret the results.

TROUBLE SHOOTING CHARTS

INTRODUCTION

The charts on the following pages are typical of those appearing in most technical manuals. They can assist you in listing the possible causes of trouble when you begin the diagnosis and testing steps.

Once you have located the cause of the problem, the chart can be used again to find a possible remedy.

The chart shown here is general in coverage, however, the technical manual for each machine will give more detailed and more specific causes and remedies.

HYDRAULIC OIL CONDITION

OIL MILKY OR DIRTY
Water in oil (milky).
Filter failures (dirty).
Metal particles (mechanical failure).

OIL DISCOLORED OR HAS BURNED ODOR
Kinked pipes.
Plugged oil lines.
Wrong oil viscosity.
Internal leaks.

SYSTEM INOPERATIVE

NO OIL IN SYSTEM
Fill to full mark. Check system for leaks.

OIL LOW IN RESERVOIR
Check level and fill to full mark. Check system for leaks.

OIL OF WRONG VISCOSITY
Refer to specifications for proper viscosity.

FILTER DIRTY OR PLUGGED
Drain oil and replace filters. Try to find source of contamination.

RESTRICTION IN SYSTEM
Oil lines could be dirty or have inner walls that are collapsing to cut off oil supply. Clean or replace lines. Clean orifices.

AIR LEAKS IN PUMP SUCTION LINE
Repair or replace lines.

DIRT IN PUMP
Clean and repair pump. If necessary, drain and flush hydraulic system. Try to find source of contamination.

BADLY WORN PUMP
Repair or replace pump. Check for problems causing pump wear such as misalignment or contaminated oil.

BADLY WORN COMPONENTS
Examine and test valves, motors, cylinders, etc. for external and internal leaks. If wear is abnormal, try to locate the cause.

OIL LEAK IN PRESSURE LINES
Tighten fittings or replace defective lines. Examine mating surfaces on couplers for irregularities.

COMPONENTS NOT PROPERLY ADJUSTED
Refer to machine technical manual for proper adjustment of components.

RELIEF VALVE DEFECTIVE
Test relief valves to make sure they are opening at their rated pressure. Examine seals for damage that could cause leaks. Clean relief valves and check for broken springs, etc.

PUMP ROTATING IN WRONG DIRECTION
Reverse to prevent damage.

OPERATING SYSTEM UNDER EXCESSIVE LOAD
Check specifications of unit for load limits.

HOSES ATTACHED IMPROPERLY
Attach properly and tighten securely.

SLIPPING OR BROKEN PUMP DRIVE
Replace couplers or belts if necessary. Align them and adjust tension.

PUMP NOT OPERATING
Check for shut-off device on pump or pump drives.

SYSTEM OPERATES ERRATICALLY

AIR IN SYSTEM
Examine suction side of system for leaks. Make sure oil level is correct. (Oil leak on the pressure side of the system could account for loss of oil.)

COLD OIL
Viscosity of oil may be too high at start of warm-up period. Allow oil to warm up to operating temperature before using hydraulic functions.

COMPONENTS STICKING OR BINDING
Check for dirt or gummy deposits. If dirt is caused by contamination, try to find the source. Check for worn or bent parts.

PUMP DAMAGED
Check for broken or worn parts. Determine cause of pump damage.

DIRT IN RELIEF VALVES
Clean relief valves.

RESTRICTION IN FILTER OR SUCTION LINE
Suction line could be dirty or have inner walls that are collapsing to cut off oil supply. Clean or replace suction line. Also, check filter line for restrictions.

VALVE OR REGULATORS PLUGGED
Clean dirt from components. Clean orifices. Check source of dirt and correct.

OIL LEAK IN PRESSURE LINES
Tighten fittings or replace defective line. Examine mating surfaces on couplers for irregularities.

COMPONENTS NOT PROPERLY ADJUSTED
Refer to machine technical manual for proper adjustment of components.

SYSTEM OPERATES SLOWLY

COLD OIL
Allow oil to warm up before operating machine.

OIL VISCOSITY TOO HEAVY
Use oil recommended by the manufacturer.

INSUFFICIENT ENGINE SPEED
Refer to operator's manual for recommended speed. If machine has a governor, it may need adjustment.

LOW OIL SUPPLY
Check reservoir and add oil if necessary. Check system for leaks that could cause loss of oil.

ADJUSTABLE ORIFICE RESTRICTED TOO MUCH
Back out orifice and adjust it. Check machine specifications for proper setting.

AIR IN SYSTEM
Check suction side of the system for leaks.

BADLY WORN PUMP
Repair or replace pump. Check for problems causing pump wear such as misalignment or contaminated oil.

RESTRICTION IN SUCTION LINE OR FILTER
Suction line could be dirty or have inner walls that are collapsing to cut off oil supply. Clean or replace suction line. Examine filter for plugging.

RELIEF VALVES NOT PROPERLY SET OR LEAKING
Test relief valves to make sure they are opening at their rated pressure. Examine valves for damaged seats that could leak.

BADLY WORN COMPONENTS
Examine and test valves, motors, cylinders, etc. for external and internal leaks. If wear is abnormal, try to locate the cause.

SYSTEM OPERATES TOO FAST

ADJUSTABLE ORIFICE INSTALLED BACKWARD OR NOT INSTALLED
Install orifice parts correctly and adjust.

OBSTRUCTION OR DIRT UNDER SEAT OF ORIFICE
Remove foreign material. Readjust orifice.

OVERHEATING OF OIL IN SYSTEM

OPERATOR HOLDS CONTROL VALVES IN POWER POSITION TOO LONG, CAUSING RELIEF VALVE TO OPEN
Return control lever to neutral position when not in use.

USING INCORRECT OIL
Use oil recommended by manufacturer. Be sure oil viscosity is correct.

LOW OIL LEVEL
Fill reservoir with clean oil. Look for leaks.

DIRTY OIL
Drain and refill with clean oil. Look for source of contamination.

ENGINE RUNNING TOO FAST
Reset governor or reduce throttle.

INCORRECT RELIEF VALVE PRESSURE
Check pressure and clean or replace relief valves.

INTERNAL COMPONENT OIL LEAKAGE
Examine and test valves, cylinders, motors, etc. for external and internal leaks. If wear is abnormal, try to locate cause.

RESTRICTION IN PUMP SUCTION LINE
Clean or replace.

DENTED, OBSTRUCTED OR UNDERSIZED OIL LINES
Replace defective or undersized oil lines. Remove obstructions.

OIL COOLER MALFUNCTIONING
Clean or repair.

CONTROL VALVE STUCK IN PARTIALLY OR FULL OPEN POSITION
Free all spools so that they return to neutral position.

HEAT NOT RADIATING PROPERLY
Clean dirt and mud from reservoir, oil lines, coolers, and other components.

AUTOMATIC UNLOADING CONTROL INOPERATIVE (IF EQUIPPED)
Repair valve.

INLET SCREEN PLUGGED
Clean screen.

BROKEN OR DAMAGED PUMP PARTS
Repair pump. Look for cause of damage like contamination or too much pressure.

STICKING OR BINDING PARTS
Repair binding parts. Clean parts and change oil if necessary.

PUMP MAKES NOISE

LOW OIL LEVEL
Fill reservoir. Look for leaks.

WATER IN OIL
Drain and replace oil.

WRONG KIND OF OIL BEING USED
Use oil recommended by manufacturer.

AIR LEAK IN LINE FROM RESERVOIR TO PUMP
Tighten or replace suction line.

KINK OR DENT IN OIL LINES (RESTRICTS FLOW)
Replace oil lines.

WORN SEAL AROUND PUMP SHAFT
Clean sealing area and replace seal. Check oil for contamination or pump for misalignment.

FOAMING OF OIL IN SYSTEM

LOW OIL LEVEL
Fill reservoir. Check system for leaks.

OIL VISCOSITY TOO HIGH
Change to lighter oil.

PUMP SPEED TOO FAST
Operate pump at recommended speed.

SUCTION LINE PLUGGED OR PINCHED
Clean or replace line between reservoir and pump.

SLUDGE AND DIRT IN PUMP
Disassemble and inspect pump and lines. Clean hydraulic system. Determine cause of dirt.

RESERVOIR AIR VENT PLUGGED
Remove breather cap, flush, and clean air vent.

AIR IN OIL
Tighten or replace suction line. Check system for leaks. Replace pump shaft seal.

WORN OR SCORED PUMP BEARINGS OR SHAFTS
Replace worn parts or complete pump if parts badly worn or scored. Determine cause of scoring.

PUMP LEAKS OIL

DAMAGED SEAL AROUND DRIVE SHAFT
Tighten packing or replace seal. Trouble may be caused by contaminated oil. Check oil for abrasives and clean entire hydraulic system. Try to locate source of contamination. Check the pump drive shaft. Misalignment could cause the seal to wear. If shaft is not aligned, check the pump for other damage.

LOOSE OR BROKEN PUMP PARTS
Make sure all bolts and fittings are tight. check gaskets. Examine pump castings for cracks. If pump is cracked, look for a cause like too much pressure or hoses that are attached incorrectly.

LOAD DROPS WITH CONTROL VALVE IN NEUTRAL POSITION

LEAKING OR BROKEN OIL LINES FROM CONTROL VALVE TO CYLINDER
Check for leaks. Tighten or replace lines. Examine mating surfaces on couplers for irregularities.

OIL LEAKING PAST CYLINDER PACKINGS OR O-RINGS
Replace worn parts. If wear is caused by contamination, clean hydraulic system and determine the source.

OIL LEAKING PAST CONTROL VALVE OR RELIEF VALVES
Clean or replace valves. Wear may be caused by contamination. Clean system and determine source of contamination.

CONTROL LEVER NOT CENTERING WHEN RELEASED
Check linkage for binding. Make sure valve is properly adjusted and has no broken or binding parts.

CONTROL VALVE STICKS OR WORKS HARD

MISALIGNMENT OR SEIZING OF CONTROL LINKAGE
Correct misalignment. Lubricate linkage joints.

TIE BOLTS TOO TIGHT (ON VALVE STACKS)
Use manufacturer's recommendation to adjust tie bolt torque.

VALVE BROKEN OR SCORED INTERNALLY
Repair broken or scored parts. Locate source of contamination that caused scoring.

CONTROL VALVE LEAKS OIL

TIE BOLTS TOO LOOSE (ON VALVE STACKS)
Use manufacturer's recommendation to adjust tie bolt torque.

WORN OR DAMAGED O-RINGS
Replace O-rings (especially between valve stacks). If contamination has caused O-rings to wear, clean system and look for source of contamination.

BROKEN VALVE PARTS
If valve is cracked, look for a cause like too much pressure or hoses that are attached incorrectly.

CYLINDERS LEAK OIL

DAMAGED CYLINDER BARREL
Replace cylinder barrel. Correct cause of barrel damage.

ROD SEAL LEAKING
Replace seal. If contamination has caused seal to wear, look for source. Wear may be caused by external as well as internal contaminants. Check piston rod for scratches or misalignment.

LOOSE PARTS
Tighten parts until leakage has stopped.

PISTON ROD DAMAGED
Check rod for nicks or scratches that could cause seal damage or allow oil leakage. Replace defective rods.

CYLINDER LOWERS WHEN CONTROL VALVE IN IN "SLOW RAISE" POSITION

DAMAGED CHECK VALVE IN LIFT CIRCUIT
Repair or replace check valve.

LEAKING CYLINDER PACKING
Replace packing. Check oil for contamination that could cause wear. Check alignment of cylinder.

LEAKING LINES OR FITTINGS TO CYLINDER
Check and tighten. Examine mating surfaces on couplers for irregularities.

POWER STEERING DOES NOT WORK, STEERS HARD, OR IS SLOW

AIR IN SYSTEM
Bleed system. Check for air leaks.

INTERNAL LEAKAGE IN SYSTEM
Components may not be adjusted properly. Parts may be worn or broken. Check for cause of wear.

SYSTEM NOT PROPERLY TIMED
Time according to manufacturer's instructions.

WORN OR DAMAGED BEARINGS
Check and replace bearings in steering components.

INSUFFICIENT PRESSURE
Check pump and relief valves. Contamination could cause valves to leak or pump to wear.

POWER BRAKES MALFUNCTION

HEAVY OIL OR IMPROPER BRAKE FLUID
Warm up fluid or change to one of lighter viscosity. Use proper oil or brake fluid (see machine operator's manual).

Note: Many brake circuits use brake fluid instead of hydraulic oil. DO NOT MIX.

AIR IN SYSTEM
Bleed brake system. Find out where air is coming from.

CONTAMINATED OIL
This may cause components to wear or jam. Clean and repair system and check for cause of contamination.

BRAKE PEDAL RETURN RESTRICTED
Clean dirt from moving parts. Check linkage for damage.

ACCUMULATOR NOT WORKING (IF EQUIPPED)
Check accumulator precharge. If accumulator is defective, repair or replace it.

TEST YOURSELF

QUESTIONS

1. Give the seven basic steps for good trouble shooting.
2. During which of the seven steps should you begin replacing parts?
3. True or false? "Test the system flow first so that you have a guide for readings on the pump flow tests."
4. (Fill in the blanks with "circuit" or "system".) "As a general rule, faulty __________ relief valves will affect pressure readings on all tests, while faulty _________ relief valves will affect only some readings."
5. (Fill in the blanks with "fixed" or "variable") "The ________ displacement pump is generally used on closed-center systems and the _________ displacement pumps used on the open-center systems."
6. (Fill in the blanks with "constant" or "varied") "In a 'load-sensitive' system, the flow is _______ and the pressure is __________."
7. What three senses are used during an Operational Checkout?
8. What are the four types of measuring meters and gauges required for testing and adjusting the components of a hydraulic system?

APPENDIX

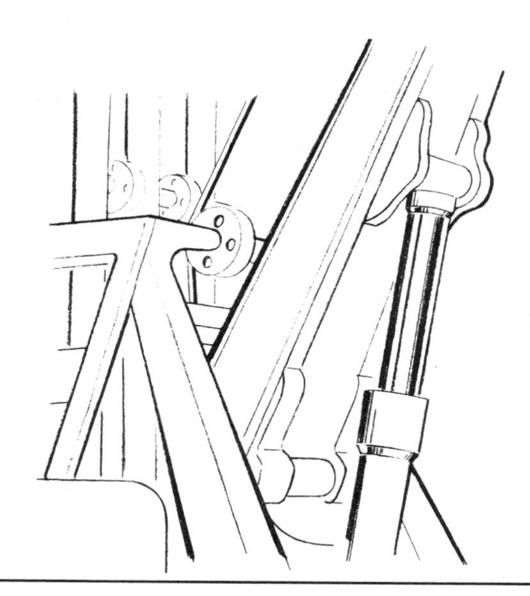

Definitions of Terms and Symbols

A

ACCUMULATOR

A container which stores fluids under pressure as a source of hydraulic power. It may also be used as a shock absorber.

ACTUATOR

A device which converts hydraulic power into mechanical force and motion. (Examples: hydraulic cylinders and motors.)

ARTICULATE

To rotate from side to side.

B

BLEED

The process by which air is removed from a hydraulic system.

BOTTOM

To reach the maximum travel of a component. As in a fully retracted cylinder.

BYPASS

A secondary passage for fluid flow.

C

CAM

A rotary member with a raised portion to provide motion to, or be rotated by a roller, pin, cylinder or valve.

CAM LOBE MOTOR

A hydraulic radial piston motor in which rotational force is created by the outward movement of pistons against the lobes of a stationary cam.

CAVITATION

A phenomenon in which vacuum bubbles occur in a hydraulic system. The violent collapse of these bubbles can cause erosion of metal parts, noise and vibration.

CHARGE SYSTEM

A low-pressure hydraulic system for supplying oil to the main system hydraulic pump.

CLOSED-CENTER SYSTEM

A hydraulic system in which the control valves are closed during neutral, stopping the inlet oil flow.

CLOSED-LOOP SYSTEM

A hydraulic system in which the pump and actuator are connected directly, without the use of a directional valve, and on which the operating oil does not normally return to the reservoir.

COMPONENT

One assembly in a hydraulic system.

CONTROLLER

A microprocessor that controls electro-hydraulic valve functions.

COOLER (Oil)

A heat exchanger which removes heat from a fluid (See "Heat Exchanger").

COUPLER

A device to connect two hoses or lines, or to connect hoses to valve receptacles.

CUSHION

A device built into the end of a cylinder that restricts outlet flow and thereby slows down the piston at the end of its travel.

CYCLE

A single complete operation of a component, which begins and ends in a neutral position.

CYLINDER

An "actuator." A device for converting fluid power into limited linear or circular motion.

Double-Acting Cylinder - A cylinder in which fluid force can be applied to the movable element in either direction.

Followup Cylinder - Non-powered cylinder that signals the position of components or powered pistons in the system.

Piston-Type Cylinder - A cylinder that uses a sliding piston in a housing to produce straight movement.

Rotary Cylinder - A cylinder in which fluid force is applied to a movable vane to produce rotary motion.

Single-Acting Cylinder - A cylinder in which fluid force is applied to only one side of a piston to move the element in only one direction.

Slave Cylinder - See Followup Cylinder.

Vane-Type Cylinder - A cylinder that uses a turning vane in a circular housing to produce rotary movement. See Rotary Cylinder.

D

DISPLACEMENT

The volume of oil displaced by one complete stroke or revolution (of a pump, motor, or cylinder).

DRAG LINK

A mechanical feedback linkage or a connection which backs up a hydraulically actuated function.

DRIFT

Movement of a cylinder or motor due to internal leakage past components in the hydraulic system.

E

EFFICIENCY

The percentage of actual performance of a hydraulic component compared to the theoretical performance.

ENERGY

The mechanical work put into generating flow, pressure and speed in the hydraulic oil that can be converted back to a mechanical output.

Heat Energy - That energy which is lost in a system when pressure drops without accomplishing work turns to heat. Example: internal leakage or line restriction.

Hydraulic Energy - The result of mechanical work put into generating oil flow with pressure that can be converted back to a mechanical output.

Kinetic Energy - The result of mechanical work put into generating velocity in hydraulic fluid as would be used in a torque converter.

Potential Energy - The work put into storing pressurized oil as in loading an accumulator.

F

FILTER (OIL)

A device that removes contaminants from a fluid.

FLOW METER

A testing device that measures flow rate.

FLOW RATE

The volume of fluid passing a point in a given time.

FLUID POWER

Energy transmitted and controlled through use of a pressurized fluid.

FORCE

A push or pull acting upon a body. In a hydraulic cylinder, it is the product of the fluid pressure, multiplied by the piston area. It is measured in pounds (newtons).

FRICTION

The resistance to fluid flow in a hydraulic system. An energy loss in terms of power output.

G

GEROTOR

A form of internal gear pump which uses lobes instead of gear teeth.

H

HEAT EXCHANGER

A device that transfers heat through a conducting wall from one fluid to another. See "Cooler, (Oil)".

HORSEPOWER

English measure of work produced per unit of time.

HOSE

A flexible line usually made of reinforced rubber used to route oil from one component to another.

HYDRAULICS

The engineering science of liquid pressure and flow.

Hydrodynamics - The engineering science of the energy of liquid momentum (as used in a torque converter).

Hydrostatics - The engineering science of the energy of liquids pressurized flow. All systems discussed in this manual are hydrostatic.

I

INERT GAS

A non-explosive gas.

L

LINE

A tube, pipe, or hose for conducting a fluid.

LOAD

The resistance to flow in an actuator as it is moved.

M

MANIFOLD

A fluid conductor that provides many ports.

MOTOR - (Hydraulic)

A device for converting fluid energy into rotary mechanical force and motion.

O

OPEN-CENTER SYSTEM

A hydraulic system in which the control valves are open to continuous oil flow when in neutral

ORIFICE

A restricted passage in a hydraulic circuit. Usually a small drilled hole to limit flow or to create a pressure differential in a circuit.

O-RING

A static and/or dynamic seal for curved or circular mating surfaces.

OUT-OF-STROKE

The condition of a variable displacement pump when it is maintaining a pressure but not delivering oil to the system.

P

PACKING

Any material, or device, that seals by compression. Common types are U-packings, V packings, "Cup" packings, and O-rings.

PIPE

A line whose outside diameter is standardized for threading.

PISTON

A cylindrical part which moves or reciprocates in a cylinder to transfer hydraulic energy back to mechanical movement.

PORT

An opening for a fluid passage in a component.

POUR POINT

The lowest temperature at which a fluid remains liquid enough to pour.

POWER BEYOND

An adapting sleeve that allows working oil to be sent to another component.

POWER LIFT

Any device for hydraulic and/or mechanically raising an implement. On modern machines, it is often used to refer to the three-point-hitch.

PRESSURE

Force of a fluid per unit area, usually expressed in pounds per square inch (psi) or kilo pascals (kPa).

Back Pressure - A pressure on the return side of a function.

Breakout Pressure - The minimum pressure which starts moving an actuator.

Charge Pressure - The pressure of the oil pumped to a system hydraulic pump.

Cracking Pressure - The pressure at which a relief valve, etc., begins to open and pass fluid.

Differential Pressure - The difference in pressure between any two points in a system or a component (Also called a "pressure drop").

Full-Flow Pressure - The pressure at which a valve will pass the full flow of the system.

Operating Pressure - The pressure at which a system is normally operated.

Pilot Pressure - Pressure in a circuit used to actuate or control a component.

Rated Pressure - The manufacturer's recommended operating pressure for a component or the pressure at which a component will provide specific results.

Static Pressure - The pressure in a fluid at rest (A form of "potential energy").

Suction Pressure - The absolute pressure (Vacuum) of the fluid at the inlet side of the pump.

Surge Pressure - A momentary pressure increase caused from a rapidly accelerated column of oil.

System Pressure - The pressure that overcomes the total resistances in a system. It includes all losses as well as useful work.

Working Pressure - The pressure that overcomes the resistance of the working device.

PULSATION

Repeated small fluctuation of pressure in a circuit.

PUMP

A device that converts mechanical force into hydraulic fluid power.

Fixed-Displacement Pump - A pump in which the output per cycle cannot be varied.

Variable-Displacement Pump - A pump in which the output per cycle can be varied.

R

REGENERATIVE CIRCUIT

A circuit in which oil returning from the cylinder rod end is returned to the piston end. It can be used to speed up the action of a cylinder when very little force is required.

REMOTE

A hydraulic function, such as a cylinder, which is separate from its supply source. Usually connected to the source by flexible hoses.

RESERVOIR

A container for keeping a supply of working fluid in a hydraulic system.

RESTRICTION

A reduced cross-sectional area in a line or passage, which causes a pressure drop.

ROCKSHAFT

A rotating shaft and arm device to transfer a mechanical force. Usually used when referring to the hydraulically operated three-point hitch.

S

SEAT

The sealing surface on which a valve rests.

SERVO

A low pressure control mechanism.

SOLENOID

An electro-magnetic device used to position a hydraulic valve.

STARVATION

A lack of oil in vital areas of a system usually the pump inlet.

STRAINER

A coarse filter.

STROKE

The length of travel of a piston in a cylinder or the changing of the displacement of a variable delivery pump.

SURGE

A momentary rise of pressure in a hydraulic circuit.

SWASH PLATE

A tilted plane surface in an axial piston pump used to create reciprocating movements to pistons.

SYMBOLS, SCHEMATIC

Used on drawings as a shorthand representation of hydraulic system components.

SYSTEM

A series of components or circuits connected to each other. Usually includes everything using the oil of a pump's output.

T

THERMAL EXPANSION

The increase in fluid volume due to temperature increase.

TORQUE

Turning effort usually measured in pounds inch (lb-in), pounds foot (lbs-ft) or newton meters (N-m).

TUBE

A metal line sized by its outside diameter.

V

VALVE

A device that controls the pressure, direction, or the flow rate of fluid.

Bypass Regulator Valve - A valve that maintains a constant flow to a circuit by dumping the excess oil.

Check Valve - A valve that permits oil flow in only one direction.

Closed-Center Valve - A valve in which inlet ports are closed in the neutral position, stopping flow from pump.

Directional Control Valve - A valve that directs oil through selected passages.

Electro-Hydraulic Valve - A valve that is opened and closed by a solenoid.

Flow Control Valve - A valve that controls the rate of flow. Also called a "volume control valve".

Flow Divider Valve - A valve that divides the flow from one source into two or more branches (Includes "priority" and "proportional" types).

Needle Valve - A valve with a tapered point which moves in relation to a seat to regulate the flow rate.

Open-Center Valve - A valve in which the inlet and outlet ports are open in the neutral position, allowing continuous pump flow through the valve.

Pilot Valve - A valve that is used to operate another valve or control.

Pilot Operated Valve - A valve that is actuated by oil from a pilot valve.

Poppet Valve - A valve that rests on a seat. It can be manually or hydraulically opened to pass oil in only one direction.

Pressure Control Valve - A valve whose function is to control pressure (Includes relief, pressure reducing, sequencing, and unloading valves).

Pressure Reducing Valve - A pressure control valve which controls outlet pressure.

Pressure Sequence Valve - A pressure control valve which directs flow in a preset sequence.

Priority Valve - A valve that insures a supply of oil to one function before allowing flow to other functions.

Priority Flow Divider Valve - A valve that directs a constant oil flow to one circuit and dumps excess flow into another circuit.

Proportional Flow Divider Valve - A valve that divides the oil proportionally and directs it to all circuits at all times.

Relief Valve - A valve that limits the pressure in a system.

Rotary Directional Valve - A cylindrical shaped valve that is turned to open and close hydraulic passages.

Selector Valve - A valve that selects one of two or more circuits into which it directs oil flow.

Shuttle Valve - A free floating spool valve that moves in its bore to control oil flow based on the pressures acting against it.

Shutoff Valve - A valve that is opened to allow full flow, or closed to block or restrict oil flow.

Spool Directional Valve - A valve designed as a spool, which slides in a bore, opening and closing passages.

Stroke Control Valve - A device to control the output of a variable-displacement pump.

Thermal Relief Valve - A valve that limits the pressure in a system caused by heat expansion of oil.

Two-, Three-, Four-, or Six-Way Valve - A valve spool sections having 2, 3, 4, or 6 connecting ports.

Unloading Valve - A valve that dumps excess pump flow at a low pressure when the system requirements are met.

Volume Control Valve - A valve that controls flow rate. Includes flow control valves, flow divider valves, and bypass flow regulators.

VALVE STACK

A series of control valves in a stack with common end plates and a common oil inlet and outlet.

VELOCITY

The speed at which fluid travels. Usually given as feet or millimeters per second.

VENT

An air breathing opening in a fluid reservoir or cylinder.

VISCOSITY

The measure of resistance of a fluid to flow.

VOLUME

The measure of a container's capacity, or the amount of fluid flow per unit of time. Usually given as gallons per minute (gpm) or liters per minute (L/m).

ABBREVIATIONS

ANSI — American National Standards Institute

ASAE — American Society of Agricultural Engineers (Sets standards for many hydraulic components for agricultural use.)

bar — Metric unit of measure for pressure

C — degrees Celsius (temperature)

F — degrees Fahrenheit (temperature)

gpm — gallons per minute...fluid flow

hp — horsepower

I.D. - inside diameter

ISO — International Organization for Standardization (Establishes many standards for worldwide use.)

kg/cm2 — kilograms per square centimeter (metric unit for pressure)

kPa — kilo Pascals, metric unit of measure for pressure

kW — kilowatts (metric unit of measure for power)

lb-ft — pounds-foot...torque or turning effort

lb-in. — pound-inch...torque or turning effort

L/m — liters per minute

Nm — newton meters (metric unit of measure for torque)

O.D. — outside diameter

psi — pounds per square inch (pressure)

rpm — revolutions per minute

SAE —· Society of Automotive Engineers (sets standards for many hydraulic components)

CONVERSION CHART

Multiply	By	To Get or Multiply	By	To Get
AREA				
Square Inch (in.^{2})	645	Square Millimeter (mm^2)	0.00155	Square Inch
Square Inch (in.^{2})	6.45	Square Centimeter (cm^2)	0.155	Square Inch
Square Foot (ft^2)	0.0929	Square Meter (m^2)	10.76	Square Foot
Square Yard (yd^2)	0.836	Square Meter (m^2)	1.196	Square Yard
Acre	0.4047	Hectare (ha) or 1 0,000m^2	2.471	Acre
Square Mile (Mi^2)	2.59	Square Kilometer (Km^2)	0.386	Square Mile
ENERGY				
British Thermal Unit (BTU)	1.055	Kilojoule (kJ)	0.948	British Thermal Unit
Calorie	4.10	Joule (J)	0.239	Calorie
FLOW				
Gallons per minute (gpm)	3.785	Liter per Minute (Urn)	0.246	Gallons per minute
FORCE				
Pound (lb force)	4.448	Newton (N)	0.225	Pound
LENGTH				
Inch (in.)	25.4	Millimeter (mm)	0.03937	Inch
Foot (ft)	0.3048	Meter (m)	3.281	Foot
Yard (yd)	0.9144	Meter (m)	1.094	Yard
Mile (mi)	1.608	Kilometer (km)	0.621	Mile
MASS (WEIGHT)				
Ounce (oz)	28.3	Gram (g)	0.035	Ounce
Pound (lb)	0.4535	Kilogram (kg)	2.205	Pound
Ton(20001b)	0.9071	Metric Ton	1.102	Ton
POWER				
Horsepower (hp)	0.7457	Kilowatt (kW)	1.34	Horsepower
Pound-Foot per Second (lb-f/sec)	1.36	Watt (W)	0.735	Pound-Foot per Second
PRESSURE				
Pounds per Square Inch (psi)	6.89475	Kilopascal (kPa)	0.145	Pounds per Square Inch
Pounds per Square Inch (psi)	0.06895	Bar	14.5	Pounds per Square Inch
Inches of Mercury (in. Hg)	3.38	Kilopascal (kPa)	0.30	Inches of Mercury
Inches of Water (in. H^{2}0)	0.249	Kilopascal (kPa)	4.0	Inches of Water
STRESS				
Pounds per Square Inch (psi)	0.006895	Megapascal (MPa or N/mm2)	145.	Pounds per Square Inch
TORQUE				
Pound-Inch (lb-in)	0.113	Newton-Meter (N•m)	8.85	Pound-inch
Pound-Foot (lb-ft)	1.3568	Newton-Meter (N•m)	0.737	Pound-Foot

CONVERSION CHART

Multiply	By	To Get or Multiply	By	To Get
VOLUME				
Cubic Inch ($in.^3$)	16.4	Cubic Centimeter (cm^3)	0.061	Cubic Inch
Cubic Inch ($in.^3$)	0.016	Liter (L)	61.	Cubic Inch
Fluid Ounce (fl.oz.)	29.6	Milliliter (ml) or cubic centimeter	0.034	Fluid Ounce
Cubic Foot (ft3)	0.0283	Cubic Meter (m^3)	35.31	Cubic Foot
Cubic Yard (yd3)	0.7646	Cubic Meter (m^3)	1.31	Cubic Yard
Bushel (bu)	0.03524	Cubic Meter (m^3)	28.38	Bushel
Bushel (bu)	35.24	Liter (L)	0.29	Bushel
Quart (qt)	0.946	Liter (L)	1.06	Quart
Gallon (gal)	3.79	Liter (L)	0.26	Gallon

TEMPERATURE

Fahrenheit to Celsius
$(T_F - 32)/1.8 = T_C$

Celsius to Fahrenheit
$1.8\,(T_C) + 32 = T_F$

UNIFIED INCH BOLT AND CAP SCREW TORQUE VALUES

SAE Grade and Head Markings	NO MARK	1 or 2[b]	5	5.1	5.2	8	8.2
SAE Grade and Head Markings	NO MARK	2	5			8	

Size	Grade 1				Grade 2b				Grade 5, 5.1, or 5.2				Grade 8 or 8.2			
	Lubricated[a]		Dry[a]		Lubricated[a]		Dry[a]		Lubricated[a]		Dry[a]		Lubricated[a]		Dry[a]	
	N•m	lb-ft	N•m	lb-ft	N•m	lb-ft	N•m	lb-ft	N•m	lb-ft	N•m	lb-ft	N•m	lb-ft	N•m	lb-ft
1/4	3.7	2.8	4.7	3.5	6	4.5	7.5	5.5	9.5	7	12	9	13.5	10	17	12.5
5/16	7.7	5.5	10	7	12	9	15	11	20	15	25	18	28	21	35	26
3/8	14	10	17	13	22	16	27	20	35	26	44	33	50	36	63	46
7/16	22	16	28	20	35	26	44	32	55	41	70	52	80	58	100	75
1/2	33	25	42	31	53	39	67	50	85	63	110	80	120	90	150	115
9/16	48	36	60	45	75	56	95	70	125	90	155	115	175	130	225	160
5/8	67	50	85	62	105	78	135	100	170	125	215	160	215	160	300	225
3/4	120	87	150	110	190	140	240	175	300	225	375	280	425	310	550	400
7/8	190	140	240	175	190	140	240	175	490	360	625	450	700	500	875	650
1	290	210	360	270	290	210	360	270	725	540	925	675	1050	750	1300	975
1-1/8	470	300	510	375	470	300	510	375	900	675	1150	850	1450	1075	1850	1350
1-1/4	570	425	725	530	570	425	725	530	1300	950	1650	1200	2050	1500	2600	1950
1-3/8	750	550	950	700	750	550	950	700	1700	1250	2150	1550	2700	2000	3400	255
1-1/2	1000	725	1250	925	990	725	1250	930	2250	1650	2850	2100	3600	2650	4550	3360

DO NOT use these values if a different torque value or tightening procedure is listed for a specific application. The torque values listed are for general use only. Check the tightness of the cap screws periodically.

Shear bolts are designed to fail under predetermined loads. Always replace the shear bolts with an identical grade.

Fasteners should be replaced with the same or higher grade. If higher grade fasteners are used, these should only be tightened to the strength of the original.

Make sure the fastener threads are clean and you properly start thread engagement. This will prevent themfrom failing when tightening.

Tighten the plastic insert or crimped steel-type lock nuts to approximately 50 percent of amount shown in chart. Tighten the toothed or serrated-type lock nuts to full torque

[a] *"Lubricated" means coated with a lubricant such as engine oil, or fasteners with phosphate and oil coatings. "Dry" means plain or zinc plated without any lubrication.*

[b] *Grade 2 applies for hex cap screws (not hex bolts) up to 152 mm (6-in.) long. Grade 1 applies for hex cap screws over 152 mm (6-in.) long, and for all other types of bolts and screws of any length.*

METRIC BOLT AND CAP SCREW TORQUE VALUES

Property Class and Head Markings	4.8	8.8 / 9.8	10.9	12.9
Property Class and Head Markings	5	10	10	12

Size	Class 4.8				Class 8.8 or 9.8				Class 10.9				Class 12.9			
	Lubricated[a]		Dry[a]		Lubricated[a]		Dry[a]		Lubricated[a]		Dry[a]		Lubricated[a]		Dry[a]	
	N•m	lb-ft	N•m	lb-ft	N•m	lb-ft	N•m	lb-ft	N•m	lb-ft	N•m	lb-ft	N•m	lb-ft	N•m	lb-ft
M6	4.8	3.5	6	4.5	9	6.5	11	8.5	13	9.5	17	12	15	11.5	19	14.5
M8	12	8.5	15	11	22	16	28	20	32	24	40	30	37	28	47	35
M10	23	17	29	21	43	32	55	40	63	47	80	60	75	55	95	70
M12	40	29	50	37	75	55	95	70	110	80	140	105	130	95	165	120
M14	63	47	80	60	120	88	150	110	175	130	225	165	205	150	260	190
M16	100	70	125	92	190	140	240	175	275	200	350	225	320	240	400	300
M18	135	100	175	125	260	195	330	250	375	275	475	350	440	325	560	410
M20	190	140	240	180	375	275	475	350	530	400	675	500	625	460	800	580
M22	260	190	330	250	510	375	650	475	725	540	925	675	850	625	1075	800
M24	330	250	425	310	650	475	825	600	925	675	1150	850	1075	800	1350	1000
M27	490	360	625	450	950	700	1200	875	1350	1000	1700	1250	1600	1150	2000	1500
M30	675	490	850	625	1300	950	1650	1200	1850	1350	2300	1700	2150	1600	2700	2000
M33	900	675	1150	850	1750	1300	220	1650	2500	1850	3150	2350	2900	2150	3700	2750
M36	1150	850	1450	1075	2250	1650	2850	2100	3200	2350	4050	3000	3750	2750	4750	3500

CAUTION: Use only metric tools on metric hardware. Other tools may not fit properly. They may slip and, cause injury.

DO NOT use these values if a different torque value or tightening procedure is listed for a specific application. The torque values listed are for general use only. Check the tightness of the cap screws periodically.

Shear bolts are designed to fail under predetermined loads. Always replace the shear bolts with an identical grade.

Fasteners should be replaced with the same or higher grade. If higher grade fasteners are used, these should only be tightened to the strength of the original.

Make sure the fastener threads are clean and you properly start thread engagement. This will prevent them from failing when tightening.

Tighten the plastic insert or crimped steel-type lock nuts to approximately 50 percent of amount shown in chart. Tighten the toothed or serrated-type lock nuts to full torque value.,

[a] "Lubricated" means coated with a lubricant such as engine oil, or fasteners with phosphate and oil coatings. "Dry" means plain or zinc plated without any lubrication.

ANSWERS TO ' TEST YOURSELF ' QUESTIONS

CHAPTER 1

1. False - A pump creates flow, pressure is caused by resistance to flow.

2. b. drop - An orifice always causes a pressure drop as oil flows through it.

3. Force = piston area times the oil pressure.

4. The four essential elements for a system are:

   a. The PUMP - to push the fluid through the system.

   b. The CYLINDER (MOTOR) - to convert fluid movement into work.

   c. The LINES - to connect the components..

   d. The RESERVOIR - to store the fluid.

   (Most systems also require Valves to control fluid pressure and flow.)

5. False - The pump converts mechanical to hydraulic power.

6. c. Power - Power equals work per time unit.

7. b. the control valve - The neutral flow of oil determines the type of system. (See # 8)

8. In neutral, the open-center valve provides a passage for pump flow to return to the reservoir. The closed-center valve blocks the flow of oil from the pump in neutral.

9. Oil is pushed into the pump by the atmospheric pressure.

10. In the load sensing system both the flow and pressure vary.

ANSWERS TO ' TEST YOURSELF ' QUESTIONS

CHAPTER 2

1. The safety alert symbol indicates "Be careful Safety hazard."

2. The three safety words are; Caution — Warning — Danger.

3. Always use a piece of cardboard to locate a high-pressure leak.

4. Hydraulic accumulators store Energy.

5. A crush point is a place where two objects move toward each other, or where one object moves toward a stationary object.

6. Before servicing

   a. Lower all equipment to the ground.

   b. Stop engine and remove key.

   c. Disconnect the battery ground strap.

   d. Hang a "Do Not Operate" sign, or tag in the opera tor station.

ANSWERS TO ' TEST YOURSELF ' QUESTIONS

CHAPTER 3

1. ISO symbols are internationally recognized; cross language barriers; they simplify design, fabrication, analysis, and function of circuits; and they show connections, flow paths, and functions of components

2. The difference is indicated by the placement of the arrowheads inside the circle. Arrowheads pointing out mean that it is a pump. Arrowheads pointing in to the center of the circle mean that it is a motor.

3. The arrow indicates that the pump or motor is a variable displacement type.

4. False. Four squares represent a four-position valve.

5. Three — one for each position.

6. Only one because the valve can only be shown in one valving position.

7. Nonactivated.

8. True.

9. Hydraulic oil reservoir or tank.

10. International Standards Organization.

ANSWERS TO ' TEST YOURSELF ' QUESTIONS

CHAPTER 4

1. A hydraulic pump converts **mechanical** force into **hydraulic** force.

2. True.

3. False. They have no way of preventing backfeeding of oil.

4. Gear, vane and piston pumps are generally used.

5. An external gear pump has a gear on a gear and these gears rotate in the opposite direction of each other. The internal gear pump has an internal gear running within an external gear, both gears rotating in the same direction.

6. A pump cavitates when less oil is allowed into the inlet than the pump is capable of pumping. This is usually the result of a restricted inlet line. This allows vacuum spaces to develop in the incoming fluid. When these spaces collapse they erode the parts which they contact.

7. In "axial" pumps, the pistons are parallel with the axis of the rotating parts (cylinder block). In "radial" pumps, the pistons are perpendicular to the rotating parts (usually a cam)

8. The axial piston pump is the most easily adopted to the load sensing systems.

9. The "rotating group" of an axial piston pump consists of the piston block, the pistons with slippers and a spring mechanism to load the slippers against the swashplate.

10. Human error is the No. 1 cause of pump failures. Specifically, improper condition of the hydraulic fluid is the most frequent.

11. In standby, the pump will pump enough to satisfy controlled internal leakage but the outlet flow will be **zero.**

12. Doubling the operating pressure will decrease the life of the pump bearings and thus the pump to **1/8 of its normal service life.**

ANSWERS TO ' TEST YOURSELF ' QUESTIONS

CHAPTER 5

1. Pressure control valves, directional control valves, and volume control valves.

2. Pressure control valves a. Limit pressure, b. Reduce pressure, c. Unload a pump and d. Set pressure for secondary circuit(s).

3. **Direct acting** and **pilot operated.**

4. False - The two pressures are closer on the pilot operated valve.

5. False - Pressure comes from the inlet.

6. The spool valve is the most used of all direction control valves.

7. The two types of valves are the **open center** and **closed center.**

8. In neutral, pump flow passes through the open center valve, but is blocked in the closed center spool valve.

9. Return port.

10. True - The flow will be affected by the pressure drop across the valve.

11. False - The flow will remain the same regardless of the pressures.

12. The fixed amount reduction senses both the inlet and outlet.

ANSWERS TO ' TEST YOURSELF ' QUESTIONS

CHAPTER 6

1. False. Cylinders convert fluid power to mechanical power.

2. First blank — "straight." Second blank — "rotary."

3. First blank — "double." Second blank — "single.

4. One side is filled with hydraulic oil; the other side is filled with air.

5. The rod fills an area of the piston that is not exposed to pressure oil. So the stroke produced on this side is less powerful, but faster (since less oil is needed to move the piston).

6. It slows down the cylinder movement at the end of its stroke by restricting discharge oil from the cylinder.

ANSWERS TO ' TEST YOURSELF ' QUESTIONS

CHAPTER 7

1. A hydraulic motor converts fluid force into mechanical force.

2. The motor works in reverse of a pump. The pump draws in fluid and pushes it out, while the motor has fluid forced in and exhausts it out. In other words, the pump drives fluid, while the motor is driven by fluid.

3. Gear, vane, and piston types.

4. Torque is a measure of the turning force (twist) exerted at the motor drive shaft.

5. False. Though very similar, port sizes, special bearings, internal oil passages and other parts make each unique to its purpose. To avoid trouble, use pumps as pumps and motors as motors.

6. First blank — "rotating" carrier- Second blank — "fixed" cam ring

ANSWERS TO ' TEST YOURSELF ' QUESTIONS

CHAPTER 8

1. "precharge" — this determines the operating range of the accumulator and accumulator capacity at the normal working pressures.

2. If hydraulic oil and oxygen mix under pressure, an explosion can occur (dieseling). Air contains moisture, which condenses to cause rust which in turn damages seals and other parts. ***Always use dry nitrogen that is an inert gas.***

3. Relieve all hydraulic pressure in the accumulator.

4. The four major uses are: 1) Store energy, 2) Absorb shocks, 3) Build pressure gradually, 4) Maintain constant pressure.

ANSWERS TO ' TEST YOURSELF ' QUESTIONS

CHAPTER 9

1. A full-flow system filters all the oil for each cycle. A bypass system filters only a small part of the oil.

2. It opens to prevent the filter from rupturing by allowing oil to flow around the filter when it becomes plugged.

3. **Micron** rating and the **Beta** rating

4. The beta rating is determined by performance tests while the micron rating is the theoretical particle size the filter will stop.

5. Filter tests include: 1.) Pore size test, 2.) Rupture test, 3.) Cold weather performance test, 4.) Burst test, 5.) Initial restriction test, 6.) Material compatibility test and 7.) Flow fatigue test.

ANSWERS TO ' TEST YOURSELF ' QUESTIONS

CHAPTER 10

1. The reservoir helps to keep oil clean, it cools the oil, and it separates air from the oil.

2. The pressurized reservoir prevents air and dirt from entering the hydraulic system by eliminating the constant breathing. In some cases it provides pressure to the pump to prevent suction side cavitation.

3. Air and water.

ANSWERS TO ' TEST YOURSELF ' QUESTIONS

CHAPTER 11

1. False.

2. False.

3. Seamless cold-rolled steel and welded cold-rolled steel.

4. Replace all brackets and clamps to their original position.

5. Skive fittings require the removal of the rubber outer layers of a hose before the fitting is installed.

6. 45 degree S.A.E and 37 degree J.I.C. cone seats used on flared hydraulic fittings.

7. Quick couplers are <u>very susceptible</u> to dust and dirt.

ANSWERS TO ' TEST YOURSELF ' QUESTIONS

CHAPTER 12

1. First blank "static"; second blank "dynamic"

2. O-Rings are the most common seals.

3. True, dynamic O-rings must be allowed to roll slightly to lubricate themselves, resulting in slight leakage.

4. False, it is always best to replace any seals that are disturbed or accessible in the repair process.

5. False, any tool that is not guided does not assure that the seal will not be twisted. Use only the proper driving tool.

ANSWERS TO ' TEST YOURSELF ' QUESTIONS

CHAPTER 13

1. Good hydraulic fluids are a highly refined fluid with proper additives to: transmit the power applied to it, provide lubrication of moving parts, be stable for long periods of time, protect parts from rust and corrosion, resist foaming and oxidation, readily separate from air, water and other contaminants, be compatible with parts and seals, maintain proper viscosity, and be affordable.

2. Thinner.

3. Thicker.

4. Oxidation changes oil's chemical balance allowing formation of acids and sludge.

5. Oils and filters are recommended and change periods established based on manufacturers testing to provide the best service life from a machine.

ANSWERS TO ' TEST YOURSELF ' QUESTIONS

CHAPTER 14

1. True.

2. False - Operate control valves.

3. False - Always use cardboard to avoid injury.

4. False - These fluids can break down oil and damage seals.

5. Hotter.

6. Heat.

7. Internal leak.

8. a. - 3.
   b. - 1.
   c. - 2.

ANSWERS TO ' TEST YOURSELF ' QUESTIONS

CHAPTER 15

1. 1) Know the system, 2) Ask the operator, 3) Operate the machine, 4) Inspect the machine, 5) List the possible causes, 6) Reach a conclusion, 7) Test your conclusion.

2. During none of these steps. Do the 7 steps before you start repairing the system.

3. False. Test the pump first. It is the generating force for the whole system and affects the operation of all other system components. System leakage is compared to the pump test results.

4. First blank — "system." Second blank — "circuit."

5. First blank — "variable." Second blank — "fixed."

6. First blank — "varied." Second blank — "varied."

7. Look, Listen and Feel.

8. Pressure gauges, thermometer, tachometer, and flow meter.